ADVANCED DISTANCE SAMPLING

Advanced Distance Sampling

S. T. BUCKLAND
University of St Andrews

D. R. ANDERSON
Colorado Cooperative Fish and Wildlife Research Unit

K. P. BURNHAM
Colorado Cooperative Fish and Wildlife Research Unit

J. L. LAAKE
National Marine Mammal Laboratory, Seattle

D. L. BORCHERS
University of St Andrews

L. THOMAS
University of St Andrews

OXFORD
UNIVERSITY PRESS

OXFORD

UNIVERSITY PRESS

Great Clarendon Street, Oxford OX2 6DP

Oxford University Press is a department of the University of Oxford.
It furthers the University's objective of excellence in research, scholarship,
and education by publishing worldwide in

Oxford New York

Auckland Bangkok Buenos Aires Cape Town Chennai
Dar es Salaam Delhi Hong Kong Istanbul Karachi Kolkata
Kuala Lumpur Madrid Melbourne Mexico City Mumbai Nairobi
São Paulo Shanghai Taipei Tokyo Toronto

Oxford is a registered trade mark of Oxford University Press
in the UK and in certain other countries

Published in the United States
by Oxford University Press Inc., New York

A catalogue record for this title is available from British Library

Library of Congress Cataloging in Publication Data
(Data available)
ISBN 0 19 850783 6 (hbk)

10 9 8 7 6 5 4 3 2 1

Typeset by Newgen Imaging Systems (P) Ltd., Chennai, India
Printed in Great Britain
on acid-free paper by
Biddles Ltd., King's Lynn

Preface

Standard uses of distance sampling to estimate the density or abundance of biological populations were addressed in the companion volume to this book, *Introduction to Distance Sampling* (Buckland *et al.* 2001). In this book, we look at advanced topics, most of which have been developed in the last decade. Some of these topics reflect new methodologies to take advantage of new technologies. Others build on recent developments in statistical theory to improve on standard distance sampling methods. Relative to *Introduction to Distance Sampling*, the content is more technical.

Unlike the introductory book, this book has different authors by chapter, and the authors of the introductory book are the editors of this one. We have sought to produce a book that reads more like a multi-authored text rather than an edited volume. We have attempted to standardize notation throughout the book, but given the large amount of notation, and the interaction with different fields of statistics in different chapters, some notation applies only locally. We use $g(y)$ to indicate the probability of detection at distance y from a line or point when $g(0) = 1$, but when we generalize to methods in which probability of detection on the line or point is uncertain, we denote this probability by $p(y)$. This allows us to write $g(y) = p(y)/p(0)$, where $p(0) \leq 1$ and $g(0) = 1$. Where we need to distinguish between the three, we denote perpendicular distance from a line by x, distance from a point by r, and generic distance from a line or point by y. Other notation and conventions are given where they are needed.

We deliberately did not invite a range of experts to contribute to this book; rather, we decided on the topics we wished to cover, and invited authors to write chapters accordingly. Thus the book does not provide a snapshot of distance sampling research around the world; rather, it reflects our own priorities for making new methods available in the Distance software. These priorities are determined by our own judgements on what the user community needs. The methods of the introductory book were largely implemented in version 3.5 of Distance (Thomas *et al.* 1998). Version 4.1 (Thomas *et al.* 2003) encompasses everything that was in version 3.5, together with some of the methods of this book. We plan to include most of the other methods in future versions.

A short introductory chapter is followed by a chapter providing a general formulation for distance sampling, in which the likelihood function plays a central role. Chapter 2 establishes a theoretical framework that is used in several of the later chapters.

Chapter 3 develops a framework for modelling the detection function as a function of covariates, by specifying a model for the scale parameter of the detection function. Thus the effective width of search becomes a function of covariates associated with an animal or animal cluster. These methods are included in Distance 4.1, and are already in wide use.

In Chapter 4, we consider spatial models for distance sampling data. This is an area which generates much interest from the user community. Further development of methodology and implementation of the methods of Chapter 4 in Distance both have very high priority for the team working on Distance.

Chapter 4 addresses the problem of modelling spatial trend. In Chapter 5, we consider the issues in estimating temporal trend, starting with simple regression and considering progressively more sophisticated models, the most advanced of which involves embedding process models for the population dynamics in the analysis.

A basic assumption of standard distance sampling is that animals on the line or at the point are detected with certainty. In some studies, such as surveys of marine mammals or animals that burrow, this assumption cannot be made. Extensions of distance sampling methods to accommodate such studies are addressed in Chapter 6. At the time of writing, these methods are being implemented in Distance.

Survey design is generally done with pencil and paper. In many surveys, design is straightforward, and this approach is adequate. However, in line transect sampling especially, apparently innocuous decisions can lead to designs in which some parts of the study region are much more likely to be sampled than others. This leads to biased estimates of abundance when animal density varies by location, because standard methods assume that all locations are equally likely to be sampled. In Chapter 7, we develop auto-mated design algorithms, which allow easy computer-generation of better designs. We also address how to estimate probability of sampling by loca-tion when this varies, and show how to estimate abundance in this case. Distance 4.1 has GIS functionality, and allows the user to select from a range of automated design algorithms. A future priority is to add simula-tion capabilities, so that populations that mimic the population of interest can be simulated, and different designs generated, evaluated, and compared for efficiency. This will provide an invaluable tool for users who need to find efficient designs to take full advantage of limited resources.

Another approach to efficient survey design is to use adaptive methods. In Chapter 8, we develop adaptive distance sampling methods. Much of this chapter is concerned with conventional adaptive methods, but we also consider fixed-effort line transect methods; that is, methods for which the total length of transect line is predetermined. In shipboard surveys, for example, the ship is typically available for a fixed number of days, so that adaptive methods in which the total line length is not known in advance are unworkable. Adaptive distance sampling methods are not currently available in Distance, and given the relative complexity of analysis, are unlikely to see much use until software is available.

Many populations are not amenable to sightings survey methods, perhaps because animals are difficult to detect, even when very close to the observer. Trapping methods are often more feasible for such populations. Passive distance sampling methods (Chapter 9) combine trapping methods with distance sampling, so that the animals themselves record their distance from the line or point, by entering a trap at a known distance. The methods are typically used for reptiles, small mammals, and some species of insect such as ground beetles. Analyses can be conducted in Distance.

Standard distance sampling relies on a combination of model-based and design-based methods. Abundance within the surveyed strips or circles is estimated by modelling the detection function, and so is model-based, whereas extrapolation to the wider survey region relies on design-based methods. In Chapter 10, we state the theoretical foundation for distance sampling, and provide a framework for testing the performance of standard methods.

The final chapter, Chapter 11, covers a range of topics, none of which justify a full chapter of their own. Topics covered are distance sampling in three dimensions, full likelihood examples for conventional distance sampling, line transect surveys with random line length, models for the search process in sightings surveys, combined mark-recapture and distance sampling surveys, combined removal methods and distance sampling surveys, point transect sampling of cues, migration counts, measurement error models, the theory underlying indirect surveys of animal signs (usually dung or nests), goodness-of-fit tests, and pooling robustness.

The editors are:

- *Stephen Buckland*   Steve is Professor of Statistics at the University of St Andrews, and is also Director of the Centre for Research into Ecological and Environmental Modelling. His research interests in distance sampling date from 1980. Much of his 20-year-old FORTRAN code can still be found lurking deep in the bowels of Distance. He has been an instructor on annual distance sampling training workshops for 10 years. He acknowledges the support from the University of St Andrews in the preparation of this book.

- *David Anderson*   David retired in May, 2003 after 37 years of federal service. He is still a professor in the Department of Fishery and Wildlife Biology at Colorado State University. His interests in line transect sampling and analysis methods date back to 1966, when he was surveying populations of duck nests in southern Colorado. He acknowledges the support and freedom he was afforded in his research by the US Geological Survey, Biological Resources Division.

- *Kenneth Burnham*   Ken is a research statistician with the Colorado Cooperative Fish and Wildlife Research Unit at Colorado State University. His research focuses on data-based inference about animal populations, especially from capture–recapture and distance sampling, and also the issue of model selection for such inferences. His formal interest in distance sampling dates from 1976. He too is grateful to his employer, the US Geological Survey, for support and freedom in his research.

- *Jeffrey Laake*   Jeff is both a statistician and research biologist for the National Marine Mammal Laboratory of the Alaska Fisheries Science Center, US National Marine Fisheries Service because he could never decide which field he preferred. His primary research interests are distance sampling, mark-recapture, and population dynamics focusing particularly on marine mammals. He wrote the original program TRANSECT with Ken and the original versions of DISTANCE with Steve. When he is not at his desk programming or analyzing data, he can be found in the field conducting line transect surveys or capturing and marking sea lions. His input to this book was funded by the US National Marine Fisheries Service (NOAA Fisheries).

- *David Borchers*   David is an applied statistician with 15 years experience estimating animal abundance, much of that using distance sampling. His initial bewilderment at the variety of abundance estimation methods out there has led to a particular interest in developing ideas that span more than one method. Together with Jeff Laake, he wrote the double-platform analysis engine of program Distance. He has been an instructor on distance sampling training workshops for 10 years and is head of the Research Unit for Wildlife Population Assessment at the University of St Andrews. His contribution to the book was funded by the University of St Andrews.

- *Len Thomas*   Len is a research fellow within the Centre for Research into Ecological and Environmental Modelling. He is a relative newcomer to distance sampling, having only started work on the methods in 1997 when he came to St Andrews. He coordinated development of the Windows versions of Distance with sponsorship from the UK Engineering and Physical Sciences Research Council, the UK Biotechnology and Biological Sciences Research Council, the US National Marine Fisheries Service, the US National Park Service and the Wildlife Conservation Society.

Each editor also coauthored at least two chapters. Additional authors are:

- *Jonathan Bishop*   Jon is working on a Ph.D. at St Andrews, funded by the UK Engineering and Physical Sciences Research Council and supervised by Steve Buckland and Ken Newman. The topic is development of methodologies for fitting population dynamics models to survey data. Jon was an undergraduate at St Andrews, where he completed a project on goodness of fit testing in distance sampling. He was then hired to incorporate his methods into version 4.1 of Distance. A summary of his work is included in the section on goodness of fit tests and q–q plots in Chapter 11.

- *Rachel Fewster* Rachel conducted her Ph.D. research on spatio-temporal models in ornithology at St Andrews under the supervision of Steve Buckland. On completion, she took up a post in the Department of Statistics at Auckland, where she is now a lecturer. Despite her time at St Andrews, her active involvement in distance sampling is relatively recent. Her interest in developing a better theoretical underpinning for distance sampling methods has resulted in Chapter 10.

- *Alan Franklin* Alan's Ph.D. was on the population dynamics of northern spotted owls. His Ph.D. program was at Colorado State University, supervised by David Anderson. He is now Research Scientist affiliated with the Colorado Cooperative Fish and Wildlife Research Unit, and continues to work on modelling population dynamics, both of spotted owls and of a range of other bird species. He is also working with Paul Lukacs and David Anderson to develop the passive distance sampling methods of Chapter 9.

- *Sharon Hedley* Sharon developed the spatial modelling methods of Chapter 4 as part of her Ph.D. at St Andrews, supervised by Steve Buckland and David Borchers. The work was funded by the International Whaling Commission. On completion of her thesis in 2000, Sharon took up a National Research Council research associateship at the Southwest Fisheries Science Center, La Jolla, CA. She returned in 2001, to work in the Research Unit for Wildlife Population Assessment (RUWPA) at St Andrews, where the spatial modelling methods are being applied and further developed.

- *Paul Lukacs* Paul is a Ph.D. student at Colorado State University, under Ken Burnham's supervision. The subject of his research is the use of genotypes as a means of 'marking' animals in mark-recapture studies. For his masters degree, he developed trapping web methods, under the supervision of David Anderson and Ken Burnham. This included development of a software package WebSim as an aid to the design of trapping web surveys. Much of that research, and of more recent work on trapping line transects, is summarized in Chapter 9.

- *Fernanda Marques* Fernanda studied for her Ph.D. at St Andrews, supervised by Steve Buckland. She was funded by the Inter-American Tropical Tuna Commission, to develop methods of analysis for line transect data from platforms of opportunity. She developed the covariate methods of Chapter 3 as part of her Ph.D., and programmed the methods into Distance. Fernanda returned to Brazil on completion of her thesis in 2001, and is currently a post-doc at Fundação Universidade Federal do Rio Grande, funded by the Brazilian National Council for Scientific and Technological Development (CNPq).

- *Tiago Marques* Tiago developed an interest in distance sampling while working on his masters degree under the supervision of Dinis Pestana at the University of Lisbon. He recently started on his Ph.D. on distance sampling at St Andrews, funded by Fundação para a Ciência e Tecnologia (FSE, III Quadro Comunitário de Apoio) and supervised by Steve Buckland and Dinis Pestana. He and David Borchers drafted the measurement error section of Chapter 11.

- *Jonathan Pollard* John registered for a part-time Ph.D. at St Andrews, under the supervision of Steve Buckland, because of a strong personal interest in applying his mathematical knowledge to wildlife populations. He is a software advisor at Unisys, who funded his Ph.D. His research topic was adaptive distance sampling, and the work of his thesis is summarized in Chapter 8. It is a testament to his determination that he completed his thesis in 2002, despite the demands of his work and the arrival of a son and a daughter.

- *Samantha Strindberg* Samantha came to St Andrews to work for the Research Unit for Wildlife Population Assessment in 1994. She subsequently embarked on a Ph.D. on optimized automated survey design in wildlife population assessment, funded by the University of St Andrews and supervised by Steve Buckland. Chapter 7 draws on information originally presented in her thesis. The automated survey design algorithms and GIS functionality in the Distance 4 software are due in no small measure to Samantha's work. On completing her thesis in 2001, she went on to take up a position as a quantitative conservation scientist with the Living Landscapes Program of the Wildlife Conservation Society in New York.

We thank all of our coauthors for their contributions and commitment to the distance sampling team. We also thank Tiago Marques for his helpful comments on drafts of the book.

January 2004

S. T. Buckland
D. R. Anderson
K. P. Burnham
J. L. Laake
D. L. Borchers
L. Thomas

Contents

6 Methods for incomplete detection at distance zero 108

J. L. Laake and D. L. Borchers

7 Design of distance sampling surveys and Geographic Information Systems

S. Strindberg, S. T. Buckland, and L. Thomas

8 Adaptive distance sampling surveys

J. H. Pollard and S. T. Buckland

1

Introduction to advanced distance sampling

S. T. Buckland and D. R. Anderson

Distance sampling, primarily line transect and point transect sampling, has had a relatively short history. The earliest attempts to use distances to detected animals to estimate abundance date back to the 1930s, and the first line transect estimator with a rigorous mathematical basis was due to Hayne (1949). Nearly 20 years later, Gates *et al.* (1968) and Eberhardt (1968) made important contributions to the development of line transect sampling methodology. Neither the radial distance model of Hayne (1949) nor the negative exponential model of Gates *et al.* (1968) is based on plausible assumptions about the detection process. Eberhardt's (1968) work was more conceptual, and attempted to provide a class of models that were robust to differing detection processes. None of these early methods are now recommended. Three papers in the early 1970s prompted Burnham and Anderson (1976) to develop the general theory needed for reliable estimation. The first of these papers was Anderson and Pospahala (1970), who used polynomials to fit the distance data, but who did not provide underlying theory. The field experiments of Robinette *et al.* (1974) were important in providing data sets with known abundance, on which estimation methods could be tested. The third paper was by Sen *et al.* (1974), which gave an erroneous formulation. Burnham and Anderson (1976) corrected this formulation, and provided a general framework for both parametric and nonparametric methods, applied to data that were either grouped or ungrouped, and truncated or untruncated. The first comprehensive treatment of the topic was by Burnham *et al.* (1980). Point transect sampling (or variable circular plots) was conceptualized in the early 1970s for songbird surveys, although the initial work was not published until 1980 (Reynolds *et al.* 1980), by which time several papers using the technique had already been published. The method is still largely restricted to avian studies (Rosenstock *et al.* 2002), although other applications are now starting to appear in the literature. Reviews of these historical developments are given by Buckland *et al.* (2000, 2001).

Given its short history, it is perhaps surprising that distance sampling is the most widely used technique for estimating abundance of wild animal populations. The use of mark-recapture in fisheries dates as far back as Walton (1653), although work by Petersen (1896) seems to have led to its use for estimating abundance (see Buckland *et al.* 2000). The use of harvest models based on the 'catch equations' to estimate abundance was first documented by Baranov (1918); catch-per-unit-effort by Hjort and Ottestad (1933); change-in-ratio by Kelker (1940); and removal methods by Moran (1951). The success of distance sampling perhaps stems from the fact that it provides more robust estimation of abundance more cheaply than methods based on catching animals, at least for populations for which the key assumptions hold to a good approximation. Further, the effects on abundance estimates when key assumptions fail tend to be more readily understood than for most competitive methods. Distance sampling represents a suite of methods which are extensions of complete counts of sample plots (called sample counts here). Their advantage over sample counts is that not all animals within the sampled plots (strips in the case of line transect sampling and circles for point transect sampling) need be counted, so that the approach can usually achieve a given level of precision at lower cost than a comparable method based on sample counts. Relative to mark-recapture, the modelling element in distance sampling is straightforward. The detection function model seldom requires more than four parameters, and one is often sufficient. Only a handful of contending models need be considered. By contrast, mark-recapture models commonly require more than 20 parameters—possibly substantially more—and there are many possible models for a given data set.

The methods that are currently considered standard are largely as set out by Buckland *et al.* (1993a). For an updated introduction to standard methodology, the reader is referred to the companion volume to this book (Buckland *et al.* 2001). Here, we concentrate on more advanced methodologies, some of which are reviewed by Buckland *et al.* (2002). A few of the more advanced sections from Buckland *et al.* (1993a), especially from chapter 6 of that book, are updated and reproduced here.

For a general methodological framework for distance sampling and other methods of estimating abundance of closed populations, the reader is referred to Borchers *et al.* (2002). That book is intended as an advanced student text, and provides a basis for many of the developments outlined here. Williams *et al.* (2002) cover a range of methods for estimating animal abundance, and show how these methods are used in the management of animal populations.

In Chapter 2, a general likelihood framework is presented. The three components of estimation for standard methodology, encounter rate, effective area surveyed, and mean cluster size (when objects occur in clusters), are integrated into a single Horvitz–Thompson-like estimator, in which the inclusion probabilities are estimated

$$\widehat{N} = \sum_{i=1}^{n} \frac{s_i}{\widehat{P}_i}, \qquad (1.1)$$

where $\widehat{N}$ is estimated population size, n is the number of animal clusters detected in the covered area, s_i is the size of the ith detected cluster, and $\widehat{P}_i$ is the estimated inclusion probability for that cluster. This probability has two components: a coverage probability P_c, which is determined by design, and given by $P_c = a/A$, where A is the size of the study region and a is the covered area (the total area of surveyed strips or circles); and an estimated detection probability $\widehat{P}_a$, given that a cluster is in the covered area. If we choose to estimate a single $\widehat{P}_a$ for all clusters, then $\widehat{P}_i = P_c \times \widehat{P}_a$ and the effective area surveyed is $a \times P_a$, estimated as $a \times \widehat{P}_a$. Alternatively, if we obtain cluster-specific estimates, then $\widehat{P}_i = P_c \times \widehat{P}_{ai}$ for cluster i.

If objects occur singly, then $s_i = 1$ for every detection, and

$$\widehat{N} = \sum_{i=1}^{n} \frac{1}{\widehat{P}_i}. \qquad (1.2)$$

In the case of clustered populations, this equation gives the estimated number of clusters in the study area.

Full likelihood and conditional likelihood methods are described in Chapter 2. Because the distribution of covariates in the study population or area is generally unknown and not easily estimated, inference is generally based on conditional likelihood methods, for which we condition on the values of covariates observed.

Covariate models for the detection function (Chapter 3) potentially yield more efficient estimates of abundance, and eliminate the bias that may arise, for example, when abundance estimates by stratum are required, but data are pooled across strata for estimating the detection function. They also offer the potential for more reliable estimates of trend in abundance when surveys are conducted from platforms of opportunity, such as ferries or fishing vessels at sea, although they do not address the problems that arise because the region is not randomly sampled in such surveys.

Spatial line transect models (Chapter 4) allow a surface to be fitted, representing animal density throughout the study region. This in turn allows estimation of abundance for any subset of the area, by integrating over the relevant section of the surface. It also allows abundance to be related to spatial covariates, so that managers can assess the importance of habitat and environment to the population of interest. Spatial models also potentially reduce bias in abundance estimates from platforms of opportunity survey data, in which survey effort is non-random.

Spatial models estimate variation in density or abundance through the region. Wildlife managers are often more interested in modelling temporal trends, to identify whether management action is required, or to assess the effects of such action. Methods for trend estimation addressed in Chapter 5 fall into two categories: empirical estimation of trend from a series of abundance estimates; and fitting of a population dynamics model to the time series of estimates. The second approach has the advantages that estimated trends are consistent with biological reality, and the effects of management actions that affect survival or productivity can be modelled and predicted.

For some populations, such as whales or porpoise, or burrowing animals such as rabbits or tortoise, one of the key assumptions that any animal on the line or at the point will be detected may be violated. Double-platform methods (Chapter 6) allow distance sampling methodology to be combined with mark-recapture methods, so that this assumption is no longer required. The second 'platform' may be a standard sightings platform on a ship or aircraft, or it may comprise an independent method of locating a subset of the surveyed animals, such as a radio-tagging experiment. In the latter case, individual animals are marked, or are identified from natural markings, whereas in the former case, no marking takes place. Instead, it is necessary to develop field methods so that duplicate detections (animals detected by both platforms) can be identified.

Automated design algorithms, linked with Geographic Information Systems (GIS) functionality, are covered in Chapter 7. They allow quick and easy generation of survey designs, and enable different designs to be compared for efficiency and accuracy of the subsequent abundance estimates, using simulation. For complex surveys in which coverage probability is not uniform, they also allow estimation of coverage probability by location. This in turn allows valid abundance estimation.

For populations that typically have an aggregated spatial distribution, adaptive distance sampling surveys (Chapter 8) potentially yield more precise estimates of abundance. They also give more detections than can conventional surveys with the same overall effort, which can be an important advantage for scarce species, for which sample size may otherwise be inadequate for modelling the detection function.

In most distance sampling surveys, one or more observers actively try to detect objects, usually animals. In passive distance sampling methods (Chapter 9), animals are not actively searched for. Instead, detections are made, for example, by using traps, or devices to secure a sample of hair or feathers, or remote systems such as cameras. A detection occurs when an animal enters a trap or a sensed area at a known distance from a central point (trapping web) or line (trapping line transect). The density of traps or sensed areas is greater near the centre point or line. Trapping webs (Anderson *et al.* 1983) have their roots in point transect sampling theory. Trapping transects (Chapter 9) make similar use of line transect sampling theory. These methods can be particularly useful for species that do not generally meet the assumptions of standard distance sampling, perhaps because they hide and are undetectable for much of the time (e.g. reptiles) or because they are too small to be reliably detected by line transect observers (e.g. beetles).

Standard distance sampling methods blend model-based statistical methods (to model detectability within the surveyed strips or circles) with design-based statistical methods (to estimate the number of animals outside the surveyed strips). (Spatial line transect models, discussed in Chapter 4, replace the design-based element by a spatial model of animal density.) In Chapter 10, we give a rigorous basis for the composite approach, show why it leads to robust estimation of animal abundance, and explore the limitations of this robustness.

Other advanced topics are covered in Chapter 11: three-dimensional distance sampling methods; full likelihood methods for conventional distance sampling; line transect surveys with random line length; models for the search process in sightings surveys; combined mark-recapture and distance sampling surveys; combined removal methods and distance sampling surveys; point transect sampling of cues; migration counts; measurement error models; theory for indirect surveys of animal signs (usually dung or nests); quantile–quantile plots, the Kolmogorov–Smirnov test and Cramér–von Mises tests; and pooling robustness.

Most of the above advances have been implemented, or will shortly be implemented, in version 4 of the software distance (Thomas *et al.* 2003). Version 4 also incorporates the standard methods of Version 3.5 (Thomas *et al.* 1998). Version 3.5 is the companion software for Buckland *et al.* (2001), and Version 4 is the companion software for this book.

2

General formulation for distance sampling

D. L. Borchers and K. P. Burnham

2.1 Introduction

Full likelihood functions for distance sampling methods involve assuming probability models for the encounter rate or animal distribution, for the detection process, for cluster size (if animals are detected in clusters), and possibly for other data. Conventional distance sampling (CDS) methods avoid assuming a probability model for encounter rate by using an empirical estimator of its variance and getting confidence intervals assuming density D to be log-normally distributed. Similarly, a point estimator and sampling variance of mean cluster size, $E[s]$ can be obtained in a regression framework, so no probability model $\pi(s)$ for cluster size s in the population need be assumed. Probability models and likelihood inference are conventionally used only for the distance part of the data.

In principle, analysis of distance data could be based on a full likelihood for all the data. Advantages of this approach include the facts that estimators of density and abundance (N) then enjoy the properties of maximum likelihood estimators (MLEs) (asymptotic efficiency, for example) and profile likelihood confidence intervals can be calculated instead of assuming log-normality. Disadvantages include the facts that more probability models need to be specified, and specifying plausible models for some components may be difficult.

In this chapter, we develop a general likelihood and estimation framework that places most distance sampling methods in a common statistical context. Using this, we give an overview of conventional methods and the advanced methods that are developed later in the book. The likelihoods and estimation methods are expanded upon in the chapters that follow.

We start by developing the framework for CDS.

2.2   CDS revisited

Consider a line or point transect survey in which a region out to a distance w from the lines or points is searched and this searched region has an area a. (We refer to this region as the 'covered region'.) If the area of the whole survey region is A, the fraction of the survey region that was covered by the survey is

$$P_c = \frac{a}{A}. \tag{2.1}$$

Suppose that lines or points were located randomly and independently of the animals. Then on average over many surveys, any given animal would fall in the covered region with probability P_c. We refer to P_c as the 'coverage probability'.

If we detected n animals in total and all animals in the covered region were seen, we could reasonably estimate the number of animals in the region, N, by

$$\widehat{N} = \frac{n}{P_c}. \tag{2.2}$$

For example, if $P_c = \frac{1}{2}$, then we would estimate that there are twice as many animals there as were seen ($\widehat{N} = n/(1/2) = 2n$).

There are two reasons we might miss animals on the survey: because they are not in the covered region, or because we fail to detect them even though they are in the covered region. In distance sampling surveys, some of the animals in the covered region are typically not detected. We use P_a to denote the probability that any particular animal which is in the covered region is detected: on average, a fraction P_a of animals in the covered region is detected.

If the probability that an animal is in the covered region and the probability that we detect an animal in the covered region are independent and the same for all animals, then we detect each animal with probability $(P_a \times P_c)$. So if we detected n animals on the survey, we could reasonably estimate the number of animals in the survey region by

$$\widehat{N} = \frac{n}{P_c P_a}. \tag{2.3}$$

For example, if we covered half the survey region ($P_c = \frac{1}{2}$) and half the animals in the covered region were detected on average ($P_a = \frac{1}{2}$), then we would estimate that there are four times as many animals there as were seen ($\widehat{N} = n/(\frac{1}{2} \times \frac{1}{2}) = 4n$).

While the coverage probability P_c is usually known by design, in most distance sampling applications P_a is unknown; its estimation using distance data is central to distance sampling inference. Suppose for the moment that we have an estimate $\widehat{P}_a$ of P_a, obtained by some means which we will discuss later. We could then estimate N by

$$\widehat{N} = \frac{n}{P_c \widehat{P}_a}. \tag{2.4}$$

Conventional distance sampling estimators of abundance have this form, as we illustrate below. The advanced distance sampling methods described in this book involve more general methods for obtaining or estimating P_c and for estimating P_a. Before we consider the generalizations, we show that the conventional line transect and point transect estimators are in fact of the form of eqn (2.4).

2.2.1 *Conventional line transect estimator*

In the case of line transect surveys with random transect placement and total line length L, the area of the covered region (the 'covered area') is $a = 2wL$, where w is the half-width of the strips. The coverage probability can therefore be written as

$$P_c = \frac{2wL}{A}. \tag{2.5}$$

Suppose on average you miss half the animals in the covered region ($P_a = \frac{1}{2}$). In effect, the survey has covered only half the strip width, that is, an area $2 \times \frac{1}{2}w \times L$. The 'effective strip half-width' (often abbreviated to 'effective strip width' or just 'esw') is $\mu = \frac{1}{2}w$ in this case. More generally, you can think of a line transect survey as effectively covering strips with total area $2\mu L$ (for some $\mu \leq w$). The probability that an animal in the covered region is detected can then be written as

$$P_a = \frac{2\mu L}{2wL} = \frac{\mu}{w}. \tag{2.6}$$

In practice we do not know μ; it is estimated from the distance data gathered on the survey (see later). Using an estimate $\hat{\mu}$ in place of the unknown μ and substituting the above two equations into eqn (2.4), we get the familiar line transect estimator of abundance:

$$\widehat{N} = \frac{n}{(2wL/A)(\hat{\mu}/w)} = \frac{nA}{2\hat{\mu}L}, \tag{2.7}$$

which is also often written as

$$\widehat{N} = \frac{n\hat{f}(0)A}{2L},$$

(2.8)

where $\hat{f}(0) = 1/\hat{\mu}$.

2.2.2 *Conventional point transect estimator*

In the case of point transect surveys with random point placement and a total of k points, the covered area is $a = k\pi w^2$. The coverage probability can therefore be written as

$$P_c = \frac{k\pi w^2}{A}$$

(2.9)

and the probability that an animal in the covered region is detected can be written as

$$\widehat{P}_a = \frac{k\pi\hat{\rho}^2}{k\pi w^2} = \frac{\hat{\rho}^2}{w^2},$$

(2.10)

where $\hat{\rho}$ is an estimator of the effective radius of the circular plots comprising the covered region, and w is the actual radius of each plot. Substituting the above two equations into eqn (2.4), we get the usual point transect estimator of abundance:

$$\widehat{N} = \frac{n}{(k\pi w^2/A)(\hat{\rho}^2/w^2)} = \frac{nA}{k\pi\hat{\rho}^2},$$

(2.11)

which is also often written as

$$\widehat{N} = \frac{nA}{k\hat{\nu}} \quad \text{or} \quad \widehat{N} = \frac{n\hat{h}(0)A}{2\pi k},$$

(2.12)

where $\hat{\nu} = \pi\hat{\rho}^2$ is the effective search area per point and $\hat{h}(0) = 2\pi/\hat{\nu}$.

2.3 Horvitz–Thompson: a versatile estimator

Horvitz and Thompson (1952) developed an estimator that has proved to be amazingly versatile. Its general form is

$$\hat{\tau} = \sum_{i=1}^{n} \frac{z_i}{P_i},$$

(2.13)

where z_i is the value of the response of interest for the ith unit, $\hat{\tau}$ is an estimate of the sum of all the z_i in the population, and P_i is the probability that the ith unit appears in the sample (its 'inclusion probability'). The estimator was developed for cases in which the inclusion probabilities are known, and in this case it is unbiased.

2.3.1 *Animals that occur as individuals*

For the case in which animals are the sampling unit, we obtain an estimate of abundance by letting $z_i = 1$ for all N animals in the population. The following is an intuitive rationale for the estimator. If a particular animal has probability $P_i = \frac{1}{2}$ of appearing in the sample, then on average it will appear in half of all possible samples. Thus if we count $2 = 1/(1/2)$ every time the animal does appear in the sample, its average contribution to $\hat{\tau}$ per sample over all possible samples will be one. More generally, if we count $1/P_i$ in every sample for which animal i is detected, then *on average* over all possible samples, every animal in the population will be counted once per sample. Formal proofs that the estimator is unbiased can be found for example in Thompson (2002).

Distance sampling estimators can be viewed as Horvitz–Thompson estimators in which we must estimate the unknown true inclusion probabilities. We refer to this sort of estimator as a Horvitz–Thompson-like estimator. In contrast to Horvitz–Thompson estimators proper, Horvitz–Thompson-like estimators are not necessarily unbiased. Nevertheless, this form of estimator turns out to be a versatile general form for distance sampling and we use it repeatedly in the chapters that follow.

Using an estimator $\widehat{P}_i$ in place of the unknown P_i, the Horvitz–Thompson-like estimator of abundance is

$$\widehat{N} = \hat{\tau} = \sum_{i=1}^{n} \frac{1}{\widehat{P}_i}. \tag{2.14}$$

There is no unique way to estimate P_i; it depends on what additional information we have and what assumptions we can or cannot make. The simplest case is when all the inclusion probabilities are considered to be the same: $P_i = P$ for all i. In this case, eqn (2.14) reduces to

$$\widehat{N} = \frac{n}{\widehat{P}}. \tag{2.15}$$

2.3.2 *Animals that occur in clusters*

If animals occur in clusters, we could let $z_i = s_i$ be the observed cluster size of the ith cluster and estimate individual abundance by

$$\widehat{N} = \sum_{i=1}^{n} \frac{z_i}{\widehat{P}_i} = \sum_{i=1}^{n} \frac{s_i}{\widehat{P}_i}. \tag{2.16}$$

An alternative approach, more commonly used in distance sampling, is to multiply the estimator of cluster abundance by an estimate of mean cluster size, $\widehat{E}[s]$:

$$\widehat{N} = \sum_{i=1}^{n} \frac{1}{\widehat{P}_i} \widehat{E}[s]. \tag{2.17}$$

2.3.3 *CDS estimators*

The conventional line transect estimator of eqn (2.7) is readily seen to be a Horvitz–Thompson-like estimator of the form of eqn (2.15) with estimated P:

$$\widehat{P} = P_c \widehat{P}_a = \left(\frac{2wL}{A} \right) \left(\frac{\hat{\mu}}{w} \right). \tag{2.18}$$

Similarly, the conventional point transect estimator of eqn (2.11) is also a Horvitz–Thompson-like estimator of the form of eqn (2.15) with estimated P:

$$\widehat{P} = P_c \widehat{P}_a = \left(\frac{k\pi w^2}{A} \right) \left(\frac{\hat{\rho}^2}{w^2} \right). \tag{2.19}$$

2.4 Maximum likelihood estimation

The description above begs the question of how to estimate the unknown inclusion probabilities. They can be factorized into a probability of the animal (cluster) being in the covered region (P_c) and a probability that it is detected, given that it is in the covered region (P_a). The coverage probability P_c is conventionally treated as known (it is determined by the survey design), but P_a must be estimated from the observed distances. In this book we focus on estimation by maximum likelihood.

See Borchers *et al.* (2002: 14–19) for a simple description of maximum likelihood estimation. Buckland *et al.* (2001: 61–68) provide a more detailed description of maximum likelihood methods as they apply to the estimation of P_a. Most texts on statistical inference contain descriptions of maximum likelihood estimation in general terms. Severini (2000) gives a detailed introduction to the modern theory of likelihood methods.

The fundamental idea behind maximum likelihood estimation is very straightforward: given an equation quantifying the probability of what was observed (the data) as a function of some unknown parameters, the maximum likelihood estimates are those parameter values at which this function is a maximum. Given the data, they are the most likely parameter values.

In the sections below, we develop likelihood functions and MLEs for CDS methods. After that, we outline how these can be generalized to accommodate the advanced distance sampling methods covered in later chapters. For brevity, we develop the likelihood functions in terms of animals only (rather than clusters), except when we deal explicitly with estimation of individual abundance from detection of clusters.

2.4.1 'Covered' animals

We define the covered region to be the region within a distance w of the line or point transects. It is convenient to refer to animals that fall in this region as 'covered animals': there are N_c covered animals on a survey. The probability that a given animal is covered is P_c.

Each of the N animals in the survey region can be thought of as a trial and each animal falling in the covered region as a 'success'. If animals fall in the covered region independently of each other, then the number of 'successes' (which we will call N_c) is a binomial random variable with parameters N and P_c:

$$\mathcal{L}_{N_c}(N) = \binom{N}{N_c} P_c^{N_c} (1 - P_c)^{N - N_c}. \qquad (2.20)$$

If we are able to observe all animals in the covered region (i.e. $P_a = 1$), then once we have done the survey, we know N_c, and N would be the only unknown parameter in this function. The function then quantifies the likelihood of what was observed (N_c) as a function of the single unknown parameter, N; it is the likelihood function for N when N_c is observed. The maximum likelihood estimate is that value of N that maximizes this function. It is easy to show that this occurs when

$$\widehat{N} = \left[\frac{N_c}{P_c}\right], \qquad (2.21)$$

where the square brackets mean 'the integer part of'. We usually omit these brackets, and allow non-integer animals.

The likelihood eqn (2.20) is appropriate for quadrat, strip, or circular plot surveys, in which all animals in the covered region are detected. In practice, however, inference from this sort of survey is seldom conducted by maximum likelihood. One reason for this is that the likelihood contains stronger assumptions about the independence of animal locations than are reasonable in most cases. In particular, 'successes' (detections in this context) are assumed to be independent events. This requires that animals fall in the covered region independently of one another. In practice this will very seldom be the case because animals tend to either cluster together or avoid each other (if they are territorial). The variance of the counts will be higher than predicted by the binomial likelihood function $\mathcal{L}_{N_c}(N)$ if animals cluster, and lower if they avoid each other. The binomial is a poor model for these cases.

One way of dealing with populations that cluster or are territorial is to use a probability density function (pdf) that is more able to deal with this than is the binomial. The negative binomial is a flexible candidate (see below).

In practice, however, design-based methods (Chapter 10) are usually used in preference to maximum likelihood methods for these surveys. Design-based methods do not rely on any assumptions about animal distribution. The Horvitz–Thompson estimator proper was developed for design-based sampling theory and it is directly applicable here because the inclusion probability (P_c) is known (from the design). The Horvitz–Thompson estimator is identical to the MLE of eqn (2.21) in this case; the difference between the two methods lies in the estimation of variance. The Horvitz–Thompson variance estimator involves no assumptions about animal distribution, whereas the likelihood function contains strong assumptions about independence of animal locations. Except in cases where these independence assumptions hold, the Horvitz–Thompson variance estimator will be the more reliable of the two.

2.4.2 *Random detection with known probability*

Animals have different detection probabilities. In particular, distance sampling is based on models of detection probability that depend on the distance that the animal is from the observer (y). Suppose that the probability of detecting animal i at distance y_i is $g(y_i)$ and $g(0) = 1$ (i.e. all animals at distance zero are detected). We could construct a Horvitz–Thompson estimator using the n $g(y_i)$ values in the denominator, but the estimator generally has lower variance if we use in the denominator the average detection probability of all animals in the covered region.

In CDS, an estimate of this average inclusion probability ($\widehat{P} = P_c \widehat{P}_a$) is used in the Horvitz–Thompson-like estimator—see eqns (2.18) and (2.19).

We can write the average detection probability in the covered region (P_a) as

$$P_a = \int_0^w g(y)\pi(y)\,dy, \qquad (2.22)$$

where $\pi(y)$ is the pdf of distances of detected and undetected animals within distance w. In distance sampling, $\pi(y)$ is treated as known.

Suppose for the moment that detection probability depended only on distance and that we somehow knew the detection function $g(y)$ and hence knew P_a. Given the number of animals detected (n) and P_a, the number of animals in the covered region can be estimated from the following likelihood

$$\mathcal{L}_n(N_c) = \binom{N_c}{n}(P_a)^n\,(1 - P_a)^{N_c - n} \qquad (2.23)$$

from which the MLE of N_c is

$$\widehat{N}_c = \frac{n}{P_a}. \qquad (2.24)$$

Note that the likelihood $\mathcal{L}_n(N_c)$ on its own does not provide any basis for estimating the abundance (or density) of animals in the whole survey region. It provides a basis for estimating the number (or density) of animals in the covered region alone. In order to draw inferences about abundance in the whole survey region from observations within the covered region, we must either model the relationship between counts in the covered region and abundance in the whole survey region (using a likelihood of the sort dealt with in Section 2.4.1 for example), or we must invoke design-based inference methods.

For a fully likelihood-based approach to estimating N from n when P_c and P_a are known, we take the product of $\mathcal{L}_{N_c}(N)$ and $\mathcal{L}_n(N_c)$ and sum over all possible N_c to obtain the likelihood

$$\mathcal{L}_n(N) = \binom{N}{n}(P_c P_a)^n\,(1 - P_c P_a)^{N - n} \qquad (2.25)$$

(which corresponds to a binomial distribution with N trials, each with probability of success $P_c P_a$) from which the MLE of N is

$$\widehat{N} = \left[\frac{n}{P_c P_a}\right], \qquad (2.26)$$

where square brackets are used here to indicate the integer part of the expression inside them. Below, we omit the square brackets for simplicity.

(The MLE again coincides with the Horvitz–Thompson estimator since the inclusion probability is $P_c P_a$.)

2.4.2.1   *Other probability models for n and N_c*

The likelihood eqns (2.23) and (2.25) apply if we treat N_c and N as fixed parameters. If instead, we treat density (D) as a fixed parameter and retain the strong independence assumptions made to get the binomial case, a Poisson likelihood applies. In this case,

$$\mathcal{L}_{N_c}(D) = (Da)^{N_c} \frac{e^{-Da}}{N_c!} \tag{2.27}$$

and

$$\mathcal{L}_n(D) = (DaP_a)^n \frac{e^{-DaP_a}}{n!}. \tag{2.28}$$

For the most part, we develop likelihoods using the binomial form, but the Poisson form is a ready alternative. See Section 11.2 for examples of likelihoods using the Poisson distribution. The difference between the Poisson and binomial in this context is largely philosophical: the binomial treats the number of animals present as fixed, while the Poisson treats density as fixed and the number of animals in the survey region as a random variable.

The Poisson or binomial distributions for n (or N_c) are not reasonable. (For brevity we discuss only models for n below.) The negative binomial model is more tenable (although a reasonable model for $\mathcal{L}_n(N)$ may need more than two parameters). The negative binomial is given by

$$\mathcal{L}_n(\underline{\phi}) = \frac{\Gamma(\phi_1 + n)}{\Gamma(\phi_1)\Gamma(n+1)} (1 - \phi_2)^n \phi_2^{\phi_1}, \tag{2.29}$$

where $0 < \phi_1$, $0 < \phi_2 < 1$, $0 \leq n$ and

$$E[n] = \phi_1 \frac{1 - \phi_2}{\phi_2} \tag{2.30}$$

$$\text{var}[n] = \frac{E[n]}{\phi_2}. \tag{2.31}$$

$\mathcal{L}_n(\underline{\phi})$ is not used as parameterized, but rather the relationship $E[n] = DaP_a$ must be imposed on the parameters in the distribution. With a multiparameter model for n, such as the negative binomial, there is no obvious unique way to reparameterize $\mathcal{L}_n(\underline{\phi})$. We suggest it will instead be necessary to optimize the log-likelihood function subject, for example, to the constraint $E[n] = DaP_a$, where P_a is replaced by its form as a function of the parameters in the detection function $g(y)$.

2.4.3   CDS likelihoods

The likelihood functions above relate only to the encounter rate (the number of animals detected per unit length of line for line transects or the number detected per visit to a point for point transects); they do not involve data that would allow P_a to be estimated. These data are the distance data y_i $(i = 1, \ldots, n)$ and when we do not know P_a, we need an additional component in the likelihood to allow us to estimate it by maximum likelihood. This component is the likelihood conditional on n:

$$\mathcal{L}_y(\underline{\theta}) = \prod_{i=1}^{n} f(y_i) = \prod_{i=1}^{n} \frac{g(y_i)\pi(y_i)}{\int_0^w g(y)\pi(y)\, dy} = \prod_{i=1}^{n} \frac{g(y_i)\pi(y_i)}{P_a}, \quad (2.32)$$

where $\underline{\theta}$ is the vector of parameters of $g(y)$, the detection function.[1] The function $f(y)$ is the pdf of the observed y. The function $\pi(y)$ is the pdf of all y (whether or not they were observed).

When P_a is unknown, $\mathcal{L}_n(N_c)$ and $\mathcal{L}_n(N)$ depend on the unknown parameter vector $\underline{\theta}$ and we write them as $\mathcal{L}_n(N_c, \underline{\theta})$ and $\mathcal{L}_n(N, \underline{\theta})$. The likelihood for estimating N_c or density from the observed y is

$$\mathcal{L}_{n,y}(N_c, \underline{\theta}) = \mathcal{L}_n(N_c, \underline{\theta})\mathcal{L}_y(\underline{\theta})$$

$$= \binom{N_c}{n} (P_a)^n (1 - P_a)^{N_c - n} \prod_{i=1}^{n} f(y_i). \quad (2.33)$$

While we can estimate density or abundance in the covered region using this likelihood, it alone is inadequate for estimating density or abundance in the whole survey region. To do this, we can use the full likelihood:

$$\mathcal{L}_{n,y}(N, \underline{\theta}) = \mathcal{L}_n(N, \underline{\theta})\mathcal{L}_y(\underline{\theta}), \quad (2.34)$$

where $\mathcal{L}_n(N, \underline{\theta})$ is from eqn (2.25), but with $P_a = \int_0^w g(y)\pi(y)dy$ expressed as a function of $\underline{\theta}$.

Section 11.2 contains several examples of estimation using a full likelihood approach, together with results of a simulation study to compare the performance of profile likelihood, log-normal based, and standard confidence intervals.

However, because of the unrealistic independence assumptions implicit in $\mathcal{L}_{N_c}(N)$, inference is not usually conducted by maximizing the full

---

[1] We do not write $g(y)$ explicitly as a function of $\underline{\theta}$ to avoid the notation becoming more complicated than necessary.

likelihood with respect to N and $\underline{\theta}$. It is conventionally conducted in two stages:

1. The detection function parameter vector $\underline{\theta}$ is estimated from $\mathcal{L}_y(\underline{\theta})$.
2. Abundance N is estimated conditional on the estimate of $\underline{\theta}$ obtained from stage 1 using the Horvitz–Thompson-like estimator

$$\widehat{N} = \frac{n}{P_c \widehat{P}_a} = \frac{n}{P_c \int_0^w \hat{g}(y)\pi(y)\,dy}, \tag{2.35}$$

where $\hat{g}(y)$ is the detection function evaluated using the estimate of θ from stage 1 and $\pi(y)$ is the pdf of distances, which is assumed to be $\pi(y) = 1/w$ in the case of line transect surveys, and $\pi(y) = 2y/w^2$ in the case of point transect surveys.

2.4.3.1 *Interval distance data*
All the above assumes that distance data are exact. If they are gathered in J distance intervals (with cutpoints $c_0, c_1, \ldots, c_J = w$), the likelihood $\mathcal{L}_y(\underline{\theta})$ of eqn (2.32) is inappropriate. Instead, the following multinomial likelihood function is applicable:

$$\mathcal{L}_{y_G}(\underline{\theta}) = \left(\frac{n!}{\prod_{j=1}^J n_j!} \right) \prod_{j=1}^J p_j^{n_j}, \tag{2.36}$$

where $p_j = \int_{c_{j-1}}^{c_j} f(y)\,dy$ is the probability that a detected animal falls in interval j and n_j is the number of detected animals in interval j.

The full likelihood in this case is

$$\mathcal{L}_{n,y_G}(N, \underline{\theta}) = \mathcal{L}_n(N, \underline{\theta})\mathcal{L}_{y_G}(\underline{\theta}). \tag{2.37}$$

The same likelihood applies if exact distance data are grouped into J distance intervals before analysis.

2.5 Summary so far and preview of advances

2.5.1 *Summary*

You can think of distance sampling as involving the following three inference stages, each with a corresponding component of a full distance sampling likelihood function.

1. Estimation of detection probabilities and P_a: component $\mathcal{L}_y(\underline{\theta})$ (or $\mathcal{L}_{y_G}(\underline{\theta})$).

2. Estimation of the number, N_c, (or density) of animals in the covered region, given the detection probabilities and number observed, n: component $\mathcal{L}_n(N_c, \underline{\theta})$.

3. Estimation of the number, N, (or density) of animals in the survey region, given the number, N_c, (or density) in the covered region: component $\mathcal{L}_{N_c}(N, \underline{\theta})$. This inference step is conventionally design-based: the covered region is representative of the whole survey region in the sense that each point in the survey region was equally likely to be in the covered region, or (with unequal coverage probability designs) falls in the covered region with a known probability determined by the design.

Inference is usually conducted by estimating the detection function (often by maximum likelihood from stage 1 above) and then estimating density or abundance conditional on the estimated detection function, using a Horvitz–Thompson-like estimator of the following form

$$\widehat{N} = \frac{n}{P_c \widehat{P}_a}, \qquad (2.38)$$

where $\widehat{P}_a$ is the estimated mean detection probability obtained using the estimated detection function.

2.5.2 *Preview of advances*

The advanced methods covered in this book generalize one or more of the components above. They include:

1. $\mathcal{L}_y(\underline{\theta})$ (or $\mathcal{L}_{y_G}(\underline{\theta})$): methods that generalize detection functions.

   (a) Methods that incorporate covariates other than distance into the detection function and estimation process, while retaining the assumption that detection at distance zero is certain.

   (b) Methods for cases in which animals at distance zero may be missed. These methods involve mark-recapture-like detection data and covariates other than distance to estimate the probability of detecting animals at distance zero.

   (c) Methods that accommodate errors in measuring distances.

2. $\mathcal{L}_n(N_c, \underline{\theta})$ and $\mathcal{L}_{N_c}(N, \underline{\theta})$: methods that model animal density within and outside the covered region as a smooth surface. This allows estimation of density 'hot spots' and 'cold spots' and estimation of empirical relationships between explanatory variables and animal density.

3. $\mathcal{L}_{N_c}(N, \underline{\theta})$: methods that accommodate variable coverage probability designs:

   (a) Methods that accommodate survey designs with smoothly varying but different coverage probabilities in different parts of the survey region. Methods are also developed to automate survey design, and designs are developed to give equal or approximately equal coverage probabilities in irregularly shaped survey regions while minimizing off-effort time.

   (b) Adaptive distance sampling design and analysis methods; these involve coverage probabilities that depend on the history of detections at any point in the survey.

Inference with the advanced methods is seldom by maximization of the full likelihood. More often it involves maximum likelihood estimation from some component of the full likelihood and abundance or density estimation using Horvitz–Thompson-like estimators. Nevertheless, the full likelihood can be developed for most of the advanced methods and we outline both the likelihood development and the estimation process in what follows.

2.6 Advanced methods for detection function estimation

2.6.1 *Multiple covariate distance sampling*

In reality, detection probability does not depend on distance only. It may depend on the ability of the surveyor, the characteristics of the individual animals, environmental or weather conditions, and a host of other factors. However, when animals at zero distance are detected with certainty ($g(0) = 1$), then providing that the fitted detection function model is flexible enough, distance sampling estimators of abundance and density are unbiased even though all things other than distance are ignored in estimating detection probability (Section 11.12). This property, known as 'pooling robustness', is a very powerful feature of distance sampling methods.

Nevertheless, it is sometimes useful to model the detection probability as a function of variables other than distance. This might be true in the following cases:

(1) density is correlated with detection probability;

(2) a large component of the variance of the abundance estimate is due to estimation of the detection function, and this variance can be 'explained' by variables other than distance; and

(3) detection probability changes across strata but there are inadequate detections in some strata to allow separate estimation of detection probability within each stratum.

In these cases, inference might be improved by estimating a detection function $g(y, \underline{z})$, which depends on both distance y and some appropriate vector of explanatory variables $\underline{z}$. Inference for this case requires generalization of the associated likelihood functions and estimators (Chapter 3).

2.6.1.1 *Likelihood function*
The conventional detection function component of the likelihood, eqn (2.32), is easily generalized to accommodate additional explanatory variables $\underline{z}$:

$$\mathcal{L}_{y,z}(\underline{\theta}) = \prod_{i=1}^{n} f(y_i, \underline{z}_i) = \prod_{i=1}^{n} \frac{g(y_i, \underline{z}_i)\pi(y_i, \underline{z}_i)}{P_a}, \qquad (2.39)$$

where $\pi(y, \underline{z})$ is the joint pdf of y and $\underline{z}$ in the population in the covered region (both detected and undetected),

$$P_a = \int_Z \int_0^w g(y, \underline{z})\pi(y, \underline{z}) \, dy d\underline{z} \qquad (2.40)$$

and $\int_Z \ldots d\underline{z}$ means integration (or summation if $\underline{z}$ is discrete) over all values of $\underline{z}$.

However, inference from this likelihood is problematic because $\pi(y, \underline{z})$ is unknown. Recall that with CDS, $\pi(y)$ is treated as known—based on the assumption that animals are distributed uniformly in the vicinity of the observer, and independently of the observer. This is usually a reasonable assumption (unless, for example, animals respond to the observer before detection), whereas generally there is no reasonable basis for assuming any particular known pdf for the covariates $\underline{z}$.

If values of y and $\underline{z}$ in the population are independent however, we can factorize $\mathcal{L}_{y,z}(\underline{\theta})$ into two components, one of which does not involve the pdf for $\underline{z}$, so that we can base inference on this component. The factorization is as follows:

$$\begin{aligned}
\mathcal{L}_{y,z}(\underline{\theta}) &= \mathcal{L}_z(\underline{\theta})\mathcal{L}_{y|z}(\underline{\theta}) \\
&= \left[\prod_{i=1}^{n} \frac{P_a(\underline{z}_i)\pi(\underline{z}_i)}{P_a}\right]\left[\prod_{i=1}^{n} \frac{g(y_i, \underline{z}_i)\pi(y_i \mid \underline{z}_i)}{P_a(\underline{z}_i)}\right], \qquad (2.41)
\end{aligned}$$

where $\pi(y \mid \underline{z})$ is the conditional pdf of y given $\underline{z}$, and

$$P_a(\underline{z}_i) = \int_0^w g(y, \underline{z}_i)\pi(y \mid \underline{z}_i) \, dy. \qquad (2.42)$$

Given population independence of y and $\underline{z}$ (achieved through random line placement), then $\pi(y \mid \underline{z}_i) = \pi(y)$ and $\mathcal{L}_{y|z}(\underline{\theta})$ involves only the pdf of y

which can be treated as known, in the same way as it is with conventional distance sampling. In this case, the detection function parameter vector $\underline{\theta}$ and $g(y, \underline{z})$ can be estimated from $\mathcal{L}_{y|z}(\underline{\theta})$ with no additional assumptions about the distribution of $\underline{z}$.

For interval distance data, a likelihood $\mathcal{L}_{y_G|z}(\underline{\theta})$ can be developed to replace $\mathcal{L}_{y|z}(\underline{\theta})$ in eqn (2.41).

2.6.1.2 *Estimation*
Maximum likelihood estimation of N from the full likelihood function

$$\mathcal{L}_{n,y,z}(N, \underline{\theta}) = \mathcal{L}_n(N, \underline{\theta}) \mathcal{L}_z(\underline{\theta}) \mathcal{L}_{y|z}(\underline{\theta}) \qquad (2.43)$$

necessarily involves some assumptions about $\pi(\underline{z})$. In order to estimate N without making these assumptions, a Horvitz–Thompson-like estimator can be used in preference to a full likelihood approach. However, P_a cannot be used in the inclusion probability because it involves $\pi(\underline{z})$: $P_a = \int_Z P_a(\underline{z}) \pi(\underline{z}) d\underline{z}$, where $P_a(\underline{z})$ is obtained from eqn (2.42). This obstacle is easily overcome by using $P_a(\underline{z}_i)$ in the inclusion probability—that is, we use the inclusion probability for individual i averaged over y only, rather than the mean inclusion probability for all animals in the covered region, averaged over y and $\underline{z}$. Assuming coverage probability P_c is constant, the abundance estimator is

$$\widehat{N} = \frac{1}{P_c} \sum_{i=1}^{n} \frac{1}{\widehat{P}_a(\underline{z}_i)}, \qquad (2.44)$$

where $\widehat{P}_a(\underline{z}_i)$ is found by evaluating eqn (2.42) using the estimated detection function parameter vector $\widehat{\underline{\theta}}$.

When animals are detected in clusters, and n clusters are detected, eqn (2.44) estimates cluster abundance. Individual abundance can be estimated by multiplying this $\widehat{N}$ by an estimate of mean cluster size in the population, $\widehat{E}[s]$, or by using cluster size, s_i instead of one in the numerator. In the latter case,

$$\widehat{E}[s] = \frac{\sum_{i=1}^{n} s_i / \widehat{P}_a(\underline{z}_i)}{\sum_{i=1}^{n} 1 / \widehat{P}_a(\underline{z}_i)}. \qquad (2.45)$$

2.6.2 *Mark-recapture distance sampling*

A key assumption of CDS methods is that detection of animals at distance zero is certain. This assumption is necessary because the distance data tells you only about relative detectability at different distances (the shape of the detection function), not about absolute detectability at any distance (the detection function intercept). When the assumption fails, the degree to which CDS estimators underestimate abundance is in direct proportion to the true detection probability at distance zero. In some applications, the probability of detection at distance zero can be 0.3 or lower, making CDS estimators negatively biased by a factor of three or more. In these cases, estimation of detection probability at zero distance is important.

The most successful methods for dealing with uncertain detection at zero distance use surveys that gather mark-recapture-like data. Two (or more) independent observers are used and each can be thought of as 'marking' animals for the other. Providing animals detected by both observers can be identified, surveys of this kind yield capture history data in addition to distance data. With two independent observers, the possible observed capture histories are

$\underline{\omega} = (1, 0)$: detection by observer 1 but not observer 2;
$\underline{\omega} = (0, 1)$: detection by observer 2 but not observer 1;
$\underline{\omega} = (1, 1)$: detection by both observers (a 'duplicate' detection).

We denote the capture history of the ith detected animal $\underline{\omega}_i = (\omega_{i1}, \omega_{i2})$, where $\omega_{ij} = 1$ if observer j detected the animal, and $\omega_{ij} = 0$ otherwise $(j = 1, 2)$.

2.6.2.1 *Likelihood function*
A new likelihood component is required to deal with the capture history data. It is constructed using the conditional probability of observing capture history $\underline{\omega}$, given that the animal was observed by at least one of the observers, as follows:

$$\mathcal{L}_\omega(\underline{\theta}) = \prod_{i=1}^{n} \frac{p_{\underline{\omega}_i}(y_i, \underline{z}_i)}{p_\cdot(y_i, \underline{z}_i)}, \tag{2.46}$$

where $p_{\underline{\omega}_i}(y_i, \underline{z}_i)$ is the probability of observing the capture history $\underline{\omega}_i$ for animal i and $p_\cdot(y_i, \underline{z}_i)$ is the probability that animal i is detected by at least one of the observers.

Note that

1. We use $p_\cdot(y, \underline{z})$ instead of $g(y, \underline{z})$ here in order to distinguish detection function models that have certain detection at the line or point $(g(0, \underline{z}) = 1)$ from those that do not $(p_\cdot(0, \underline{z}) \leq 1)$.

2. We model detection probability as a function of both y and other explanatory variables, $\underline{z}$. Incorporating explanatory variables is more important with mark-recapture-like data than with CDS data because mark-recapture estimators of capture probability (detection probability in our terms) and abundance are notoriously biased if variables

that affect detection probability are neglected—a phenomenon called 'unmodelled heterogeneity'.

A likelihood analogous to $\mathcal{L}_\omega(\underline{\theta})$ can be developed for the case in which distances are grouped into intervals.

The full likelihood function when capture history data are included is

$$\mathcal{L}_{n,y,z,\omega}(N,\underline{\theta}) = \mathcal{L}_n(N,\underline{\theta})\mathcal{L}_z(\underline{\theta})\mathcal{L}_{y|z}(\underline{\theta})\mathcal{L}_\omega(\underline{\theta}). \tag{2.47}$$

Inferences about components of $\underline{\theta}$ relating to the intercept of the detection function can be based solely on the mark-recapture component $\mathcal{L}_\omega(\underline{\theta})$ of this likelihood.

2.6.2.2 *Estimation*

It is possible to estimate N and $\underline{\theta}$ simultaneously by maximizing $\mathcal{L}_{n,y,z,\omega}(N,\underline{\theta})$ with respect to both, and this is sometimes done. This approach does, however, require that $\pi(y,\underline{z})$, the joint pdf of y and $\underline{z}$, be specified—at least up to some unknown parameters.

An alternative is to use a Horvitz–Thompson-like estimator. The estimator

$$\widehat{N} = \frac{1}{P_c}\sum_{i=1}^{n}\frac{1}{\widehat{P}_a(\underline{z}_i)} = \frac{1}{P_c}\sum_{i=1}^{n}\frac{1}{\int_0^w \hat{p}(y,\underline{z}_i)\pi(y)\,dy}, \tag{2.48}$$

where $\hat{p}(y,\underline{z}_i)$ is the estimated probability that at least one of the observers detects an animal at y with variables $\underline{z}_i$, has less small-sample bias and higher precision than a Horvitz–Thompson-like estimator which uses $\hat{p}(y_i,\underline{z}_i)$ in the denominator.

If animals are detected in clusters and n clusters are detected, eqn (2.48) is an estimator of cluster abundance. It can be converted into an estimator of animal abundance either by multiplying it by an estimate of mean cluster size in the population or by using cluster size s_i instead of one in the numerator.

2.6.3 *Estimation when $\pi(y)$ is unknown*

In CDS, the detection function $g(y)$ and the pdf of distances $\pi(y)$ are completely confounded. The pdf of observed distances, $f(y)$, is proportional to the product of these two functions. Only if we know $\pi(y)$ can we extract the shape of $g(y)$ from $f(y)$. Put another way: without knowledge of $\pi(y)$, there is no way we can interpret a decline (or rise) in the observed number of detections with distance as a decline (or rise) in the detection probability $g(y)$—it might equally well be due to a decline (or rise) in $\pi(y)$. When animals are attracted to the observer before detection, for example, we draw incorrect inferences about $g(y)$ (we infer that it falls off with distance faster than it really does) because $\pi(y)$ is not what we assume it to be (it has higher density near the observer).

Knowledge of $\pi(y)$ is central to reliable inference from CDS methods.

This is not the case with mark-recapture distance sampling (MRDs) methods because the detection function $p.(y)$ or $p.(y, \underline{z})$ can be estimated from the likelihood component $\mathcal{L}_\omega(\underline{\theta})$ alone, and this does not involve $\pi(y)$. As a consequence, these methods need not treat $\pi(y)$ as known. In fact, given a functional form for $\pi(y)$ with unknown parameter vector $\underline{\phi}_y$, the distribution of animals with respect to the observer $(\pi(y))$ can be estimated from the likelihood components $\mathcal{L}_{y|z}(\theta, \underline{\phi}_y) \times \mathcal{L}_\omega(\underline{\theta})$. Alternatively, it can be estimated from the full likelihood.

Mark-recapture distance surveys can therefore be used to check whether animals are distributed as assumed in CDS methods (i.e. uniformly in space in the vicinity of the observer) and/or to estimate their distribution when it is unknown. This allows the likelihood components $\mathcal{L}_{y|z}(\theta, \underline{\phi}_y) \times \mathcal{L}_\omega(\underline{\theta})$ to be used for migration surveys of cetaceans, for example, where the distribution of distances is known to be different from that normally assumed in distance sampling (few animals pass in shallow water very close to a shore-based observer) but its exact shape is unknown (Section 11.8).

2.7 Estimating animal density surfaces

It is often the case that animals are not uniformly distributed through the survey region. Survey regions typically span regions of suitable habitat in which animal density tends to be high, and less suitable habitat when animal density tends to be lower.

It is also often the case that there is interest in the relationships between habitat variables and density, and in identifying regions of high and low density. CDS methods allow inferences of this sort only to the resolution of strata (some strata will have higher density than others). To get higher resolution, more strata are required, and the number of strata that can be used is always constrained by sample size. Fitting a smooth density surface to distance data is a way of obtaining high-resolution estimates of density and drawing inferences about the empirical relationships between habitat variables and density without using many parameters.

Two forms of density surface estimator have been developed. The first involves a generalization of the likelihoods $\mathcal{L}_n(N)$ and $\mathcal{L}_n(N_c)$ of eqns (2.25) and (2.23); the second involves a generalization of the corresponding likelihoods in which distances between detections rather than number of detections is treated as the response.

Both methods are model-based methods: inferences about density outside the covered region are based on the estimated density surface, not on the survey design. The stepwise approach to inference using a Horvitz–Thompson-like estimator to estimate abundance given estimated detection probability, as described in Section 2.5.1, is therefore not an option.

2.7.1   The count method

Let (u, v) be Cartesian coordinates used to locate points within the survey region and suppose that animal density varies as a smooth function $D(u, v)$ over the survey region; we assume that $D(u, v)$ has an unknown parameter vector $\underline{\phi}$ associated with it. The pdf of animal locations in the survey region can then be written as

$$\pi(u, v) = \frac{D(u, v)}{\iint D(u, v)\, dudv},\tag{2.49}$$

where integration is over the whole survey region.

Assuming that animal location is independent of $\underline{z}$, the mean probability of detecting an animal with explanatory variables $\underline{z}$ in this case is

$$P_a(\underline{z}) = \iint p(u, v, \underline{z})\pi(u, v)\, dudv,\tag{2.50}$$

where $p(u, v, \underline{z})$ is the detection function expressed in terms of (u, v)—something that can easily be done once the line or point transects have been placed. $P_a(\underline{z})$ depends on the unknown parameter vector $\underline{\phi}$, although this is not shown explicitly above.

The likelihood functions $\mathcal{L}_n(N)$ and $\mathcal{L}_n(N_c)$ now involve $P_a(\underline{z})$ of eqn (2.50) and hence the additional density surface parameter vector $\underline{\phi}$. Maximizing the associated full likelihood with respect to N, $\underline{\theta}$, and $\underline{\phi}$ yields an estimate of the density surface $D(u, v)$. If explanatory variables relating to habitat are included in addition to (u, v), estimates of the empirical relationships between these variables and density are obtained.

In practice it can be easier to estimate $\underline{\phi}$ by methods other than maximum likelihood. One such approach involves estimating N_c at each point transect or in each of many small segments of line transect (either by maximizing $\mathcal{L}_n(N_c)$ or by a Horvitz–Thompson-like estimator, conditional on $\hat{\underline{\theta}}$) and then using generalized linear or additive regression models to fit a flexible $D(u, v)$ surface.

2.7.2   The waiting distance method

This approach is based on the fact that if the number of events (detections in our terms) per unit distance is a Poisson random variable with parameter λ, the distances between events are exponentially distributed random variables with parameter λ. Fitting a density surface when count is the response can be awkward because counts are necessarily over some covered area, but explanatory variables may change over this area and it is not always clear what value of the explanatory variables in the area should be used. Working in terms of waiting distances is convenient because events (detections) occur at points and there is no ambiguity about explanatory variable values at points. See Chapter 4 for details.

2.7.3 *Cluster size surface estimation*

Care has to be taken when estimating a density surface for individuals from detections of clusters. If mean cluster size changes in space, then estimating a cluster abundance surface and multiplying it by a single estimate of mean cluster size in the survey region will result in a biased individual density surface (too high in regions where mean cluster size is small and too low in regions where mean cluster size is large). One way of dealing with this is to model the mean cluster size surface. Suppose that $\underline{z}$ is a scalar s, representing cluster size. If $\pi(\underline{z}) \equiv \pi(s)$ is parameterized as a function of location, (u, v), with some unknown parameter vector $\underline{\phi}_s$, an estimate of $\pi(s)$ can be obtained by maximizing the full likelihood with respect to all parameters. Then $\hat{\pi}(s)$ is an estimated mean cluster size surface. Alternatively, if estimation is being performed using the count method, estimating N_c at each point or in each subsection of a line, N_c can be estimated using observed cluster size s_i in the numerator of the Horvitz–Thompson-like estimator. In this case, $\widehat{N}_c$ is an estimate of individual abundance and the density surface fitted to the $\widehat{N}_c$ is that for individuals.

2.8   Survey design

The coverage probability P_c is central to estimation of N. Without it, we can estimate the abundance in the covered region, N_c, but we would have no reasonable basis for estimating N from $\widehat{N}_c$ (stage 3 in Section 2.5.1).

In all of the above, coverage probability has been assumed to be equal throughout the survey region. This can quite often be difficult to achieve, because of constraints on use of the survey platform and/or because the survey region has a very irregular shape. If coverage probability is not the same everywhere but is treated as equal in analysis, abundance estimates may be biased (sometimes severely biased). The ability to deal with unequal coverage probabilities is essential in these circumstances.

2.8.1 *Likelihood-based inference*

Suppose that instead of being equal to P_c everywhere, coverage probabilities could be different for every location in the survey region. In this case, the binomial (or Poisson) likelihood function components for n are inappropriate because the probability that an animal is in the covered region is different for different animals—it depends on where they are in the survey region. However, for any but very simple functions (such as a stratified design with different coverage probability between strata but constant coverage probability within strata), the appropriate likelihood is quite complicated and can be very difficult to evaluate and maximize.

2.8.2 *Design-based inference*

Distance sampling methods conventionally infer $\widehat{N}$ from $\widehat{N}_c$ using design-based methods, not likelihood-based methods. When P_c is a function of location, design-based methods of estimating N given $\widehat{N}_c$ are no more difficult than for the equal coverage probability case, provided that P_c can be calculated for every detected animal.

Given a survey design (a set of probabilistic rules determining where the covered region falls) with unequal coverage probabilities, it is relatively easy to evaluate coverage probability for any point by simulation. You simulate many realizations of the design and calculate the proportion of these on which the point in question falls within the covered region.

Suppose we have the coverage probabilities for detected animals, $P_c(u_i, v_i)$, where (u_i, v_i) are the coordinates for the location of detected animal i, and estimated mean detection probabilities $\widehat{P}_a(\underline{z}_i)$ ($i = 1, \ldots, n$). Then abundance can be estimated by

$$\widehat{N} = \sum_{i=1}^{n} \frac{1}{P_c(u_i, v_i)\widehat{P}_a(\underline{z}_i)}.$$ (2.51)

In the case of detection of clusters, an estimate of individual abundance can be obtained by using observed size s_i in place of one in the numerator, or by multiplying the above estimate by an estimate of mean cluster size in the population.

These issues are addressed in Chapter 7.

2.8.3 *Adaptive distance sampling*

When sampling is adaptive (i.e. the design can change as new detections are made), developing a full likelihood is challenging and of questionable utility. Adaptive methods were developed in a sampling theory context using either Horvitz–Thompson or Hansen–Hurwitz estimators. Adaptive distance sampling methods modify these estimators for application with line and point transect data (Chapter 8).

Development of these methods was motivated by efficiency considerations: when surveying a rare species known to aggregate, it seems inefficient not to spend more time and effort sampling the area surrounding detections because it is likely that other animals will be in the vicinity, and every detection of a rare species is valuable. However, use of conventional design-based inference methods can result in large bias in this case because the coverage probability of animals detected during the additional sampling triggered by a detection is higher than for other animals. The methods of Thompson and Seber (1996) have been modified somewhat for application to distance sampling surveys and have been found to increase precision substantially in the case of markedly clustered populations. When populations are less clustered and/or sample size from conventional designs is higher, the gains of adaptive distance sampling are smaller.

2.9 Model selection

Estimation by maximum likelihood allows you to use the powerful set of model selection methods associated with maximum likelihood methods. These include likelihood ratio (LR) tests, Akaike's information criterion (AIC; Akaike 1973, 1985) and related criteria; see Buckland *et al.* (2001: 68–71) for more details.

If inference is based on an estimated detection function obtained from the likelihood function $\mathcal{L}_y(\underline{\theta})$ or $\mathcal{L}_{y_G}(\underline{\theta})$ only and the distribution of distances $\pi(y)$ is treated as known, model selection involves only the selection of the detection function, as described in Buckland *et al.* (2001). If likelihood function components other than that for the detection function are used in inference, model selection involves all these components. In particular, likelihood functions could include one or more of the following models,[2] each with its own set of parameters to be estimated:

$\pi(y)$: a model for the distribution of distances to animals,
$\pi(\underline{z})$: a model for characteristics of the animals or animal clusters (size, for example),
$\pi(u, v)$: a model for the distribution of animals in the survey region (where (u, v) represents location),
$\pi(y_o \,|\, y)$: a distance estimation error model (where y_o is observed distance and y true distance),
$\pi(v)$: a model for the availability process; availability is symbolized by v here. (see Chapter 6 for more on availability.)

LR tests, AIC and related criteria can be used for selecting each and all of these models.

---

[2] Other formulations are possible; in particular, Cooke and Leaper (unpublished) developed a more general formulation than that described here.

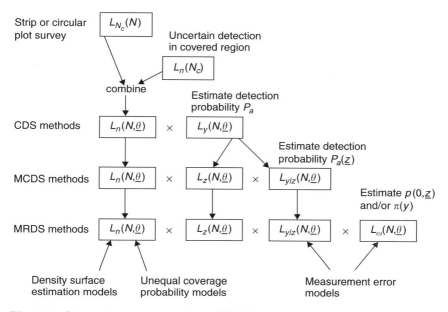

Fig. 2.1. Schematic representation of likelihood components for conventional and advanced distance sampling methods.

2.10 Summary

This chapter describes how full likelihood functions that apply to conventional distance sampling methods can be generalized to accommodate the various advanced methods described in later chapters of this book. A schematic summary of the generalizations is shown in Fig. 2.1.

Reasons to conduct inference based on the full likelihood include:

1. The resulting estimators of N enjoy all the properties of MLEs, including asymptotic unbiasedness and efficiency.
2. Well-developed likelihood-based theory for profile likelihood intervals[3] for N and for model selection (such as AIC) are then available.
3. Full likelihoods are a necessary part of a Bayesian approach to distance sampling.

Disadvantages of the full likelihood approach include:

1. The full likelihood functions are often based on unrealistic assumptions about independence of animal locations.

[3] See Section 11.2 for a description of profile likelihood confidence intervals.

2. Full likelihood functions require more intensive modelling of the probability mechanisms governing the state of the population (where they are, cluster sizes and other characteristics) and we may have little information to decide what models are appropriate.

3. Full likelihood functions can be difficult to evaluate and maximize.

Horvitz–Thompson-like estimators provide a versatile means of estimating abundance and density without having to resort to full likelihood maximization. Maximization of likelihood function components relating to the detection function provides estimates of detection probability for each detected animal, and coverage probabilities can be evaluated by simulation or, with equal coverage probability designs, analytically. A general form of the Horvitz–Thompson-like estimator for distance sampling has the form

$$\widehat{N} = \sum_{i=1}^{n} \frac{s_i}{P_c(u_i, v_i)\widehat{P}_a(\underline{z}_i)}, \qquad (2.52)$$

where $P_c(u_i, v_i)$ is the coverage probability of a detected animal located at Cartesian coordinates (u_i, v_i) in the survey region, $\widehat{P}_a(\underline{z}_i)$ is the estimated detection probability for an animal or cluster with characteristics $\underline{z}_i$, and s_i is the cluster size of the ith detected unit. If individuals are the detection unit, $s_i \equiv 1$. The s_i for detected clusters are generally assumed to be recorded without error. If, for example, estimates of s_i were made by more than one observer, a model could be fitted to these estimates, and a predicted $\hat{s}_i$ obtained for each detection, giving

$$\widehat{N} = \sum_{i=1}^{n} \frac{\hat{s}_i}{P_c(\underline{u}_i)\widehat{P}_a(\underline{z}_i)}. \qquad (2.53)$$

In CDS, we use a common estimate of mean cluster size, and $\widehat{N}$ is obtained by substituting $\hat{s}_i = \widehat{E}[s]$ in eqn (2.53).

If Bayesian methods are to be fully developed for distance sampling, they will require full likelihoods to augment priors on the parameters. In either case of a full likelihood or a Bayesian approach, there is a need for numerical optimization and integration methods, possibly on objective functions with many parameters.

3
Covariate models for the detection function
F. F. C. Marques and S. T. Buckland

3.1 Introduction

In standard line transect sampling, we assume that probability of detection $g(y)$ of an animal is solely a function of its distance y from the line. We then use models for the detection function that are 'pooling robust'; that is, they yield reliable estimates of density when heterogeneity in the detection probabilities due to covariates other than distance is ignored (Section 11.12). Provided $g(0) = 1$, the models in distance have this property. However, if separate estimates of density D are required for different strata, and the sightings data are pooled across strata, then heterogeneity causes bias in these stratum-specific estimates. (Recall that $\widehat{D} = n\hat{f}(0)/2L$ for unclustered populations, where L is total line length.)

Bias in stratum-specific estimates due to heterogeneity can be eliminated by estimating $f(0)$ separately by stratum. However, independent estimation of $f(0)$ by stratum requires an adequate sample size in each stratum. If this is not possible, then a better option is to model $g(y)$, and hence $f(0)$, as a function of covariates.

Beavers and Ramsey (1998) proposed regressing the logarithm of the observed perpendicular distances on the covariates and adjusting these distances to average covariate values using the estimated regression parameters. The main advantage of this approach is that it allows standard software to be used for the estimation of $f(0)$, by analyzing the adjusted distances (e.g. Fancy 1997). Alternatively, the covariates can be directly incorporated into the estimation procedure via a multivariate detection function (Ramsey *et al.* 1987). In this chapter, we summarize the work of Marques (2001) and of Marques and Buckland (2003), which extends the approach of Ramsey *et al.* to the 'key + series adjustment' methodology of Buckland (1992a), and we extend the work to point transect sampling.

3.2   A conditional likelihood framework for distance sampling

Direct incorporation of covariates into the standard distance sampling detection function requires estimation of the joint density of the observed perpendicular distances x (line transect sampling) or sighting distances r (point transect sampling) and associated covariates $\underline{z}$ ($\underline{z} = z_1, \ldots, z_q$). Using y to represent distance x or r, we have

$$f(y, \underline{z}) = \frac{g(y, \underline{z})\pi(y, \underline{z})}{\iint g(y, \underline{z})\pi(y, \underline{z})\, dy d\underline{z}}, \tag{3.1}$$

where $g(y, \underline{z})$ is a multivariate detection function representing the probability of detecting an animal at distance y from the line or point and with covariates $\underline{z}$, and $\pi(y, \underline{z})$ is the joint density of y and $\underline{z}$ in the population. Assuming that y and $\underline{z}$ are independent (which holds under random line or point placement), then $\pi(y, \underline{z}) = \pi(y)\pi(\underline{z})$, where $\pi(y)$ and $\pi(\underline{z})$ denote the densities of the y and $\underline{z}$ respectively, so that we have:

$$f(y, \underline{z}) = \frac{g(y, \underline{z})\pi(y)\pi(\underline{z})}{\iint g(y, \underline{z})\pi(y)\pi(\underline{z})\, dy d\underline{z}}. \tag{3.2}$$

The density $\pi(\underline{z})$ is usually not known, and so must be either estimated or factored out.

For univariate z, Chen (1996) proposed a bivariate density estimator based on the product of two Gaussian kernels. The density $f(x, z)$ is then directly estimated from the data, without the need to assume any parametric form for it. A similar method based on a single Gaussian kernel has been proposed by Mack and Quang (1998). Both methods were developed with the primary aim of estimating mean cluster size and its effect on detectability. Although they seem to perform relatively well, they suffer from a few disadvantages. One difficulty is the choice of bandwidth h. For the estimators of both Chen (1996) and Mack and Quang (1998), $h = \hat{\sigma}n^{-\delta}$, where $\hat{\sigma}$ is the sample standard deviation of observed distances, n corresponds to the total number of detected animals, and δ is a constant. The choice of value for δ depends on the criterion being used to derive unbiased distributional properties. As the rate of shrinkage of the bandwidth varies according to the criterion being used, confidence intervals may be biased (Mack and Quang 1998). Related to this, density estimates tend to be sensitive to the choice of bandwidth (Buckland 1992a). Second, as the kernel method is based on local averaging of observations, it is more likely to produce biased estimates when the detection function is not very smooth near $y = 0$, or when small distances tend to be rounded to zero (Buckland 1992a). Finally, at least one of the methods (Mack and

Quang 1998) requires relatively large sample sizes ($n \geq 70$) for unbiased estimation of mean cluster size.

An alternative approach that does not require knowledge of $\pi(\underline{z})$ is maximum likelihood estimation of the conditional density of y, given the observed $\underline{z}$. Following the derivation of Borchers (1996), we have:

$$f(\underline{z}) = \int f(y, \underline{z}) \, dy = \frac{\pi(\underline{z}) \int g(y, \underline{z}) \pi(y) \, dy}{\iint g(y, \underline{z}) \pi(y) \pi(\underline{z}) \, dy d\underline{z}} \qquad (3.3)$$

and

$$f(y \mid \underline{z}) = \frac{f(y, \underline{z})}{f(\underline{z})} = \frac{g(y, \underline{z}) \pi(y)}{\int g(y, \underline{z}) \pi(y) \, dy} \qquad (3.4)$$

so that the conditional likelihood is given by:

$$\mathcal{L}(\underline{\theta}; \underline{y}, \underline{z}) = \prod_{i=1}^{n} f(y_i \mid \underline{z}_i) = \prod_{i=1}^{n} \frac{g(y_i, \underline{z}_i) \pi(y_i)}{\int g(y, \underline{z}_i) \pi(y) \, dy}. \qquad (3.5)$$

3.3 Line transect sampling

3.3.1 *The conditional likelihood*

We first develop the methodology for line transect sampling (so that y is the perpendicular distance from the line, x), before considering point transect sampling more briefly. Given random line placement, $\pi(x) = 1/w$, $0 \leq x \leq w$, and with $y = x$, eqn (3.2) is reduced to:

$$f(x, \underline{z}) = \frac{g(x, \underline{z}) \pi(\underline{z})}{\iint g(x, \underline{z}) \pi(\underline{z}) \, dx d\underline{z}}. \qquad (3.6)$$

Similarly, substitution of $\pi(x) = 1/w$ into eqns (3.4) and (3.5) gives the conditional distribution of x, given covariates $\underline{z}$ as

$$f(x \mid \underline{z}) = \frac{f(x, \underline{z})}{f(\underline{z})} = \frac{g(x, \underline{z})}{\mu(\underline{z})}, \qquad (3.7)$$

where $\mu(\underline{z}) = \int g(x, \underline{z}) \, dx$, and the conditional likelihood as

$$\mathcal{L}(\underline{\theta}; \underline{x}, \underline{z}) = \prod_{i=1}^{n} f(x_i \mid \underline{z}_i) = \prod_{i=1}^{n} \frac{g(x_i, \underline{z}_i)}{\mu(\underline{z}_i)}. \qquad (3.8)$$

Several authors have used this conditional approach to develop estimators for $f(0)$ (Drummer and McDonald 1987; Ramsey *et al.* 1987;

Quang 1991). Quang (1991) modified the Fourier series model of Crain *et al.* (1979) to include an additional variable other than perpendicular distance. All other estimators assume that the covariates enter the detection function via the scale parameter. Empirical evidence (Otto and Pollock 1990) suggest that this is a reasonable approach, as long as detection on the line is certain (i.e. $g(0, \underline{z}) = 1$). In the case where covariates affect probability of detection on the line, methods that allow this probability to be less than one should be used instead. The methods developed here to incorporate covariates into the scale of the detection function are nevertheless useful for surveys in which detection probability on the line is less than one (Chapter 6).

Ramsey *et al.* (1987) formulated an estimator for line or point transect surveys based on the density of detection areas a_i, where for point transect sampling $a_i = \pi r_i^2$, where r_i is the distance from the point to the ith detected animal. For line transect sampling, modelling of detected areas is equivalent to the modelling of the observed perpendicular distances themselves. Thus, following the notation used throughout this chapter, and using the observed perpendicular distances in place of detected areas, the estimator of Ramsey *et al.* has the form $g(x, \underline{z}) = G\left(x/\mu(\underline{z})\right)$. Here $G(\cdot)$ is a detection function with properties such that the conditional density of the observed perpendicular distances given the associated covariates, $f(x \mid \underline{z})$, is given by:

$$f(x \mid \underline{z}) = \frac{G\left(x/\mu(\underline{z})\right)}{\mu(\underline{z})}. \tag{3.9}$$

Thus the estimator of Ramsey *et al.* is equivalent to the conditional expression from eqn (3.7). By assuming that $\log_e(\mu(\underline{z})) = \beta_0 + \sum_{j=1}^{q} \beta_j z_j$, and that $f(x \mid \underline{z})$ follows an exponential power series distribution with a single shape (power) parameter γ, then $f(x \mid \underline{z}) = \exp[-\{x/\mu(\underline{z})\}^\gamma]$ is estimated by maximizing the log-likelihood with respect to the $q + 1$ covariate parameters and the shape parameter γ, conditional on the observed values of the covariates.

The bivariate density estimator proposed by Drummer and McDonald (1987) is based on the assumption that $\int g(x, z)dx = c \cdot \int g(x)dx$, with $c = z^\beta$. As in the formulation of Ramsey *et al.*, this approach is a special case of eqn (3.7).

It can easily be seen that both $\mu(\underline{z})$ and c, denoted in the equation below by σ, have the form (Borchers 1996):

$$\sigma = \exp\left(\beta_0 + \sum_{j=1}^{q} \beta_j z_j^*\right) \tag{3.10}$$

with $z_j^* = z_j$ for $\mu(\underline{z})$ of Ramsey *et al.*, and $\beta_0 = 0$, $z_j^* = \log_e(z_j)$ and $q = 1$ for Drummer and McDonald's c. Palka (1993) applied Drummer and McDonald's method parameterized as in eqn (3.10) to fit a bivariate detection function based on the hazard-rate model of Hayes and Buckland (1983) and estimate the effect of school size on the detectability of harbour porpoise.

Quang (1991) applied the conditional likelihood methods described above using a series representation for $f(x \,|\, \underline{z})$, while Drummer and McDonald (1987) and Ramsey *et al.* (1987) used parametric functional forms for the detection function $g(x, z)$. Buckland (1992a,b) developed a unifying framework where the two approaches are combined to model $f(0)$ in the case where the probability of detection is a function of the perpendicular distances alone. Buckland's approach involves the fitting of a known parametric form (the 'key function') to the data, with additional adjustment terms used when necessary to improve the fit. This semiparametric approach is very flexible, and so is 'model robust' in the sense of Burnham *et al.* (1980). In addition, as model parameters are estimated by maximum likelihood, likelihood-based model selection criteria can be employed.

We propose a generalization of Buckland's approach in which covariates are incorporated into the estimation of the detection probabilities via the scale parameter σ of eqn (3.10). In this formulation, the covariates are assumed to affect the rate at which detectability decreases as a function of distance. Depending on the choice of standardization of distance in the adjustment terms, covariates need not influence the shape of the detection function. Unlike kernel-based methods, this approach does not involve local fitting, and so it is more robust to rounding of measurements. Also, confidence interval estimation is not sensitive to the parameterization used. Finally, the method allows for the inclusion of more than one covariate (other than the perpendicular distance), and it is easy to implement.

3.3.2 *Incorporating covariates into semiparametric models for the detection function*

Let there be a set of transect lines of total length L placed over a survey region of area A according to some survey design. An observer travels along each transect and records the perpendicular distance x_i and covariate values $\underline{z}_i$ ($\underline{z}_i = z_{1i}, \ldots, z_{qi}$) associated with each detected animal (or cluster of animals) i ($i = 1, \ldots, n$). Only animals located up to distance w from the line are recorded. The usual assumptions of line transect methodology (Buckland *et al.* 2001) are made. Of primary importance are the assumptions that (i) all animals located on or near the line are detected with certainty (i.e. $g(0, \underline{z}) = 1$); (ii) animals are detected prior to any responsive movement; and (iii) measurements are made without errors.

Following the 'key function' formulation of Buckland (1992a), and using the result from eqn (3.7), the conditional density $f(x \mid \underline{z})$ is given by:

$$f(x \mid \underline{z}) = \frac{k(x, \underline{z})}{\int_0^w k(x, \underline{z}) \left[1 + \sum_{m=1}^M \alpha_m p_m(x_s)\right] dx}$$

$$\times \left[1 + \sum_{m=1}^M \alpha_m p_m(x_s)\right] = \frac{k(x, \underline{z})}{\mu(\underline{z})} \left[1 + \sum_{m=1}^M \alpha_m p_m(x_s)\right]. \quad (3.11)$$

Here $k(x, \underline{z})$ is a parametric function (e.g. half-normal or hazard-rate), $p_m(\cdot)$ is an adjustment term (cosine, simple polynomial, or Hermite polynomial) of order m $(m = 1, \ldots, M)$, α_m is the coefficient for adjustment term m, and x_s is a standardized x value required to avoid numerical difficulties (Buckland 1992a) and usually taken to be x/w or x/σ, where σ is the scale term (see below). The integral $\mu(\underline{z}) = \int_0^w k(x, \underline{z}) \left[1 + \sum_{m=1}^M \alpha_m p_m(x_s)\right] dx$, in which w denotes the truncation distance, is a normalizing function of $\underline{z}$ and the parameters, required to ensure that $f(x \mid \underline{z})$ integrates to unity. We consider just the half-normal and hazard-rate key functions; other key functions available in Distance (Thomas et al. 2003) either do not allow the inclusion of covariates (uniform key) or have an implausible shape close to $x = 0$ (exponential key), and hence are not considered here.

Assume that the covariates affect detectability via the scale term σ, where for animal i, σ has the form:

$$\sigma_i = \exp\left(\beta_0 + \sum_{j=1}^q \beta_j z_{ij}\right), \quad (3.12)$$

and where the covariate effects are multiplicative as previously assumed by other authors (cf. eqn (3.10)).

Parameter estimates are obtained by maximizing the conditional log-likelihood

$$l = \log_e[\mathcal{L}(\underline{\theta}; \underline{x}, \underline{z})] = \log_e\left[\prod_{i=1}^n f(x_i \mid \underline{z}_i)\right] = \sum_{i=1}^n \log_e[f(x_i \mid \underline{z}_i)] \quad (3.13)$$

with respect to the parameter vector $\underline{\theta}$, where $\underline{\theta} = \theta_1, \ldots, \theta_{J+q+1+M}$; J, $q + 1$, and M refer to the number of shape parameters of the key function, the number of scale parameters, and the number of adjustment terms, respectively. Note that $J = 0$ for the half-normal key $(\underline{\theta} = (\beta_0, \beta_1, \ldots, \beta_q, \alpha_1, \ldots, \alpha_M)')$, whereas for the hazard-rate key function, $J = 1$ (the 'power' parameter) $(\underline{\theta} = (\gamma, \beta_0, \beta_1, \ldots, \beta_q, \alpha_1, \ldots, \alpha_M)')$.

Defining

$$t(x, \underline{z}) = k(x, \underline{z}) \left[1 + \sum_{m=1}^{M} \alpha_m p_m(x_s) \right], \qquad (3.14)$$

we have

$$l = \sum_{i=1}^{n} \log_e \left[\frac{t(x_i, \underline{z}_i)}{\mu(\underline{z}_i)} \right]$$

$$= \sum_{i=1}^{n} \log_e \left[t(x_i, \underline{z}_i) \right] - \sum_{i=1}^{n} \log_e \left[\mu(\underline{z}_i) \right] \qquad (3.15)$$

and

$$\frac{\partial l}{\partial \theta_j} = \sum_{i=1}^{n} \frac{\partial \log_e [t(x_i, \underline{z}_i)]}{\partial \theta_j} - \sum_{i=1}^{n} \frac{\partial \log_e [\mu(\underline{z}_i)]}{\partial \theta_j}$$

$$= \sum_{i=1}^{n} \frac{1}{t(x_i, \underline{z}_i)} \frac{\partial t(x_i, \underline{z}_i)}{\partial \theta_j} - \sum_{i=1}^{n} \frac{1}{\mu(\underline{z}_i)} \frac{\partial \mu(\underline{z}_i)}{\partial \theta_j}. \qquad (3.16)$$

For the parameters of the scale term, it is convenient to rewrite the equation above as:

$$\frac{\partial l}{\partial \theta_j} = \sum_{i=1}^{n} \frac{1}{t(x_i, \underline{z}_i)} \frac{\partial t(x_i, \underline{z}_i)}{\partial \sigma_i} \frac{\partial \sigma_i}{\partial \theta_j} - \sum_{i=1}^{n} \frac{1}{\mu(\underline{z}_i)} \frac{\partial \mu(\underline{z}_i)}{\partial \sigma_i} \frac{\partial \sigma_i}{\partial \theta_j} \qquad (3.17)$$

for $j = J + 1, \ldots, J + q + 1$, so that now the only additional computation required beyond that required for the standard estimation approach are the partial derivatives of the σ_i with respect to θ_j (i.e. β_j from eqn (3.12)).

We use the algorithm described by Buckland (1992a,b), extended to include the covariate parameters, to obtain maximum likelihood estimates of the model parameters. Selection of the number of adjustment terms to be included in the model can be carried out based on either likelihood ratio tests or Akaike's information criterion (AIC; Akaike 1973). Variance estimates are obtained using the Hessian matrix using the final parameter estimates (Buckland 1992a).

Note that eqn (3.11) ensures that covariates affect the detection function only by adjusting its scale, provided either no adjustment terms are fitted or the standardization $x_{si} = x_i/\sigma_i$ is used in the adjustment terms, where σ_i is given by eqn (3.12). If the standardization $x_s = x/w$ is used instead, then the effect of the adjustment on the detection function is independent of the covariate values. Thus for example, if search effort was conducted

in such a way that probability of detection was consistently higher at a
given distance than the model without adjustments would predict, then
standardization using $x_s = x/w$ would be appropriate, whereas if the same
shape of detection function was appropriate for all animals, but with different
scales, standardization using $x_{si} = x_i/\sigma_i$ would be necessary. Note
that, if no adjustment terms are required, this issue does not arise.

3.3.3   Abundance estimation

3.3.3.1   Cluster abundance

In this section, we consider how to estimate cluster abundance when
animals occur in clusters. The formulae also apply for animal abundance
when animals occur singly (i.e. all clusters are of size one).

The standard univariate (i.e. based on perpendicular distances alone)
line transect estimator of abundance is given by

$$\widehat{N}_s = A \cdot \widehat{D}_s = A \cdot \frac{n \cdot \hat{f}(0)}{2L}, \tag{3.18}$$

where $\widehat{N}_s$ is an estimator of cluster abundance N_s, $\widehat{D}_s$ is an estimator of
the density of clusters D_s, and n is the total number of detections.

To derive an estimator for the case where we have a multivariate conditional
density $f(0\,|\,\underline{z})$, it is convenient to view the above abundance
estimator as a Horvitz–Thompson estimator (Horvitz and Thompson 1952),
but with the inclusion probabilities having been estimated. Hence, in the
context of line transect sampling, define $P_a(\underline{z}_i)$ to be the probability that
cluster i is detected, given that it is within the strip of half-width w, and
given the observed values of $\underline{z}_i$:

$$P_a(\underline{z}_i) = E\left[g(x_i, \underline{z}_i)\,|\,\underline{z}_i\right] = \int_0^w g(x, \underline{z}_i)\pi(x\,|\,\underline{z}_i)\,dx, \tag{3.19}$$

where $g(x, \underline{z}_i) = f(x\,|\,\underline{z}_i)/f(0\,|\,\underline{z}_i)$. Assuming random line placement, then
x and $\underline{z}$ in the population are independent, and $\pi(x\,|\,\underline{z}) = \pi(x) = 1/w$.
Thus:

$$P_a(\underline{z}_i) = \int_0^w g(x, \underline{z}_i)\pi(x)\,dx = \frac{1}{w}\int_0^w g(x, \underline{z}_i)\,dx = \frac{1}{w}\frac{1}{f(0\,|\,\underline{z}_i)}. \tag{3.20}$$

For the standard Horvitz–Thompson estimator, the $P_a(\underline{z}_i)$ are assumed
to be known. However, here we must estimate the $f(0\,|\,\underline{z}_i)$ and hence the
$P_a(\underline{z}_i)$. Hence we obtain the 'Horvitz–Thompson-like' estimator of N_{cs}, the

total number of clusters within the covered strips (Borchers 1996):

$$\widehat{N}_{cs} = \sum_{i=1}^{n} \frac{1}{\widehat{P}_a(\underline{z}_i)} = w \sum_{i=1}^{n} \hat{f}(0 \,|\, \underline{z}_i) \qquad (3.21)$$

with the corresponding Horvitz–Thompson-like estimator of cluster abundance N_s in the survey region (Borchers 1996; Borchers et $al.$ 1998a):

$$\widehat{N}_s = \frac{\widehat{N}_{cs}}{P_c} = \frac{A}{2Lw} \widehat{N}_{cs}$$

$$= \frac{A}{2L} \sum_{i=1}^{n} \hat{f}(0 \,|\, \underline{z}_i). \qquad (3.22)$$

Horvitz–Thompson estimators (in which inclusion probabilities are known constants) are unbiased (Thompson 2002). Thus, when we replace $f(0 \,|\, \underline{z})$ by its estimator $\hat{f}(0 \,|\, \underline{z})$, we obtain an asymptotically unbiased estimate of N_s, provided the estimates of $f(0 \,|\, \underline{z}_i)$ are asymptotically unbiased.

Under the assumption that detections are independent, an estimator for the variance of $\widehat{N}_{cs}$, conditional on the $\widehat{P}_a(\underline{z}_i)$, or equivalently, given $\hat{\underline{\theta}}$, is given by (Borchers 1996):

$$\widehat{\text{var}}(\widehat{N}_{cs} \,|\, \hat{\underline{\theta}}) = w^2 \sum_{i=1}^{n} \hat{f}(0 \,|\, \underline{z}_i)^2 - \widehat{N}_{cs}. \qquad (3.23)$$

Using standard results for conditional variances (Seber 1982: 9) we have:

$$\text{var}(\widehat{N}_{cs}) = E_{\hat{\underline{\theta}}} \left[\text{var}\left(\widehat{N}_{cs} \,|\, \hat{\underline{\theta}} \right) \right] + \text{var}_{\hat{\underline{\theta}}} \left(E \left[\widehat{N}_{cs} \,|\, \hat{\underline{\theta}} \right] \right), \qquad (3.24)$$

which can be estimated by:

$$\widehat{\text{var}}(\widehat{N}_{cs}) = w^2 \sum_{i=1}^{n} \hat{f}(0 \,|\, \underline{z}_i)^2 - \widehat{N}_{cs} + \sum_{j=1}^{r} \sum_{m=1}^{r} \frac{\partial \widehat{N}_{cs}}{\partial \hat{\theta}_j} \frac{\partial \widehat{N}_{cs}}{\partial \hat{\theta}_m} H_{jm}^{-1}(\hat{\underline{\theta}}), \quad (3.25)$$

where r in the summations refers to all $J + q + 1 + M$ parameters and $H_{jm}^{-1}(\hat{\underline{\theta}})$ denotes the jmth element of the inverse of the Hessian matrix, whose jmth element is given by (Buckland et $al.$ 2001):

$$H_{jm}(\hat{\underline{\theta}}) = \frac{1}{n} \left[\sum_{i=1}^{n} \frac{\partial \log_e [f(x_i \,|\, \underline{z}_i)]}{\partial \hat{\theta}_j} \cdot \frac{\partial \log_e [f(x_i \,|\, \underline{z}_i)]}{\partial \hat{\theta}_m} \right]. \qquad (3.26)$$

We now need the variance of $\widehat{N}_s$. This has an additional component to the variance of $\widehat{N}_{cs}$, because of the sampling error involved in extrapolating to abundance in the entire survey region from a sample of strips. Innes *et al.* (2002) suggest the following estimator:

$$\widehat{\mathrm{var}}(\widehat{N}_s) = \left(\frac{A}{2wL}\right)^2 \left\{ L \sum_{k=1}^{K} \frac{l_k \left(\frac{\widehat{N}_{csk}}{l_k} - \frac{\widehat{N}_{cs}}{L}\right)^2}{K-1} \right. $$

$$\left. + \sum_{j=1}^{r} \sum_{m=1}^{r} \frac{\partial \widehat{N}_{cs}}{\partial \hat{\theta}_j} \frac{\partial \widehat{N}_{cs}}{\partial \hat{\theta}_m} H_{jm}^{-1}(\hat{\theta}) \right\}, \qquad (3.27)$$

where l_k is the length of transect k, $k = 1, \ldots, K$, and $\widehat{N}_{csk} = w \sum_{i=1}^{n_k} \hat{f}(0 \,|\, \underline{z}_i)$ is estimated cluster abundance in the corresponding strip of half-width w and length l_k, with n_k the number of detections from line k. This estimator follows from a result for $\mathrm{var}(\widehat{N}_s)$ analogous to eqn (3.24):

$$\mathrm{var}(\widehat{N}_s) = E_{\hat{\underline{\theta}}} \left[\mathrm{var}\left(\widehat{N}_s \,|\, \hat{\underline{\theta}}\right) \right] + \mathrm{var}_{\hat{\underline{\theta}}} \left(E\left[\widehat{N}_s \,|\, \hat{\underline{\theta}}\right] \right). \qquad (3.28)$$

An advantage of viewing the estimator of N_s as a Horvitz–Thompson-like estimator is that its estimate of the variance, given by the equation above, incorporates the variance component due to estimation of the parameters of the detection function. This avoids the negative bias common to Drummer and McDonald's (1987) and Quang's (1991) estimators of precision (Borchers 1996), which effectively treat the estimated parameters as known.

Alternatively, the variance of $\widehat{N}_s$ can be estimated using the bootstrap (Efron and Tibshirani 1993). Assuming that detections from different transect lines are independent, the transects can be taken to be the sampling units, and the procedure would then be as follows. In each of B bootstrap resamples, sample transect lines, along with their corresponding detections, with replacement, until either the number of lines is equal to K or the total amount of effort (i.e. total line length) from the resampled lines approximates the original total effort. Estimates of $f(0 \,|\, z)$ can then be obtained using the methodology described in Section 3.3.2, and abundance estimates obtained using eqn (3.22). If in each resample model selection is carried out, then this approach has the advantage of incorporating model selection uncertainty into the estimate of precision (Buckland *et al.* 1997).

Note that, for the estimator of eqn (3.21), we integrated out perpendicular distance x. There is of course no necessity for this; indeed, it might seem more consistent to use the estimator:

$$\widehat{N}_{cs} = \sum_{i=1}^{n} \frac{1}{\hat{g}(x_i, \underline{z}_i)}. \tag{3.29}$$

The reason that we do not use this estimator is because it is less stable (Borchers *et al.* 2002). When $g(x_i, \underline{z}_i)$ is small, small absolute errors in its estimation lead to large errors in the above estimator, so that it has high variance and potentially high bias relative to the estimator of eqn (3.21).

3.3.3.2 *Animal abundance*

In the case where animals occur in clusters, the standard line transect estimator of abundance is given by

$$\widehat{N} = A \cdot \widehat{D} = A \cdot \frac{n \cdot \widehat{E}[s] \cdot \hat{f}(0)}{2L}, \tag{3.30}$$

where $\widehat{E}[s]$ denotes the estimated mean cluster size of the population.

A Horvitz–Thompson-like estimator of the total number of clusters within the covered strips, N_{cs}, is given by eqn (3.21). A Horvitz–Thompson-like estimator of the total number of animals within the covered area, N_c, is given by:

$$\widehat{N}_c = \sum_{i=1}^{n} \frac{s_i}{\widehat{P}_a(\underline{z}_i)} = w \sum_{i=1}^{n} s_i \cdot \hat{f}(0 \,|\, \underline{z}_i), \tag{3.31}$$

where s_i denotes the size of the ith detected cluster and $\widehat{P}_a(\underline{z}_i) = 1/\{w\hat{f}(0 \,|\, \underline{z}_i)\}$ is the estimated probability of detection for the ith detected cluster. An estimate of the overall abundance of animals in the survey region, $\widehat{N}$, is then obtained by substituting the expression above into eqn (3.22), so that:

$$\widehat{N} = \frac{A}{2Lw} \widehat{N}_c = \frac{A}{2L} \sum_{i=1}^{n} s_i \cdot \hat{f}(0 \,|\, \underline{z}_i). \tag{3.32}$$

As in the case of cluster abundance, this estimator is asymptotically unbiased provided the $\hat{f}(0 \,|\, \underline{z}_i)$ are also asymptotically unbiased.

Assuming that detections are independent, an estimate of the variance of $\widehat{N}_c$, conditional on $\hat{\underline{\theta}}$, is given by (Thompson 2002):

$$\widehat{\text{var}}(\widehat{N}_c \,|\, \hat{\underline{\theta}}) = \sum_{i=1}^{n} \left(\frac{1 - \widehat{P}_a(\underline{z}_i)}{\{\widehat{P}_a(\underline{z}_i)\}^2} \right) s_i^2$$

$$= \sum_{i=1}^{n} \left\{ w^2 \hat{f}(0\,|\,\underline{z}_i)^2 - w\hat{f}(0\,|\,\underline{z}_i) \right\} s_i^2, \qquad (3.33)$$

so that

$$\widehat{\text{var}}(\widehat{N}_c) = \left\{ \sum_{i=1}^{n} \left\{ w^2 \hat{f}(0\,|\,\underline{z}_i)^2 - w\hat{f}(0\,|\,\underline{z}_i) \right\} s_i^2 \right.$$

$$\left. + \sum_{j=1}^{r} \sum_{m=1}^{r} \frac{\partial \widehat{N}_c}{\partial \hat{\theta}_j} \frac{\partial \widehat{N}_c}{\partial \hat{\theta}_m} H_{jm}^{-1}(\hat{\underline{\theta}}) \right\}. \qquad (3.34)$$

The corresponding estimate of the variance of the total abundance of animals in the survey region is given by:

$$\widehat{\text{var}}(\widehat{N}) = \left(\frac{A}{2wL} \right)^2 \left\{ L \sum_{k=1}^{K} \frac{l_k(\widehat{N}_{ck}/l_k - \widehat{N}_c/L)^2}{K-1} \right.$$

$$\left. + \sum_{j=1}^{r} \sum_{m=1}^{r} \frac{\partial \widehat{N}_c}{\partial \hat{\theta}_j} \frac{\partial \widehat{N}_c}{\partial \hat{\theta}_m} H_{jm}^{-1}(\hat{\underline{\theta}}) \right\}, \qquad (3.35)$$

where $\widehat{N}_{ck} = w \sum_{i=1}^{n_k} s_i \hat{f}(0\,|\,\underline{z}_i)$ is the estimated number of animals in strip k.

Note that an estimate of the mean cluster size $E[s]$ is given by:

$$\widehat{E}[s] = \frac{\widehat{N}_c}{\widehat{N}_{cs}} = \frac{\sum_{i=1}^{n} s_i \cdot \hat{f}(0\,|\,\underline{z}_i)}{\sum_{i=1}^{n} \hat{f}(0\,|\,\underline{z}_i)}. \qquad (3.36)$$

An estimate of the variance may be obtained using the delta method, so that:

$$\widehat{\text{var}}(\widehat{E}[s]) = \widehat{E}[s]^2 \left\{ \frac{\widehat{\text{var}}(\widehat{N}_c)}{\widehat{N}_c^2} + \frac{\widehat{\text{var}}(\widehat{N}_{cs})}{\widehat{N}_{cs}^2} - \frac{2\,\widehat{\text{cov}}(\widehat{N}_c, \widehat{N}_{cs})}{\widehat{N}_c \widehat{N}_{cs}} \right\}, \qquad (3.37)$$

where $\widehat{\mathrm{var}}(\widehat{N}_{cs})$ and $\widehat{\mathrm{var}}(\widehat{N}_c)$ are obtained from eqns (3.23) and (3.33), and $\widehat{\mathrm{cov}}(\widehat{N}_c, \widehat{N}_{cs})$ can be approximated by:

$$\widehat{\mathrm{cov}}(\widehat{N}_c, \widehat{N}_{cs}) = \sum_{i=1}^{n} \frac{\{1 - \widehat{P}_a(\underline{z}_i)\}s_i}{\{\widehat{P}_a(\underline{z}_i)\}^2} + \sum_{j=1}^{r} \sum_{m=1}^{r} \frac{\partial \widehat{N}_c}{\partial \hat{\theta}_j} \frac{\partial \widehat{N}_{cs}}{\partial \hat{\theta}_m} H_{jm}^{-1}(\hat{\underline{\theta}})$$

$$= w^2 \sum_{i=1}^{n} s_i \hat{f}(0 \,|\, \underline{z}_i)^2 - \widehat{N}_c$$

$$+ \sum_{j=1}^{r} \sum_{m=1}^{r} \frac{\partial \widehat{N}_c}{\partial \hat{\theta}_j} \frac{\partial \widehat{N}_{cs}}{\partial \hat{\theta}_m} H_{jm}^{-1}(\hat{\underline{\theta}}). \tag{3.38}$$

Alternatively the variance of $\widehat{E}[s]$ can be obtained by bootstrapping the transect lines as described at the end of Section 3.3.3.1. At each bootstrap iteration, $\widehat{N}_c$ and $\widehat{N}_{cs}$ can be computed based on the resampled data, and an estimate of the mean cluster size can be obtained using eqn (3.36). Hence we can directly estimate the variance of $\widehat{E}[s]$ from its bootstrap estimates. Under this approach it is assumed that transect lines are independent, but detected clusters and their cluster size need not be.

3.4 Point transect sampling

In point transect sampling, distance y becomes sighting distance r. Given random point placement, we have

$$\pi(r) = \frac{2r}{w^2},$$

where r is distance of a detected animal from the point, and w the truncation distance. Hence eqn (3.2) yields

$$f(r, \underline{z}) = \frac{r\, g(r, \underline{z})\pi(\underline{z})}{\iint r\, g(r, \underline{z})\pi(\underline{z})\, drd\underline{z}}, \tag{3.39}$$

from which

$$f(\underline{z}) = \int f(r, \underline{z})\, dr = \frac{\pi(\underline{z}) \int r\, g(r, \underline{z})\, dr}{\iint r\, g(r, \underline{z})\pi(\underline{z})\, drd\underline{z}}, \tag{3.40}$$

and

$$f(r \,|\, \underline{z}) = \frac{f(r, \underline{z})}{f(\underline{z})} = \frac{r g(r, \underline{z})}{\int r g(r, \underline{z})\, dr}. \tag{3.41}$$

The conditional likelihood is then

$$\mathcal{L}(\underline{\theta}; \underline{r}, \underline{z}) = \prod_{i=1}^{n} f(r_i \,|\, \underline{z}_i) = \prod_{i=1}^{n} \frac{r_i g(r_i, \underline{z}_i)}{\int r g(r, \underline{z}_i)\, dr}. \qquad (3.42)$$

The key function formulation for the probability density function is the same as for line transect sampling, except that the line transect key (half-normal or hazard-rate) is multiplied by r.

Given random point placement and an animal that is within w of one of the points, the probability that it is detected is now

$$P_a(\underline{z}_i) = \int_0^w g(r, \underline{z}_i)\pi(r)\, dr = \frac{2}{w^2} \int_0^w r g(r, \underline{z}_i)\, dr$$

$$= \frac{2}{w^2} \frac{1}{h(0 \,|\, \underline{z}_i)}, \qquad (3.43)$$

where

$$h(0 \,|\, \underline{z}_i) = \lim_{r \to 0} f(r \,|\, \underline{z}_i)/r.$$

If animals occur in clusters, the Horvitz–Thompson-like estimator of N_{cs}, the total number of clusters within the covered circles, is now

$$\widehat{N}_{cs} = \sum_{i=1}^{n} \frac{1}{\widehat{P}_a(\underline{z}_i)} = \frac{w^2}{2} \sum_{i=1}^{n} \hat{h}(0 \,|\, \underline{z}_i), \qquad (3.44)$$

and the corresponding Horvitz–Thompson-like estimator of cluster abundance N_s in the survey region is

$$\widehat{N}_s = \frac{A}{K\pi w^2} \widehat{N}_{cs} = \frac{A}{2K\pi} \sum_{i=1}^{n} \hat{h}(0 \,|\, \underline{z}_i), \qquad (3.45)$$

where K is the number of points in the design.

Variances follow much as for line transect sampling:

$$\widehat{\mathrm{var}}(\widehat{N}_{cs} \,|\, \underline{\hat{\theta}}) = \frac{w^4}{4} \sum_{i=1}^{n} \hat{h}(0 \,|\, \underline{z}_i)^2 - \widehat{N}_{cs} \qquad (3.46)$$

$$\widehat{\mathrm{var}}(\widehat{N}_{cs}) = \left\{ \frac{w^4}{4} \sum_{i=1}^{n} \hat{h}(0 \mid \underline{z}_i)^2 - \widehat{N}_{cs} + \sum_{j=1}^{r} \sum_{m=1}^{r} \frac{\partial \widehat{N}_{cs}}{\partial \hat{\theta}_j} \frac{\partial \widehat{N}_{cs}}{\partial \hat{\theta}_m} H_{jm}^{-1}(\underline{\hat{\theta}}) \right\}$$

(3.47)

$$\widehat{\mathrm{var}}(\widehat{N}_s) = \left(\frac{A}{K\pi w^2} \right)^2 \left\{ K \sum_{k=1}^{K} \frac{\left(\widehat{N}_{csk} - \frac{\widehat{N}_{cs}}{K} \right)^2}{K-1} \right.$$

$$\left. + \sum_{j=1}^{r} \sum_{m=1}^{r} \frac{\partial \widehat{N}_{cs}}{\partial \hat{\theta}_j} \frac{\partial \widehat{N}_{cs}}{\partial \hat{\theta}_m} H_{jm}^{-1}(\hat{\theta}) \right\}, \quad (3.48)$$

where K indicates the total number of points surveyed and the Hessian matrix has jmth element

$$H_{jm}(\underline{\hat{\theta}}) = \frac{1}{n} \left[\sum_{i=1}^{n} \frac{\partial \log_e[f(r_i \mid z_i)]}{\partial \hat{\theta}_j} \cdot \frac{\partial \log_e[f(r_i \mid z_i)]}{\partial \hat{\theta}_m} \right].$$

(3.49)

If animals occur individually, the above formulae give estimated animal abundance. If animals occur in clusters, the above estimators are modified in the same way as for line transect sampling to obtain estimates of animal abundance.

3.5 Example

As part of a program to assess the impact of tuna purse seine fishing vessels on dolphin populations in the eastern tropical Pacific, observers are placed on board the fishing vessels. They collect line transect survey data, from which trends in relative abundance of the populations are estimated (Buckland *et al.* 1992). A substantial volume of data is collected annually, but, in common with most surveys conducted on board so-called platforms of opportunity, substantial heterogeneity in probability of detection, beyond that explained by distance from the transect line, is present. There is therefore the potential for reduction in bias in relative abundance indices if we model covariates (Marques 2001; Marques and Buckland 2003). We illustrate analyses of the 1982 sightings data on the northeastern offshore stock of spotted dolphins.

Because tuna often swim under dolphin schools, tuna fishermen search for dolphins. If there is a substantial school of tuna present, the purse seine net is 'set' on the dolphins. This involves encircling the dolphin school, and then releasing the dolphins, leaving the tuna in the net. Whether or not a detection led to a set was found to correlate with the school's detectability,

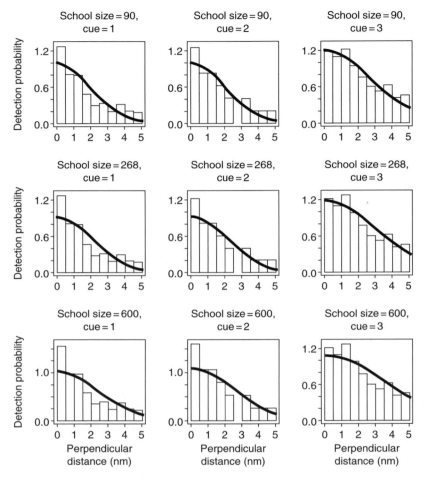

Fig. 3.1. Fitted detection functions for northeastern offshore spotted dolphins. Rows correspond to the quartiles of observed school sizes, conditional on the values of the other covariates. Columns correspond to different sighting cues: cue = 1 indicates that the sighting cue was dolphins; cue = 2 indicates splashes; cue = 3 indicates birds. Note that within a cue type, the histograms show observations across all school sizes, so that the three histograms within a column are identical.

probably because the observer is sometimes unaware that a school has been sighted, but rejected because tuna are not present. The initial sighting cue was also found to affect detectability. Birds often fly above the school, making it more detectable. A factor at three levels, corresponding to a cue of birds, splashes, or dolphins, was included in analyses. School size can vary from a few animals to several thousands, so it is unsurprising that school size contributed to the model. Other covariates such as sea state, season, and platform type were also considered for inclusion, but were not selected by AIC. We show the effects of cue type and school size on the detection function in Fig. 3.1. This plot shows that, when the cue is birds, detectability is greater as expected. It also shows that larger schools are more easily detected, although this effect is small over the range of school sizes plotted, which correspond to the lower quartile (90), median (268), and upper quartile (600) of observed school sizes.

3.6 Discussion

The methods of this chapter will generally be superior to standard distance sampling methods when there is appreciable heterogeneity in the detectability of different animals or animal clusters, provided covariates are available that correlate well with detectability. This is especially true when stratum-specific estimates of abundance are required, and sample sizes do not support separate fitting of the detection function to the data in each stratum of interest.

The methods are also useful for modelling sightings data collected from platforms of opportunity, as for our example. There is often substantial heterogeneity in detectability between different observers, observation platforms, and sighting conditions in such surveys, and if this heterogeneity is modelled, more reliable estimates of trend in abundance are likely to result (Marques and Buckland 2003).

Because the methods of this chapter are based on a Horvitz–Thompson-like estimator, in which the inclusion probabilities are estimated, abundance estimates can be sensitive to errors in the estimated probabilities. The estimator is based on $\sum 1/\widehat{P}_a(\underline{z}_i)$, which means that the sensitivity is greater for smaller $P_a(\underline{z}_i)$. As a rough guide, we recommend that the method be not used if more than say 5% of the $\widehat{P}_a(\underline{z}_i)$ are less than 0.2, or if any are less than 0.1. If these conditions are violated, the truncation distance w can be reduced. This causes some loss of precision relative to standard distance sampling without covariates.

4
Spatial distance sampling models
S. L. Hedley, S. T. Buckland, and D. L. Borchers

4.1 Introduction

Increasingly, wildlife managers wish to extract more than just an abundance estimate from their sightings surveys. They frequently need to relate animal density to spatial variables reflecting topography, habitat, and other factors that affect the animals' environment. This aids assessment of how to manage that environment, and the animals within it. There is therefore a strong demand for spatial models for analysing line transect data.

This chapter is largely based on the work of Hedley (2000), Hedley *et al.* (1999), and Hedley and Buckland (in press). This work concentrates on providing practical methods that can be conducted with standard statistical software. However, we also give some groundwork for the development of more general methods, using the theory of point processes. We expect significant methodological development in this field over the next few years.

Spatial distance sampling models allow wildlife managers to estimate abundance in any subset of a survey region, simply by numerically integrating under the relevant section of the fitted density surface. In contrast, conventional distance sampling methods restrict estimation of abundance to a set of survey blocks, defined at the design stage of a survey using stratified random or stratified systematic sampling. A spatial model for density allows total abundance to be apportioned between any set of sub-regions of interest, without falling foul of small sample sizes in sub-regions, as frequently occurs with stratified schemes.

Adopting a model-based approach for density estimation represents a departure from traditional distance sampling methodology, for which unbiased estimation relies on a survey design with randomized line or point locations. Spatial modelling does not require that the lines or points are located according to a formal survey sampling scheme. This can be useful when marine line transect surveys are conducted from 'platforms of opportunity' (ferries, merchant navy vessels, oceanographic survey vessels, etc.). Such surveys often gather large quantities of non-random data at a cost substantially lower than that of a properly designed survey. Provided the non-random transects give good spatial cover throughout the survey region, spatial models offer more reliable estimation of abundance than conventional, design-based analyses can. In the case of non-random terrestrial transects, spatial models may again be useful, but they cannot correct for bias arising, for example, when transects systematically follow geographic features such as tracks or ridges.

We first consider spatial line transect models. We then briefly address spatial point transect models, which are simpler because observations are associated with a dimensionless point rather than a one-dimensional line.

4.2   Spatial line transect models

The conventional line transect estimator of density, $\widehat{D}$, when animals occur singly, is

$$\widehat{D} = \frac{n}{2L\hat{\mu}},\tag{4.1}$$

where n is the number of detections, L is the total transect length, and $\hat{\mu}$ is the estimated effective strip half-width. This is also the estimator of cluster density when animals occur in well-defined clusters. In this chapter, we consider estimation of 'object' density, where the object might be either single animals or clusters of animals. We also note how the theory of marked point processes, in which the mark is cluster size, might be adapted to allow estimation of animal density when the objects are clusters. In the discussion, a simpler approach for clustered populations is described, in which the methods of this chapter are used to estimate cluster density, and a separate spatial model for mean cluster size is fitted to observed cluster sizes.

Point process formulations of line transect estimators were considered by Stoyan (1982) and Högmander (1991, 1995). The conventional line transect estimator given by eqn (4.1) was reformulated by Stoyan (1982) in terms of the intensity of a stationary marked point process, where the points were the locations of the objects, and individual points were marked by some probability representing the detectability at each point, based on its sighting distance. Stationarity implies that objects are uniformly distributed throughout the region of interest, which is not a plausible assumption for animal populations. Conventional line transect estimation uses design-based methods to avoid the assumption; lines are assumed to be placed at random with respect to the locations of the objects. Here, we seek a model-based solution, so that object density (or intensity) can be modelled as a function of spatial location, rather than as the average intensity of a spatial point process.

4.2.1   Deriving a likelihood

Let the function $D(u, v)$ represent the density of objects at location (u, v), using a Cartesian coordinate system. Objects' locations are modelled as an inhomogeneous Poisson process with rate $D(u, v)$. Thus if $A_1, \ldots, A_J$ are arbitrary disjoint sets within the survey area A, then the numbers of objects in each set, $N(A_j), j = 1, \ldots, J$, are independent Poisson variables, with the following expected values:

$$E[N(A_j)] = \int_{A_j} D(u, v) \, du dv, \quad j = 1, \ldots, J. \tag{4.2}$$

(Note that this formulation implies that the $N(A_j)$, and hence population size in the survey region, are random.) Suppose that the areas A_j are J non-overlapping rectangles, each of width $2w$, and of lengths $l_j, j = 1, \ldots, J$, with $\sum_j l_j = L$. Each rectangle may correspond to a surveyed (covered) strip, with the corresponding transect line down the centre of the strip, or it may correspond to a section of a surveyed strip. We term these rectangles 'segments'. Without loss of generality, we switch to a different, locally defined Cartesian coordinate system (x, y) for segment j, in which x is distance along the line from the start of the survey at $x_0 = 0$, and y is perpendicular distance from the line, with $-w \leq y \leq w$ (Fig. 4.1). Let $g(x, y)$ be the probability of detecting an object, given that it is located at (x, y).

From eqn (4.2), the expected number of objects within a rectangular segment j, starting at x_{j-1} and ending at x_j $(j = 1, \ldots, J)$, is

$$\int_{-w}^{w} \int_{x_{j-1}}^{x_j} D(x, y) \, dx dy. \tag{4.3}$$

The expected number of detections within the segment is a function of how many objects are present in the segment and of how detectable they are. Under the additional assumption that the detection process is independent of the density of objects, the detection locations represent an independent thinning of the inhomogeneous Poisson process given by all of the objects' locations. The thinned process is also an inhomogeneous Poisson process (Cressie 1991: 625–6). Thus the locations of detected objects within the segment follow an inhomogeneous Poisson process with rate $D(x, y)g(x, y)$, which can also be expressed as $D(u, v)g(u, v)$, where the point (u, v) in the global coordinate system corresponds to the point (x, y) in the local system.

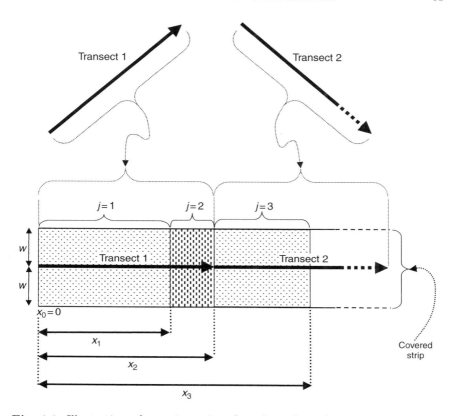

Fig. 4.1. Illustration of notation using three hypothetical segments $j = 1, 2, 3$ defined on two transect lines. Although the transects do not touch each other on the survey (top half of the figure), the coordinate system is defined such that along-transect distance x accumulates as if they did.

Let $\mu(A) = \int_A D(u, v) g(u, v) du dv$, where $\int_A$ is the integral over the entire survey region. Then the density of locations of detected objects, conditional on the number n detected, is

$$f((u_1, v_1), \ldots, (u_n, v_n) \mid n) = \frac{\prod_{i=1}^{n} D(u_i, v_i) g(u_i, v_i)}{[\mu(A)]^n} \qquad (4.4)$$

and the joint density of the number of detections and their locations is

$$f((u_1, v_1), \ldots, (u_n, v_n), n) = \begin{cases} \exp\left[-\mu(A)\right], & n = 0 \\ \exp\left[-\mu(A)\right] \prod_{i=1}^{n} D(u_i, v_i) g(u_i, v_i)/n!, & n \geq 1. \end{cases}$$
$$(4.5)$$

Note that the product $D(u, v)g(u, v)$ is zero except within the J surveyed segments, because $g(u, v) = 0$ outside the segments by assumption. The process of interest is $D(u, v)$, not the product $D(u, v)g(u, v)$. Estimation of this process from observations on the thinned process is not straightforward. The assumption that all objects on the line are detected gives $g(x, 0) = 1$, but this is not sufficient to separate D from g. The scales over which the two functions D and g operate tend to be very different. Transects are often long, so that D may vary appreciably along the line, whereas w is typically small, so that if D is modelled as a smooth function, it shows little variation across the width of the segment. By contrast, g falls rapidly with distance from the line, but at given y, is unlikely to vary appreciably as x varies. We exploit this to separate out estimation of the two functions below.

4.2.2 *A likelihood based on inter-detection distances*

We call the along-transect distance surveyed between two successive detections the 'waiting distance'. We expect waiting distances to be short in areas of high density, and we can exploit this information to estimate density. Define $l_i = x_i - x_{i-1}$ to be the along-trackline distances between successive detections, with $i = 2, \ldots, n$ (Fig. 4.2). (If successive detections are recorded from different lines, the accumulated effort is recorded; a discontinuity in $D(x, 0)$ when switching from one line to the next is accommodated.) If we denote the x-value at the start of survey effort as $x_0 = 0$ and the x-value corresponding to the end of survey effort as x_{n+1}, then l_1 is the distance

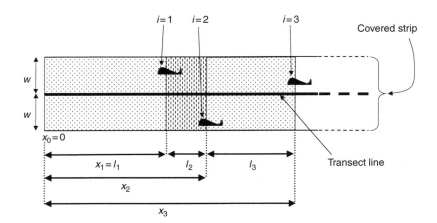

Fig. 4.2. Illustration of notation for the waiting distance model, using three hypothetical whale detections at distances x_1, x_2 and x_3 along a transect line.

surveyed until the first detection and l_{n+1} is the distance surveyed after the last detection.

We make the following assumptions:

(1) $D(x, y) \equiv D(x)$ independent of y;
(2) $g(x, y) \equiv g(y)$ independent of x;
(3) $g(0) = 1$;
(4) $g(-y) = g(y)$ for $0 \leq y \leq w$; and
(5) x and y are independent.

Denote the random variable representing distance between detections i and $i-1$ by L_i, and let l_i denote a value this random variable takes. Noting that l_{n+1} is a right-censored observation, we can write the joint density of observations as

$$f(l_1, \ldots, l_{n+1}, y_1, \ldots, y_n) = \left[\prod_{i=1}^{n} f_{l|x}(l_i \mid x_{i-1}) \right]$$
$$\times P(L_{n+1} > l_{n+1} \mid x_n) \times \left[\prod_{i=1}^{n} f_y(y_i) \right]. \quad (4.6)$$

Conventional line transect theory yields $f_y(y_i) = g(y_i)/2\mu$, $-w \leq y_i \leq w$, where $\mu = \int_0^w g(y)dy$. (Usually, the distribution is folded, so that $0 \leq y \leq w$; see Buckland *et al.* 2001: 60.)

We therefore need $f_{l|x}(l_i \mid x_{i-1})$ for $i = 1, \ldots, n+1$. The cumulative distribution function (cdf) of the random variable L_i is

$$F_{l|x}(l_i \mid x_{i-1}) = P(L_i \leq l_i \mid x_{i-1})$$
$$= 1 - P(L_i > l_i \mid x_{i-1}) \quad i = 1, \ldots, n+1. \quad (4.7)$$

Since $L_i > l_i$ if and only if there were no detections in the strip of half-width w, then

$$F_{l|x}(l_i \mid x_{i-1}) = 1 - \exp\left\{ -\int_{-w}^{w} \int_{x_{i-1}}^{x_{i-1}+l_i} D(x)g(y)\, dx dy \right\}$$
$$= 1 - \exp\left\{ -2\mu \int_{x_{i-1}}^{x_{i-1}+l_i} D(x)\, dx \right\}. \quad (4.8)$$

Differentiating, we obtain the conditional probability density function (pdf) of the waiting distance l_i as:

$$f_{l|x}(l_i \mid x_{i-1}) = 2\mu D(x_{i-1} + l_i)$$

$$\times \exp\left[-2\mu \int_{x_{i-1}}^{x_{i-1}+l_i} D(x)\,dx\right] \quad i = 1,\dots,n. \quad (4.9)$$

Note that uncertainty in n is handled implicitly in this formulation. If a sequence of l_i is generated according to an inhomogeneous Poisson process, n is determined as the largest i for which $\sum_{i'=1}^{i} l_{i'} \leq L$, the total line length.

Suppose we have a vector of spatial parameters $\underline{\beta}$ for the density surface $D(x)$ and parameters $\underline{\theta}$ for the detection function $g(y)$. Then if we write x_i for $x_{i-1} + l_i$ and with appropriate substitution into eqn (4.6), we obtain the marginal likelihood $\mathcal{L}(\underline{\beta},\underline{\theta}; \underline{l}, \underline{y})$:

$$\mathcal{L}(\underline{\beta},\underline{\theta}; \underline{l}, \underline{y}) = \left[\prod_{i=1}^{n} D(x_i)\right] \exp\left[-2\mu \sum_{i=1}^{n+1} \int_{x_{i-1}}^{x_i} D(x)\,dx\right] \left[\prod_{i=1}^{n} g(y_i)\right],$$

$$(4.10)$$

where $\underline{l}$ is the vector of observed waiting distances $l_i = x_i - x_{i-1}$, $i = 1,\dots,n+1$, $\underline{y}$ is the vector of observed perpendicular distances y_i, $i = 1,\dots,n$, and n is the number of detections.

Note that we can allow detectability to vary with x by introducing a vector of covariates $\underline{z}_e$, so that the detection function is $g(y,\underline{z}_e)$. These covariates should be 'effort-level' covariates, which are covariates associated with the line such as habitat or sea state, and which are typically recorded as the line is traversed. The methods of Chapter 3 may then be used to fit this component of the model.

4.2.3   *Clustered populations*

If animals occur in well-defined clusters, then we have a marked point process, where the mark is cluster size. If the size of cluster i is s_i, and is assumed independent of location and n, then eqn (4.5) extends to

$$f((u_1,v_1,s_1),\dots,(u_n,v_n,s_n),n) = \begin{cases} \exp\left[-\mu(A)\right], & n = 0 \\[2mm] \dfrac{\exp\left[-\mu(A)\right]}{n!} \\ \quad \times \prod_{i=1}^{n} \{D(u_i,v_i)g(u_i,v_i,s_i)\pi_s(s_i)\}, & n \geq 1 \end{cases}$$

$$(4.11)$$

where $\mu(A) = \int_A D(u,v)g(u,v,s)\pi_s(s)dudvds$, and $\pi_s(s)$ is the pdf of cluster sizes in the population. Note that cluster density is not modelled as a function of cluster size in this formulation. The assumption that $g(x,0,s) = 1$ for all x and s, together with a suitable model for $g(x,y,s) \equiv g(u,v,s)$, now potentially allows separation of D and g. Indeed, we could go further, and define a model for which $g(x,0,s) = 1$ only for larger cluster sizes s, which gives a framework for estimation when detection of smaller clusters that are on the line is not certain. In practice, rather than specify a model for $\pi_s(s)$, analysis is likely to be conducted conditional on the observed cluster sizes, as in Chapter 3. We do not pursue these ideas here.

4.3 Practical implementations of spatial line transect models

Implementation of the above methodologies is the subject of ongoing research. Sound methods for fitting general models using the likelihoods described above without convergence problems have not yet been developed. We therefore describe two approaches that enable spatial variation in density to be modelled using standard generalized linear modelling (GLM, McCullagh and Nelder 1989) or generalized additive modelling (GAM, Hastie and Tibshirani 1990) software. The first uses waiting distance data and is a simplification of the likelihood-based approach described in the previous section; the second is a simple count model, which we have found to be a practical and robust option for some data sets.

4.3.1   A waiting distance model

If object detections occurred randomly along the transect lines according to a homogeneous Poisson process, then the distances between detections would be distributed exponentially, with rate proportional to the reciprocal of the average density of detections. For an inhomogeneous Poisson process, the distances between detections would no longer be exponentially distributed. A GLM/GAM can be formulated for a situation that is intermediate between a fully spatially inhomogeneous Poisson process and a homogeneous Poisson process. By assuming that the density of objects and the expected encounter rate are constant in the interval between two successive detection locations, but may change at each location, each waiting distance can be modelled using an exponential distribution, with rate proportional to the reciprocal of the fitted density in that interval. The GAM formulation gives

$$\eta\left(E[l_i]\right) = \beta_0 + \sum_{k=1}^{K} f_k(z_{ik}), \quad i = 1, \ldots, n \qquad (4.11)$$

where the link function η is a monotonic differentiable function (the logarithmic link may often be suitable, as it ensures positive values of the mean response), K is the number of spatial covariates ($\underline{z}$) included in the model, $f_k(\cdot)$ is the one-dimensional smooth function for covariate k, and n is the number of detections. Letting $\hat{l}_i$ denote the fitted values of $E[l_i]$, then given the estimated effective strip half-width, $\hat{\mu}$, the estimated object density in the interval l_i between detections $i - 1$ and i is $[2\hat{\mu}\hat{l}_i]^{-1}$.

Object density would be expected to vary along the trackline in the interval between detections. However, the GLM/GAM framework outlined above, only allows density to change where detections occur. We therefore use an iterative procedure, in which each observed waiting distance is transformed to the corresponding distance (with respect to the cdf) had the underlying (inhomogeneous) Poisson process been homogeneous with rate equal to the estimated rate at the location of the next detection. The steps of this iterative procedure are as follows:

1. Fit a GLM or GAM to the waiting distance data, as if the effort associated with waiting distance l_i was all located at x_i rather than in the interval $(x_{i-1}, x_i]$ for $i = 1, \ldots, n$ (where $x_i = x_{i-1} + l_i$). From the fitted model, obtain the estimated density along the line, $\widehat{D}(x, 0)$. This estimated density is biased because it ignores the fact that density may vary along the line.

2. Adjust each waiting distance, l_i, as follows. From eqn (4.8) with $\mu = \int_0^w g(y)dy$ estimated by $\hat{\mu}$, we have

$$\widehat{F}(l_i \mid x_{i-1}) = 1 - \exp\left\{-2\hat{\mu}\int_{x_{i-1}}^{x_{i-1}+l_i} \widehat{D}(x, 0)\,dx\right\}.$$

If, between x_{i-1} and x_i, density were constant along the line (with estimated value $\widehat{D}(x_i, 0)$), then we would have

$$\widehat{F}(l_i \mid x_{i-1}) = 1 - \exp\left\{-2\hat{\mu}l_i\widehat{D}(x_i, 0)\right\}.$$

By equating these expressions, with l'_i substituted for l_i in the second expression, we obtain an adjusted waiting distance, where the adjustment is for variable density along the line:

$$l'_i = \frac{\int_{x_{i-1}}^{x_{i-1}+l_i} \widehat{D}(x, 0)\,dx}{\widehat{D}(x_i, 0)}, \quad i = 1, \ldots, n \qquad (4.12)$$

3. The GLM or GAM is now fitted to the adjusted waiting distances, to obtain the estimated density along the line, $\widehat{D}(x, 0)$. This should

have lower bias than the estimate from step 1, because we replace the observed waiting distances along the lines by waiting distances that have been adjusted to the estimated density at the locations of detections. Each adjusted waiting distance is therefore associated with a single location in space, allowing simple fitting of the GLM or GAM. However, $\widehat{D}(x, 0)$ was obtained by fitting to unadjusted waiting distances. We therefore recalculate the adjusted waiting distances, using the updated estimate of density along the line, and the procedure is iterated until convergence.

As depicted in Fig. 4.3, the iterative procedure equates the area under the predicted density surface between successive detections (at x_{i-1} and x_i, say) to the area of the rectangle of width l_i' and height $\widehat{D}(x_i, 0)$. We note that the procedure as described does not utilize the effort data recorded following the last detection. One solution is to modify the waiting distance data such that the distance to the first detection is redefined to be the sum of l_1 plus the right-censored distance, l_{n+1}.

4.3.2 *A count model*

Strip transect surveys are a special case of line transect surveys, in which all objects within strips of width $2w$ and total length L are counted. They

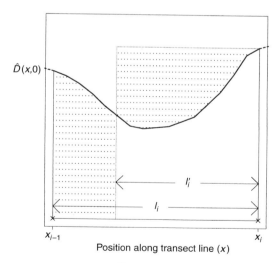

Fig. 4.3. Density on the trackline, $\widehat{D}(x, 0)$, with detections at x_{i-1} and x_i. The distance between detections is shown as l_i; the adjusted waiting distance is shown as l_i'. l_i' is such that the area of the rectangle of width l_i' and height $\widehat{D}(x_i, 0)$ equals the area under $\widehat{D}(x, 0)$ between x_{i-1} and x_i (so that the dotted regions are equal in area).

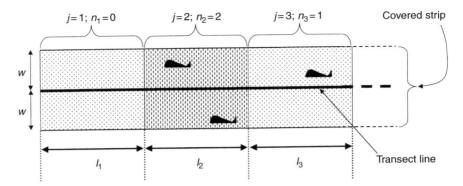

Fig. 4.4. Illustration of notation for the count model, using three hypothetical segments of lengths l_1, l_2, and l_3.

are particularly suited to species that are sufficiently common, or occur at such densities, that it is not practical to record exact locations of detected objects. Shipboard surveys of seabirds or aerial surveys of large game animals are examples. We suppose that counts of objects are made, corresponding to short line segments. In the case of aerial photographic surveys of seals on ice, for example, the size of the segments is determined by the area covered by a single photo.

Suppose that the surveyed strips are divided into T small contiguous segments. Let the length of the jth segment be denoted by l_j, and the number of objects detected within it by $n_j, j = 1, \ldots, T$ (Fig. 4.4). Now suppose that for each segment, a set of K spatial covariates is available, and let z_{jk} denote the value of the kth spatial covariate in the jth segment. The expected values of the n_j can be related to the spatial covariates using a GLM or GAM formulation. For example, for a GLM,

$$ E[n_j] = \exp\left[\log_e(2l_j w) + \beta_0 + \sum_k \beta_k z_{jk}\right], \quad j = 1, \ldots, T. \quad (4.13) $$

The logarithm of the area of each segment, $\log_e(2l_j w)$, thus enters the linear predictor as an offset, and the β_k, $k = 0, \ldots, K$, are parameters to be estimated.

To extend this formulation so that it may be used to model line transect data, we first consider the case where the estimated probability of detection of the ith object in the jth segment, unconditional on its distance y from the line, depends only on the segment and not on individual animals. We write the probability of detecting an animal on segment j, given covariates $\underline{z}_j$, as $P_a(\underline{z}_j)$. Inclusion of its estimate $\widehat{P}_a(\underline{z}_j)$ in the offset term yields a formulation suitable for spatial modelling of line transect data conditional on the $\widehat{P}_a(\underline{z}_j)$, using for example a GLM of the following general form:

$$E(n_j) = \exp\left[\log_e\{2l_jw\widehat{P}_a(\underline{z}_j)\} + \beta_0 + \sum_k \beta_k z_{jk}\right], \quad j = 1,\ldots,T.$$

$$(4.14)$$

No additional modelling difficulties arise, but the component of variance due to uncertainty in the estimation of the $P_a(\underline{z}_j)$ must be included in variance estimates of density and abundance.

If effective strip half-width μ is constant so that $P_a(\underline{z}_j) = P_a$ for all j, then eqn (4.14) becomes

$$E(n_j) = \exp\left[\log_e(2l_jw\widehat{P}_a) + \beta_0 + \sum_k \beta_k z_{jk}\right]$$

$$= \exp\left[\log_e(2l_j\hat{\mu}) + \beta_0 + \sum_k \beta_k z_{jk}\right], \quad j = 1,\ldots,T. \quad (4.15)$$

We could estimate the number of objects N_j in the jth segment using the Horvitz–Thompson-like estimator $\widehat{N}_j = n_j/\widehat{P}_a(\underline{z}_j)$. Equation (4.14) implies that $\widehat{N}_j = 2l_jw\exp[\hat{\beta}_0 + \sum_k \hat{\beta}_k z_{jk}]$. We could in principle model these $\widehat{N}_j$ instead of the n_j, although the model of eqn (4.14) is preferred, because the counts are then modelled directly, so that a Poisson model is more plausible. If we record individual covariates, we can no longer model the counts directly. Instead, the methods of Chapter 3 or 6 can be used to obtain estimates $\widehat{P}_a(\underline{z}_{ij})$ of detection probability which vary by individual object, then the $\widehat{N}_j$ are estimated by

$$\widehat{N}_j = \sum_{i=1}^{n_j} \frac{1}{\widehat{P}_a(\underline{z}_{ij})}, \quad j = 1,\ldots,T. \quad (4.16)$$

These estimates can then form the response for a spatial model, with appropriate modification of the offset in eqn (4.14):

$$E(\widehat{N}_j) = \exp\left[\log_e(2l_jw) + \beta_0 + \sum_k \beta_k z_{jk}\right], \quad j = 1,\ldots,T. \quad (4.17)$$

4.4 Spatial distribution of Antarctic minke whales

The data used in this example are sightings of minke whale pods (with typically 1–3 animals in a pod) from a line transect survey conducted in the Southern Ocean in the austral summer of 1992–3. The survey was conducted in two alternating survey modes: closing mode and independent observer (IO) mode. For illustration, we use data only from IO mode. The survey region was divided into four geographic strata, and relatively more survey effort was dedicated to the two southern strata, which are adjacent to the ice-edge, where whale densities are higher (Fig. 4.5).

Sightings data were pooled across the northern strata and southern strata for estimating effective strip half-widths μ, and in both cases, the perpendicular distance data were truncated at $w = 1.5$ nautical miles (n.miles). Resulting estimates are shown in Table 4.1. These estimates were used to apply the three methods described in this chapter. The detection function was assumed to depend only on the perpendicular distance y from the line. Three spatial covariates (latitude, longitude, and distance from the ice-edge) were available for inclusion in the models.

With $w = 1.5$ n.miles, the surveyed strips have a width of 3 n.miles, around 0.3% of the width of the survey region, so that it seems reasonable to assume that $D(x, y) = D(x)$, independent of y, for $|y| \leq w$. Direct maximization of the likelihood of eqn (4.10), which uses this simplification, was carried out, modelling $D(x)$ as a linear function of the three spatial covariates.

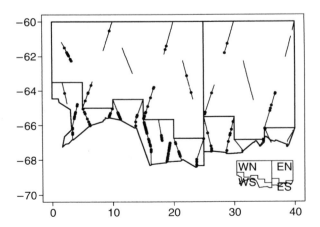

Fig. 4.5. Realised survey effort in IO mode and sightings of minke whale pods during the 1992–3 IWC/IDCR Antarctic Survey. The southern survey boundary is defined by the extent of sea ice. Subplot shows the division of the region into four strata: WN, WS, EN, and ES.

Table 4.1. Stratum estimates and coefficients of variation (% cv) of $\widehat{P}_a$, the detection probability of minke pods within a strip of half-width 1.5 n.miles. Also shown are the corresponding effective strip half-widths $\hat{\mu} = 1.5 \times \widehat{P}_a$ (in n.miles)

Pooled strata	$\widehat{P}_a$	% cv	$\hat{\mu}$
WN and EN	0.742	7.61	1.112
WS and ES	0.360	15.42	0.540

Table 4.2. Comparison of estimates of abundance of minke whale pods ($\widehat{N}$) from a conventional stratified line transect analysis, direct maximization of the likelihood, and two GAM-based approaches, one modelling the waiting distances between detections, and the other modelling the estimated number of minke whale pods in segments of length 3 n.miles. Coefficients of variation (% cv) were estimated using the parametric bootstrap

Stratum	Stratified analysis $\widehat{N}$	% cv	Likelihood maximization $\widehat{N}$	% cv	GAM waiting distances $\widehat{N}$	% cv	GAM counts $\widehat{N}$	% cv
WN	4810	40.1	4970	13.6	5810	20.8	4920	16.3
EN	1460	49.5	1440	14.5	800	31.1	760	25.0
WS	7410	25.1	7740	16.4	7820	17.1	8130	17.1
ES	640	44.3	1210	17.0	910	22.9	990	21.5
All	14320	23.0	15360	15.9	15340	16.1	14800	15.9

For the two GLM/GAM-based models, GAMs with logarithmic links were fitted using all three covariates. Latitude and longitude were included as smoothing spline terms each with 4 degrees of freedom, while the effect of distance from the ice-edge was modelled linearly. For the waiting distance model, a gamma error distribution was assumed, while for the count model, transects were divided into segments of approximately 3 n.miles, and the estimated number of minke pods in each segment was modelled assuming a variance–mean relationship corresponding to an overdispersed Poisson distribution. Pod abundance estimates for the three approaches, together with estimates from a conventional line transect analysis, are given in Table 4.2. Coefficients of variation for the spatial methods were obtained using a parametric bootstrap procedure. For each method, resampled data were generated from the original fitted model, and the original model was refitted to the resamples. The overall variance of pod abundance, var($\widehat{N}$), was estimated by combining the variance of $\hat{\mu}$ (or $\widehat{P}_a$) and the sample variance of the abundance estimates from the resamples, using the delta method (Seber 1982: 7–9). This assumes independence between the two variance components.

The spatial methods all yielded higher estimates of pod abundance than the stratified method, and the precision of the estimates was better. The spatial models gave even larger improvements in precision for estimates of abundance by stratum. This occurs because spatial models utilize data from all strata to estimate the density surface, and hence abundance, in each; stratum estimates are therefore estimated with higher precision, but also with positive correlation, so that the gain in precision in the estimate of total abundance is less. The improved precision of the spatial methods in this example is substantial, an important consideration for surveys in which ship time is expensive. A further advantage of the spatial methods is that the density surface is estimated. The three models fitted here yield the surfaces shown in Figs 4.6–4.8.

We have illustrated our methods using explanatory variables that are not ecologically meaningful. Instead, our example uses locational covariates (latitude, longitude, and distance from the ice-edge) which only serve as proxies for other covariates which might genuinely be expected to influence habitat selection, such as food resource, sea surface temperature, and upwellings. The availability of spatial line transect models will encourage researchers to identify and measure variables more directly relevant to the species of interest.

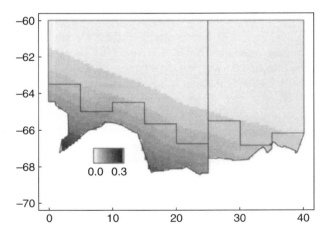

Fig. 4.6. Density of minke whale pods in the survey region, predicted by maximizing the likelihood of observed waiting distances, conditional on the estimates of effective strip half-widths given in Table 4.1.

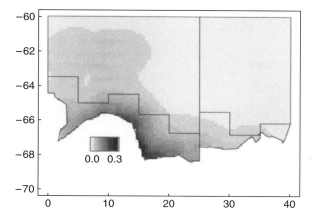

Fig. 4.7. Density of minke whale pods in the survey region, predicted from a waiting distances model. Explanatory variables were smoothing splines of latitude and longitude (each with 4 df), and a linear term for distance from the ice-edge.

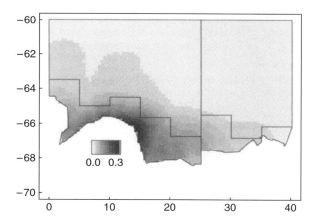

Fig. 4.8. Density of minke whale pods in the survey region, predicted from a count model. Explanatory variables were smoothing splines of latitude and longitude (each with 4 df), and a linear term for distance from the ice-edge.

4.5 Spatial point transect models

4.5.1 Deriving a likelihood

Now suppose that areas A_j are J non-overlapping circles, each of radius w, $j = 1, \ldots, J$. Without loss of generality, we switch to a locally-defined polar coordinate system (r, ϕ) for circle j, in which r is distance from the circle's centre, with $0 \leq r \leq w$. Let $g(r, \phi)$ be the probability of detecting an object, given that it is located at (r, ϕ). We assume that this probability is independent of ϕ, so that $g(r, \phi) \equiv g(r)$.

From eqn (4.2), the expected number of objects within a circle defined by $0 \le r \le w$ is

$$\int_0^w r \int_0^{2\pi} D(r, \phi) \, d\phi dr. \tag{4.18}$$

The expected number of detections within the circle is a function of how many objects are present in the circle and on how detectable they are. As for line transects, if the detection process is independent of the density of objects, the detection locations represent an independent thinning of the inhomogeneous Poisson process given by all of the objects' locations, giving rise to an inhomogeneous Poisson process with rate $D(r, \phi)g(r)$. This rate can also be expressed as $D(u, v)g(u, v)$, where the point (u, v) in the global coordinate system corresponds to the point (r, ϕ) in the local system.

Let $\mu(A) = \int_A D(u, v)g(u, v)dudv$, where $\int_A$ is the integral over the entire survey region. (In fact, $g(u, v) = 0$ unless the point (u, v) is within a distance w of a point, so this is equivalent to integrating over the covered region of size $J\pi w^2$.) Then the density of locations of detected objects, conditional on the number n detected, is given by eqn (4.4), and the joint density of the number of detections and their locations is given by eqn (4.5).

The assumption that all objects at the centre of the circle are detected gives $g(0, \phi) = 1$, but as with line transects, this is not sufficient to separate D from g. Because typically w is small relative to the scales over which D varies, or at least over scales for which variation in D is of interest, we lose little by approximating $D(r, \phi)$ by $D_j(0)$, the density at the centre $(r = 0)$ of circle j, for $r \le w$. The expected number of detections at point j becomes:

$$\int_0^w r \int_0^{2\pi} D(r, \phi)g(r) \, d\phi dr = D_j(0) \int_0^w r \, g(r) \int_0^{2\pi} d\phi dr$$

$$= D_j(0) \times 2\pi \int_0^w r \, g(r) \, dr. \tag{4.19}$$

Assuming that $g(r)$ does not depend on covariates,

$$\mu(A) = \int_A D(u, v)g(u, v) \, dudv = 2\pi \int_0^w r \, g(r) \, dr \sum_j D_j(0).$$

If we take eqn (4.4), transform from (u, v) to (r, ϕ) (noting that $du\,dv$ transforms to $r\,dr\,d\phi$), and integrate out ϕ, we obtain

$$
f(r_1, \ldots, r_n \mid n) = \frac{\prod_{i=1}^{n} 2\pi D_i(0) r_i g(r_i)}{\left[\sum_j D_j(0)\right]^n \left[2\pi \int_0^w r\, g(r)\, dr\right]^n}
$$

$$
= \frac{\prod_{i=1}^{n} D_i(0)}{\left[\sum_j D_j(0)\right]^n} \times \frac{\prod_{i=1}^{n} r_i\, g(r_i)}{\left[\int_0^w r\, g(r)\, dr\right]^n}, \qquad (4.20)
$$

where $D_i(0)$ is the density at the centre of the circle at which the ith detection was made. The second term is the conventional conditional likelihood for point transect sampling, allowing us to fit the detection function using standard methods, if we conduct inference using the likelihood conditional on n. To estimate a relative density surface D conditional on n, we simply maximize the first term with respect to the parameters of D. We can then scale this surface so that the estimated expected number of detections is equal to n.

If we wish to model detectability as a function of covariates, then the conditional likelihood of the second term of eqn (4.20) may be replaced by the corresponding likelihood from Chapter 3.

Alternatively, we can base inference on the full likelihood, which is obtained by multiplying eqn (4.20) by the Poisson density, $p(n) = \{\mu(A)\}^n \times \exp\{\mu(A)\}/n!$ (see eqn (4.5)):

$$
f(r_1, \ldots, r_n, n) = \frac{\exp\left\{2\pi \int_0^w r\, g(r)\, dr \sum_j D_j(0)\right\}}{n!}
$$

$$
\times \prod_{i=1}^{n} D_i(0) \times \prod_{i=1}^{n} r_i\, g(r_i). \qquad (4.21)
$$

4.5.2 A point transect count model

If we collected point count data in the form of complete counts out to $r = w$, then it is straightforward to model these counts using GLMs or GAMs and spatial covariates. Denote the count at point j by n_j, and let z_{jk} denote the value of the kth spatial covariate at the jth point, $j = 1, \ldots, J$, $k = 1, \ldots, K$. Then for a GLM formulation,

$$
E[n_j] = \exp\left[\log_e(\pi w^2) + \beta_0 + \sum_k \beta_k z_{jk}\right], \quad j = 1, \ldots, J. \qquad (4.22)
$$

The logarithm of the area of each circle enters the linear predictor as an offset, so that the linear predictor models animal (or cluster) density. We can readily substitute a GAM formulation to give greater flexibility.

If we model probability of detection of the ith object at the jth point as a function of covariates that vary spatially but not by individual animal, then conditioning on estimates $\widehat{P}_a(\underline{z}_j)$, a GLM formulation yields

$$E[n_j] = \exp\left[\log_e\{\pi w^2 \widehat{P}_a(\underline{z}_j)\} + \beta_0 + \sum_k \beta_k z_{jk}\right], \quad j = 1, \ldots, J. \quad (4.23)$$

The component of variance due to uncertainty in the estimation of the $P_a(\underline{z}_j)$ should be included in variance estimates of density and abundance.

If we assume $P_a(\underline{z}_j) = P_a$ for all j, then the offset becomes $\log_e(\pi\hat{\rho}^2)$, where $\hat{\rho}$ is the effective radius, estimated using conventional distance sampling theory without covariates.

From eqn (4.23), the estimated number of objects $\widehat{N}_j$ in the jth circle is $n_j/\widehat{P}_a(\underline{z}_j)$. Allowing covariates to vary by individual animal (or cluster), then the N_j are estimated by

$$\widehat{N}_j = \sum_{i=1}^{n_j} \frac{1}{\widehat{P}_a(\underline{z}_{ij})}, \quad j = 1, \ldots, J. \quad (4.24)$$

As for the line transect count model, these estimates can then form the response for a spatial model, with appropriate modification of the offset in eqn (4.23).

4.6   Discussion

We anticipate that more sophisticated statistical techniques will allow models based on the point process formulation of Section 4.2.1 to be fitted to line transect data, incorporating spatial variation in detectability and (cluster) density. For example, a Markov chain Monte Carlo approach is being developed by Fernández and Niemi (in preparation). However, simpler methods may often be sufficient for wildlife managers, and are much easier to fit. Although rather inelegant, and based on subjectively dividing the line, the count model appears robust and reliable. It appears insensitive to choice of segment length, provided that the segments are sufficiently small that expected density is unlikely to vary much within a segment. In practice, this means that detectability conditions should be as similar as possible within each segment, and for habitat-distribution studies, segment size should reflect the scale of the environment, so that habitat varies little within segments. In the example, we chose the segment length to be approximately equal to the strip width $(2w)$, so that the segments were approximately square.

We used the parametric bootstrap to estimate variances in our estimates. For both the waiting distances model and the count model, there was evidence in our example of spatial autocorrelation, after having fitted spatial trends (p-value $= 0.04$ for the waiting areas model, and 0.02 for the count model, using Monte Carlo tests). This may be true autocorrelation, caused perhaps by social behaviour of the animals, or simply variation due to covariates that have not been measured. The two cases are indistinguishable given our data, and in either case, some bias in the variance estimates can be anticipated.

The appropriate course of action to resolve the problem differs for the two cases. If we simply have the wrong covariates, the ideal would be to measure more relevant covariates, and fit a more appropriate model. This may result in reduced bias in the abundance estimates and a decrease in the variance estimates. If there is true autocorrelation, then we might seek to model it, for example, by using the approach of Gotway and Stroup (1997), who present a framework for combining generalized linear models and quasi-likelihood with geostatistical methods, defining a general variance–covariance matrix that accounts for autocorrelation. They initially fit a GLM with trend only, then estimate the autocorrelation structure from the semi-variogram of the residuals. The estimated correlation matrix is then inserted into the variance–covariance matrix when fitting the final GLM using quasi-likelihood. An alternative approach is being developed by Bravington *et al.* (in preparation). Their approach is also based on modelling large-scale variation (trend) in density distribution separately from small-scale variation (autocorrelation), where they suppose that the autocorrelation is an ephemeral quantity that is not of intrinsic interest, but which should contribute to the variance of the abundance estimate. Having modelled the autocorrelation using a multiplicative random field, the trend in density can be modelled using, for example, GAMs as an inhomogeneous Poisson process, with the likelihood (and model fitting criteria) adjusted for the residual clustering seen in the data, by integrating over possible values of the random field.

Particularly if data are only available from a single survey, it is important to consider what the objective of the analysis is, since the resulting inference is philosophically different. In modelling density in such a way that the autocorrelation is also incorporated (using, for example, the methods described above), Augustin *et al.* (1998) note that inference is then drawn not on the population that happened to be present at the time of the survey, but rather on a 'hyper-population' from which that population was drawn. This inevitably yields higher variance estimates, and such a philosophy may be of less use to wildlife managers. However, it is impossible from a single survey to distinguish between what might be a true persistent 'trend' in density and what might alternatively be modelled simply as a short-term correlation. Unravelling the spatio-temporal effects would require additional data, either from repeat surveys of the same area or from surveys in similar areas where the population under study is expected to behave (and particularly aggregate) in a similar manner. For any single survey for which trend and autocorrelation are treated as separate components of the model, inferences should be checked for robustness to the way in which the two components are separated.

Animals often occur in groups, termed clusters in the distance sampling literature, as with the minke whales of our example. In this case, the methods described here provide estimated density surfaces for clusters. In the absence of integrated modelling as hinted in Section 4.2.3, we can convert these to density surfaces for animals by independently fitting a spatial model for mean cluster size. Cluster density is then multiplied by predicted mean cluster size at any given location to estimate animal density at that location. A spatial model for cluster size can readily be fitted, because size is observed at n discrete locations, where n is the number of clusters detected. GLMs or GAMs may be used to model these observed sizes as a function of spatial covariates. In the case of the count model with individual covariates, for which cluster density is estimated by eqn (4.16), a minor modification allows us to estimate animal density more directly:

$$\widehat{N}_j = \sum_{i=1}^{n_j} \frac{s_{ij}}{\widehat{P}_a(\underline{z}_{ij})}, \quad i = 1, \ldots, T, \tag{4.25}$$

where N_j now represents the number of animals in segment j, and s_{ij} is the size of the ith detected cluster in that segment. Size bias (the bias arising from the higher probability of detection of larger clusters and hence over-sampling of those clusters) is accounted for in eqn (4.25). If size bias is suspected, then the estimated probability of detection of each cluster should be similarly included in a spatial model for cluster size, weighting each observation by the reciprocal of its estimated detection probability.

Recall that there are three categories of parameter estimated in conventional distance sampling: those relating to the encounter rate n/L, those relating to the detection function $g(y)$, and those relating to mean cluster size $E(s)$. The spatial density models described above deal primarily with estimation of parameters relating to n/L. If detection probability changes in space (due to sighting conditions changing in space, for example) and this is not modelled, changes in density will be confounded with changes in detectability. In this case, there is a danger of misinterpreting trend in detection probability as trend in density.

Similar misinterpretation is possible when animals occur in clusters and the trend in mean cluster size is not modelled. Consider the minke

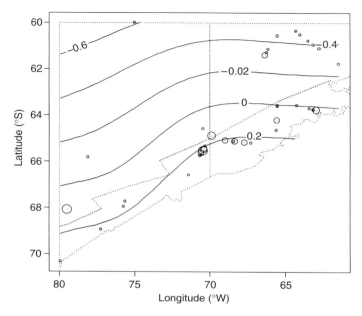

Fig. 4.9. Contour plot (solid lines) of the estimated relative log school size for the 1999–2000 IDCR/SOWER Antarctic minke whale survey. The contour of height zero is a reference contour; contours with positive heights occur in regions with estimated mean school size greater than this, while those with negative heights occur in regions with estimated mean school size less than this. The dotted lines are the stratum boundaries (the southernmost dotted line is the ice-edge). Observed school sizes (s_i) are shown as circles with radius proportional to $\log_e(s_i) + 1$.

whale example above, in which the trend in cluster density was modelled, and suppose that ultimately an estimate of the trend in individual density was required. Minke whale density tends to be high close to the ice-edge because this is their prime feeding area, and it is not unusual that they occur in larger clusters when feeding, creating a gradient in mean cluster size away from the ice-edge. An example of this sort of gradient (from a different survey) is shown in Fig. 4.9. In this example, if the cluster density estimate was converted to a whale density estimate using a single estimate of mean cluster size for the whole survey region (or even separate estimates within each stratum), the resulting individual density surface would be too low close to the ice-edge and too high far from the ice-edge. Trend in mean cluster size will have been confounded with trend in animal density.

When using the methods of this chapter to estimate density surfaces, care should be taken not to confound trend in n/L, $g(y)$, and $E(s)$. Trend in detection probability can be accommodated by modelling it as a function of relevant variables, using the methods of Chapter 3. Inclusion of location covariates like latitude and longitude should be considered even if there is no *a priori* reason that they should affect detection probability; they may be useful proxies for some unrecorded variable that does affect detection probability. As noted above, trend in mean cluster size can be modelled using GLM or GAM methods. Fig. 4.9, taken from Borchers and Burt (unpublished), is an example of a mean cluster size surface fitted to line transect data using GAMs.

5

Temporal inferences from distance sampling surveys

L. Thomas, K. P. Burnham, and S. T. Buckland

5.1 Introduction

Many distance sampling surveys are repeated on multiple occasions, with the aim of monitoring long-term changes in population density or abundance. In this chapter, we review selected methods of analysing repeated surveys, and discuss issues related to planning long-term studies. For clarity, we will assume that surveys occur annually, although the methods described here can be applied directly for surveys equally spaced in time and are readily generalized to unequal time intervals. We mostly focus on simple methods that are relatively easy to implement, although we briefly discuss more advanced approaches and those requiring further methodological development.

Distance sampling is just one type of population estimation method, and many of the issues we discuss apply to population monitoring in general. Indeed, we will illustrate some of the points we make using a non-distance sampling example: the Waterfowl Breeding Population and Habitat Survey (WBPHS). The WBPHS is a spring survey of breeding waterfowl in the north-central United States and Canada (Fig. 5.1). Each year a fixed set of aerial strip transects is surveyed, and some transects are also surveyed from the ground. The ground surveys are used to correct the aerial estimates: an example of a double-sampling survey (Thompson 2002: 139–47). We chose this example because it is the longest running extensive survey of wildlife abundance that we are aware of, with reliable annual estimates available from 1955 to the present. The data we use are from US Fish and Wildlife Service (2003); more details of the survey are given at http://migratorybirds.fws.gov.

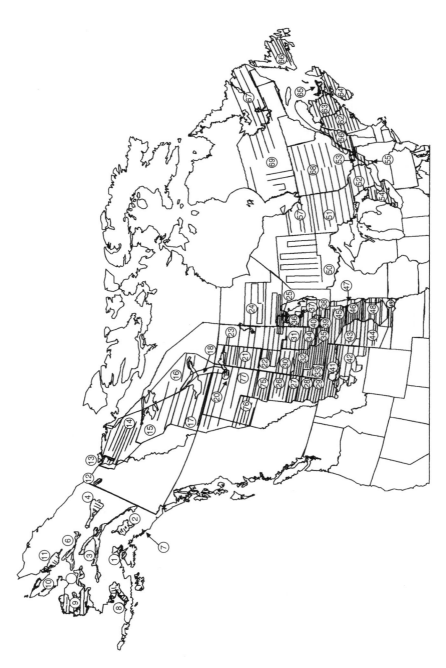

Fig. 5.1. Transects and strata for the WBPHS, reproduced from US Fish and Wildlife Service (2003). The abundance estimates for mallard in this chapter use data from strata 1–18, 20–50, and 75–7 (the 'traditional survey area').

We begin by discussing a number of concepts relating to temporal analysis of survey data before considering some analysis methods in detail. We then discuss issues relating to survey design.

5.2 Concepts

5.2.1 *Sampling and population variation*

Annual estimates of abundance for mallard (*Anas platyrhynchos*) from the WBPHS are shown in Fig. 5.2. The annual estimates vary over time, and we can quantify this total variation as

$$\widehat{\mathrm{var}}(\widehat{N}) = \frac{1}{T-1} \sum_{t=1}^{T} (\widehat{N}_t - \widehat{\bar{N}})^2, \tag{5.1}$$

where $\widehat{N}_t$ is the abundance estimate at time t, $\widehat{\bar{N}}$ is the mean of the estimates, and T is the number of time periods. Conceptually, this variation can be attributed to two sources: population variation and sampling variation.

Population variation (sometimes called *process variation*) is temporal variation in the true abundances. It can be expressed as the variance $\mathrm{var}(N)$, where

$$\mathrm{var}(N) = \frac{1}{T-1} \sum_{t=1}^{T} (N_t - \bar{N})^2 \tag{5.2}$$

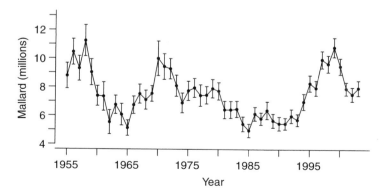

Fig. 5.2. Mallard abundance estimates and associated 95% confidence intervals from the WBPHS.

and N_t is the true abundance at time t. Because the true abundances are almost always unknown, this quantity cannot be calculated directly, but must be estimated using methods like those outlined below. Patterns of population variation are of considerable interest to theoretical ecologists (see references in Link and Nichols 1994). For example, longer time series often show more population variation (Lawton 1988; Halley 1996). In applied situations, the principle interest is in separating out various components of population variation, such as trend (see below).

Sampling variation is the variation associated with estimating the population abundances, given their true value. It can be expressed as the conditional variance $\mathrm{var}(\widehat{N}\,|\,\underline{N})$, where $\underline{N} = (N_1, \ldots, N_T)'$, and its magnitude can usually be estimated from the survey data (see below). Sampling variation relates to the way that the data were collected and analysed and is therefore not of primary interest to biologists, except as a nuisance to be minimized as far as possible.

The two sources of variance are additive (assuming that the $\widehat{N}_t$ are unbiased estimates of the N_t), in that

$$\mathrm{var}(\widehat{N}) = \mathrm{var}(N) + E\{\mathrm{var}(\widehat{N}\,|\,\underline{N})\}. \tag{5.3}$$

We can therefore estimate the population variation by subtracting the estimated sampling variation from the total variation (and setting the result to zero if it is negative). This 'naïve' estimator can often be improved upon with more complex estimators (giving better precision).

To obtain estimates of sampling variation from the survey data, we will initially assume that each annual abundance estimate depends only on true abundance in that year (i.e. there are no sampling covariances). Distance sampling survey results are usually reported as $\widehat{N}_t$ and the associated coefficient of variation $\mathrm{cv}(\widehat{N}_t)$. From this, we can obtain an estimate of sampling variance at each time point as $\widehat{\mathrm{var}}(\widehat{N}_t\,|\,N_t) = [\widehat{N}_t\,\mathrm{cv}(\widehat{N}_t)]^2$. If there is no sampling covariance, then

$$\widehat{E}\{\mathrm{var}(\widehat{N}\,|\,\underline{N})\} = \frac{1}{T}\sum_{t=1}^{T}\mathrm{var}(\widehat{N}_t\,|\,N_t), \tag{5.4}$$

so the sampling variation may be estimated as the average of the annual sampling variance estimates.

For the mallard data, the estimated total variance of $\widehat{N}$ is 2.52×10^{12}, and the average of the estimated sampling variances is 1.18×10^{11}, giving an estimated population variance of 2.41×10^{12}. In this case, the population variation is estimated to be much greater than the sampling variation, but these surveys are unusually precise (mean cv 4.45%) and the time series is unusually long. In general, sampling variation may be larger or smaller than population variation. Another example of this type of calculation is given by Link and Nichols (1994), who obtain similar formulae assuming a stochastic model for the $N_1, \ldots, N_T$. Components of variation in population time series are also discussed by White (2000) and Williams *et al.* (2002).

5.2.2 *Sampling covariance*

In many distance sampling surveys, the abundance estimate at time t depends not only on true abundance at time t, but also on abundance at the other time periods. Hence for the annual sampling variance we should write $\mathrm{var}(\widehat{N}_t \mid \underline{N})$, instead of $\mathrm{var}(\widehat{N}_t \mid N_t)$. Also, in addition to the sampling variances, there will be sampling covariances $\mathrm{cov}(\widehat{N}_{t_1}, \widehat{N}_{t_2} \mid \underline{N})$, which we expect to be positive.

Positive covariances can arise in two ways. The first is when the estimates of detection probability for each year are not independent. This will occur when the detection function is based on data pooled across years, and can occur when year is included as a covariate using the methods of Chapter 3. When sample sizes are reasonably large, it is preferable to fit a separate detection function to each year of data, to avoid confounding any changes in abundance with possible changes in detection probability. However in many cases, a limited number of observations in some or all years means that some pooling or covariate analysis must be used. Where pooling is required, pooling over adjacent years of data may limit the problem. If the focus is on estimating long-term trend, then the need for pooling is lessened because we are going to smooth (in some way) over the $\widehat{N}_t$, so they do not need to be as precise as if the focus was on a single $\widehat{N}$.

The second cause of positive sampling covariance is when encounter rates are not independent because some or all of the same transects have been surveyed in more than one year. In conventional distance sampling, we use a design-based approach to estimate variance in encounter rates (Chapter 10), so that if lines are not placed at random and independently of previous years, there will be a sampling covariance. (Similar covariances would occur in a model-based approach if our model for the distribution of animals allowed average density to vary over space, and if this variation was not independent across years.)

If the sampling covariances can be estimated, eqn (5.4) can readily be extended to accommodate them:

$$\widehat{E}\{\mathrm{var}(\widehat{N} \mid \underline{N})\} = \frac{1}{T} \sum_{t=1}^{T} \mathrm{var}(\widehat{N}_t \mid \underline{N})$$

$$- \frac{2}{T(T-1)} \sum_{t_1=2}^{T} \sum_{t_2=1}^{t_1-1} \mathrm{cov}(\widehat{N}_{t_1}, \widehat{N}_{t_2} \mid \underline{N}). \qquad (5.5)$$

In conventional distance sampling analyses, probably the easiest way to estimate the covariances is to use a nonparametric bootstrap where resamples are selected according to the survey design. By this, we mean that if a subset of transects is permanent (i.e. repeated in all years), then whenever a permanent transect is selected in a bootstrap resample for the first year, it should also be in that resample for all other years. Similar algorithms can be derived for semi-permanent transects (e.g. panel designs, Section 5.7). This strategy can also be applied to estimate variance in trend in the presence of sampling covariance, as is shown in Section 5.3.4. Other strategies are possible: Underwood (2004) considers analytic design-based estimation of variance and covariance for transect-based surveys with designs involving a subset of permanent transects.

5.2.3 *Empirical and process models*

In many temporal analyses, the main goal is to make inferences about components of population variation. Two types of analysis approach can be distinguished: empirical and process.

Empirical methods specify a statistical model for the components of population variation—for example, a linear or smooth trend plus independent fluctuations around the trend. The aim is to provide a parsimonious description of the main features of the data. Empirical methods can also be used to 'explain' these features with the inclusion of covariates such as environmental variables. A problem of empirical methods is that the estimated trend can be biologically implausible, leading to unreliable explanations and predictions. It is generally unwise to extrapolate the predictions of an empirical model beyond the years of the data.

Process modelling (also called *mechanistic modelling*) specifies an explicit model of the population dynamics for the process variation. The model is therefore more biologically realistic, and this approach will usually provide better explanations and predictions. Three drawbacks of process models are that they require strong assumptions about the dynamics of the population, they generally require long time series of data, and they can be more difficult to implement. For these reasons, we focus largely on empirical methods in this chapter, although we briefly describe process modelling in Section 5.5.

We note that the division is not absolute; for example, empirical methods based on estimating log-linear trends can also be thought of as fitting a biological model of exponential growth.

5.2.4 *Trend*

In most empirical approaches, the focus is on separating long-term changes in abundance from short-term fluctuations. Long-term changes are usually referred to as *trend* or *trajectory*, meaning 'smooth and slow movement over the long term' (Dagum and Dagum 1988). The concept of 'long' in this context is relative, and what is identified as a trend in a series of abundance estimates may be seen as part of a population cycle or irregular fluctuation when the series is extended. For example, if the WBPHS had begun in 1970, then after 15 years of data collection, we would have concluded that there were quite worrying declines in mallard numbers, with the final estimate being about half the initial one (Fig. 5.2). If it had begun in 1985, after 15 years we would have concluded that mallard populations were increasing exponentially. However, taken over the whole time series of 49 years (Fig. 5.2), it is clear that there have been large fluctuations, but no overall increase or decline in mallard populations.

Two types of trend model are commonly used: *linear* and *smooth*. Linear models have the advantage that they summarize the trend in one number and so are easy to interpret. It is easy to synthesize results from multiple studies or a multi-species study. A strong disadvantage of linear models is that they are clearly inadequate for longer time series. For example, if a population is stable for a long period of time and then starts a rapid decline, then it will be a long time before this is detected in the overall linear trend. It is a paradox of linear models that the longer the time series, the less likely we are to detect a recent rapid drop in numbers. An alternative quantitative approach for longer time series, therefore, is to fit a smooth curve to the abundance estimates. The amount of smoothing can be determined *a priori*, or using a data-driven method. A disadvantage of smooth models is that the trend estimate cannot now be summarized using a single number, although it is possible to calculate differences in point estimates from the smooth between any two time points (e.g. the beginning and end of the time series, although this choice gives relatively poor precision due to uncertainty in where the smooths should go at each endpoint). In addition, significant 'change points' can be identified (Fewster *et al.* 2000).

5.2.5 *Abundance as a fixed or random quantity*

Another important distinction between methods is whether they treat the true (but unknown) abundances as *fixed* or *random* quantities and, if random, how this randomness is modelled.

When the true abundances are assumed to be fixed quantities, inferences are conditional on these fixed values. It follows that the only uncertainty in a trend estimate comes from the sampling variation. For example, consider the case where the true abundances are known, so that the sampling variation is zero. For any given trend model, for example, a linear model, the trend is known with certainty. If we were able to re-run the study over the same time period and study area, because the abundances are fixed and there is no sampling variation, the same abundances would be recorded and therefore the same trend estimate generated. Any non-zero trend would therefore be considered statistically significant, regardless of the amount of fluctuation of the true abundances about the trend line.

Alternatively, the true abundances can be assumed to be a random realization of a stochastic process generating abundances. In this case, uncertainty in the trend estimate arises both from sampling variation and from randomness in the true abundances. Using the above example, where there is no sampling variation, the estimate of trend would be the same as in the fixed-abundance analysis, but now the trend would not be known with certainty. If we were able to re-run the study, a different realisation of the random process would result in a different set of abundances, so the trend estimate would vary.

Methods that treat abundance as random differ in which components of the total population variation they assume to be random. The empirical random-abundance methods of Sections 5.3 and 5.4 treat all variation in abundance not explained by the trend model as random. By contrast, the process methods of Section 5.5 allow us to separate out two components of population variation, *environmental* and *demographic*, and treat either or both as random. We discuss this further in Section 5.5.2.

Whether abundance should be treated as fixed or random depends on the inferences required from the analysis. An initial goal of many monitoring studies is to make inferences about the trend in abundance in the study area during the time period of the study (e.g. 'has the species declined?'), and this can be addressed with a fixed-abundance analysis. However, in many cases there is a desire to judge these trends in the context of biological variation in abundance. For example, a small decline in a species subject to large natural fluctuations in abundance may be judged less important than a small but consistent decline in a species with little natural fluctuation. In this case a random-abundance analysis is more appropriate.

Simple fixed-abundance analyses can also be used as a first step to decide whether more complex random-abundance analyses (such as those of Section 5.3.4.1) are required. Fixed-abundance methods will always have greater precision relative to the equivalent random-abundance method because they ignore all non-sampling error. Therefore if a trend estimate is not statistically significant with a fixed-abundance analysis, then it will not be with the equivalent random-abundance analysis either.

Abundances should also be considered random if predictions of future abundance are required, or if understanding of the processes underlying trends in abundance is needed. In these cases, since the questions require understanding of the underlying biological processes, it makes more sense to use process models rather than empirical models. For this reason, when we discuss process models, we focus on models where true abundance is a random quantity.

5.3 Trend estimation from global abundance estimates

In this section we discuss methods for estimating trend using study-wide estimates of abundance in each time period.

5.3.1 *Graphical exploration*

The first step in any analysis is graphical exploration of the data. A simple plot of the abundance estimates with associated estimates of sampling error or confidence intervals can reveal a great deal. For example, from Fig. 5.2 it seems clear that there is little overall trend in the 49-year mallard time series. Looking more closely, the population appears to have undergone several cycles of decline (1959–65, 1970–85) and recovery (1965–70, 1985–99) with overall numbers reaching 10–11 million birds in the best years, and falling to as low as 5 million in poor years. A more informative analysis could relate these abundances to changes in breeding habitat, continental weather patterns, or other explanatory factors. For monitoring at its most basic, however, such deeper analyses are not needed.

For short time series, this type of descriptive analysis may be all that is required. For longer series, especially where there is moderate to large sampling variation or covariation, it is often hard to assess whether the observed patterns are statistically significant from description alone. Quantitative methods are required, such as those outlined in the following subsections.

5.3.2 *Linear trend models*

5.3.2.1 *True abundances assumed random*
A simple model of trend is

$$\widehat{N}_t = \beta_0 + \beta_1 t + \delta_t + \epsilon_t, \tag{5.6}$$

where β_0 is the intercept, $\beta_1 t$ is the trend, δ_t is the sampling variation (also called *sampling error*), and ϵ_t is the non-trend population variation (often called *process error*). We will assume initially that both error terms are independent and identically distributed normal random variables. This implies no sampling covariances, and a process model for the population dynamics in which departures of the N_t from a linear trend are independent between years. We can then combine the terms and carry out standard simple linear regression, with the error term in year t equal to $\delta_t + \epsilon_t$. By using a weighted regression, variances can be allowed to vary by t or by N_t; the former will be useful when we expect sampling error to vary over time, for example, due to changing observer effort. Alternatively, we could avoid combining the two error terms by using variance components methods (Section 5.3.4.1).

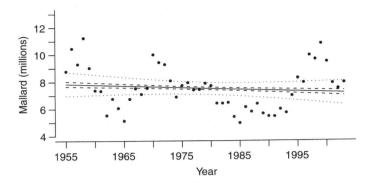

Fig. 5.3. Linear trend analysis of mallard data. Dashed and dotted lines show 95% confidence bands for points on the regression line assuming the true abundances are fixed and random respectively.

For short time series, simple linear regression can be a reasonable approach. For longer time series, the assumption of linearity is seldom plausible. Fig. 5.2 shows that, for the mallard data, the relationship between abundance and time is markedly non-linear. Alternatively, we might argue that the underlying trend is linear, but that there is a strong positive correlation between abundances in successive years. If we ignore this failure of the model, we obtain a trend estimate (slope) of $\hat{\beta}_1 = -1.37 \times 10^4$. The residual mean square (RMS) is 2.54×10^{12}—in this example, the RMS is greater than $\text{var}(\hat{N})$, underlining how little of the variation in the $\hat{N}$ is explained by the linear trend. This yields $\widehat{\text{var}}(\hat{\beta}_1) = 2.59 \times 10^8$ and 95% confidence limits for β_1 of $(-4.61 \times 10^4, +1.86 \times 10^4)$. The 95% confidence interval easily contains zero slope.

The fitted trend line is shown in Fig. 5.3, together with 95% confidence bands for points on the line. An alternative would be to display global confidence bands on the entire line, which will be wider. The difference between these bands is discussed in most standard statistics texts (e.g. Steel and Torrie 1980: 254).

An alternative linear model is

$$\log_e(\widehat{N}_t) = \beta_0 + \beta_1 t + \delta_t + \epsilon_t \tag{5.7}$$

with δ_t and ϵ_t again both independent and normally distributed. This may be more relevant in the distance sampling context since the errors now follow a log-normal distribution on the untransformed scale, which corresponds with the usual assumption about sampling error in conventional distance sampling (Buckland *et al.* 2001: 77). It is also arguably more biologically meaningful as it corresponds with the exponential growth model

$E(N_t) = N_0\lambda^t$. The parameter $\lambda = \exp(\beta_1)$ can then be interpreted as the average annual rate of change, an important quantity in population analysis (see Caswell 2001). If the two error terms are combined as discussed above, and the model is fitted using linear regression, then $\exp(\hat{\beta}_1)$ is a biased estimate of λ, but $\hat{\lambda} = \exp(\hat{\beta}_1 - 0.5\widehat{\text{var}}(\hat{\beta}_1))$ is approximately unbiased.

Another way to formulate and fit such models is using generalized linear modelling (GLM; McCullagh and Nelder 1989). In this framework, we can assume error distributions other than normal, and responses can be modelled as some function of the linear predictor. Thus a log-link function gives the following model:

$$\widehat{N}_t = \exp\left(\beta_0 + \beta_1 t\right) + \delta_t + \epsilon_t. \tag{5.8}$$

Note that this is different from eqn (5.7) as the errors are now on the same scale as the abundances, not on the log scale. An appropriate distribution for a combined error term might be the gamma distribution, since we typically assume that the sampling error in conventional distance sampling has a constant cv. When the sampling errors are expected to vary between time periods, then a weighted GLM can be used, weighting by the inverse of the estimated sampling variance. An alternative is to incorporate the distance sampling component as an offset in the GLM; this kind of approach is discussed in Chapter 4. The model then becomes (assuming line transect sampling):

$$n_t = \exp\left(\log_e(2L_t w_t \widehat{P}_{a,t}) + \beta_0 + \beta_1 t\right) + \epsilon_t, \tag{5.9}$$

where n_t is the number of animals counted at time t, L_t is the line length surveyed at time t, w_t is the truncation distance, $\widehat{P}_{a,t}$ is the estimated mean probability of detection for animals in the covered region a in year t, and $\log_e(2L_t w_t \widehat{P}_{a,t})$ is the offset term. Offsets in GLM are assumed known but $\widehat{P}_{a,t}$ is estimated, so a method is required for incorporating uncertainty in the estimation of $\widehat{P}_{a,t}$ into the trend analysis. This is most easily done using a bootstrap, in which the distance sampling and trend estimation analyses are repeated within each bootstrap resample (see Chapter 4 for further discussion).

We can generalize the above models further by adding covariates, to explain some of the variation between years that is not part of the linear trend.

5.3.2.2   *True abundances assumed fixed*
We again formulate the model of trend as

$$\widehat{N}_t = \beta_0 + \beta_1 t + \delta_t + \epsilon_t \tag{5.10}$$

with notation as before, but where $N_t = \beta_0 + \beta_1 t + \epsilon_t$ are now assumed fixed, so that we have simply decomposed the true abundances N_t into a linear trend term $\beta_1 t$ and a non-trend component $\beta_0 + \epsilon_t$.

We wish to make inferences about the trend (slope) β_1 in the context of the sampling errors δ_t, which are the only random part of the model. If we again assume the sampling errors are independently, identically, and normally distributed, then we obtain the same estimate of the trend β_1 as from simple linear regression in the previous section. We can also use the standard formulae for making inferences about uncertainty in the trend estimate (e.g. Steel and Torrie 1980: 253–8), but with two alterations. First, we use the average sampling variance in place of the RMS of the regression. Second, the degrees of freedom (df) for calculating statistics such as confidence limits should now be based on the degrees of freedom from the distance sampling density estimates, rather than the number of time periods. Since the density estimates are typically based on a large number of degrees of freedom, for simplicity we assume the degrees of freedom are infinite, leading, for example, to the use of the normal distribution rather than Student's t distribution in the formulae. With these alterations, the estimated variance of the trend estimate becomes

$$\widehat{\text{var}}(\hat{\beta}_1 \mid \underline{N}) = \frac{\widehat{\text{var}}(\hat{N} \mid \underline{N})}{\sum_{t=1}^{T} (t - \bar{t})^2}, \tag{5.11}$$

where $\bar{t}$ is the mean of the T times. Confidence intervals about the trend estimate can be calculated using

$$\hat{\beta}_1 \pm z_{\alpha/2} \sqrt{\widehat{\text{var}}(\hat{\beta}_1 \mid \underline{N})}, \tag{5.12}$$

where z_γ is the $100\gamma\%$ quantile from the standard normal distribution. Similarly, pointwise confidence bands around the trend line can be calculated using

$$\hat{\beta}_0 + \hat{\beta}_1 t \pm z_{\alpha/2} \sqrt{\widehat{\text{var}}(\hat{N}_{t,est} \mid \underline{N})} \tag{5.13}$$

and global confidence bands can be calculated using

$$\hat{\beta}_0 + \hat{\beta}_1 t \pm \sqrt{\chi^2_{(1-\alpha),2} \, \widehat{\text{var}}(\hat{N}_{t,est} \mid \underline{N})}, \tag{5.14}$$

where $\chi^2_{\gamma,\nu}$ is the $100\gamma\%$ quantile from a χ^2 distribution with ν degrees of freedom and $\widehat{\text{var}}(\hat{N}_{t,est} \mid \underline{N})$ is the estimated variance of an estimated point on the line, given by

$$\widehat{\text{var}}(\widehat{N}_{t,est} \mid \underline{N}) = \widehat{\text{var}}(\widehat{N} \mid \underline{N}) \left[\frac{1}{T} + \frac{(t - \bar{t})^2}{\sum_{t'=1}^{T} (t' - \bar{t})^2} \right]. \qquad (5.15)$$

Note that because the variances in eqns (5.11) and (5.15) do not contain the process error, they are not affected by the fit of the trend line to the data and their value will be the same regardless of the model used for trend.

Using the mallard example, we estimate the linear regression slope, β_1, as -1.37×10^4 (Fig. 5.3), the same as in the previous section. We already calculated the average sampling variance in Section 5.2.1 as 1.18×10^{11}, which gives us an estimate of $\text{var}(\hat{\beta}_1 \mid \underline{N})$ of 1.20×10^7 and so 95% confidence limits on the trend of $(-2.06 \times 10^4, -6.98 \times 10^3)$. These limits do not contain zero slope, so we conclude that there was a statistically significant decline in mallard numbers over the time period of the survey using a linear trend model. When we assumed that the N_t were random, the confidence interval included zero, so the conclusion is now very different. To put our conclusions from the fixed effect analysis into a biological context, the lower confidence limit on the trend slope of -2.06×10^4 ducks per year translates into an estimated decrease of one million ducks over the 49 years of the survey, or 10% of the initial estimated number. It is up to managers to decide whether such a decline is biologically significant or not.

An alternative approach that provides an assessment of the statistical significance of the trend is to use linear contrasts, a standard technique in analysis of variance (e.g. Steel and Torrie 1980: 177–81). The test involves calculating the weighted sum

$$\widehat{N}^* = \sum_{t=1}^{T} c_t \widehat{N}_t, \qquad (5.16)$$

where, for a linear trend, the weights are $c_t = (t - \bar{t})$. The test statistic is

$$z = \frac{\widehat{N}^*}{\sqrt{\widehat{\text{var}}(\widehat{N}^*)}}, \qquad (5.17)$$

where

$$\widehat{\text{var}}(\widehat{N}^*) = \sum_{t=1}^{T} c_t^2 \widehat{\text{var}}(\widehat{N}_t \mid N_t) \qquad (5.18)$$

and this statistic is assumed to follow a normal $(0, 1)$ distribution. This approach has the advantage of being easy to apply, and is a good candidate to provide an initial quick test for trend assuming true abundances are fixed.

Applying the above method to the mallard data, we obtain $\widehat{N}^* = -1.35 \times 10^8$ and $\widehat{\mathrm{var}}(\widehat{N}^*) = 1.29 \times 10^{15}$, which gives a test statistic of $z = -3.76$ and a corresponding p-value of 1.70×10^{-4}. Again, we conclude that there has been a statistically significant linear component to the decline in mallard numbers.

5.3.3   Smoothing

5.3.3.1   True abundances assumed random
Smooth trend models can be viewed as an extension of the linear models of the previous section. Instead of an intercept and linear trend, $\beta_0 + \beta_1 t$, we describe the trend as a smooth function of time, $s(t)$. Hence, eqn (5.6) becomes

$$\widehat{N}_t = s(t) + \delta_t + \epsilon_t \qquad (5.19)$$

while eqn (5.7) becomes

$$\log_e(\widehat{N}_t) = s(t) + \delta_t + \epsilon_t \qquad (5.20)$$

and eqn (5.8) becomes

$$\widehat{N}_t = \exp(s(t)) + \delta_t + \epsilon_t. \qquad (5.21)$$

There are numerous ways to smooth data, including polynomial regression, smoothing splines, local regression, and kernel methods (Hastie and Tibshirani 1990; Simonoff 1996; Bowman and Azzalini 1997; Loader 1999). Fewster *et al.* (2000) discuss a number of smoothing methods for estimating trend from wildlife survey data, and advocate the use of smoothing splines within the framework of generalized additive modelling (GAM). The paper gives a good introduction to GAMs, and a comprehensive treatment is given by Hastie and Tibshirani (1990). As an example, we fit eqn (5.19) to the mallard data (Fig. 5.4) using the function `gam` from the `mgcv` package (version 0.9-5; Wood 2000, 2003) in the free software R (R Development Core Team 2003). We used the default smoothing spline method (thin plate regression splines), and specified Gaussian (normal) errors and an identity link. We are again assuming that the $(\delta_t + \epsilon_t)$ are identically normally distributed, although weighted regression and other error structures are possible as discussed in the previous section.

With all smoothing methods, one must somehow specify the amount of smoothness of the function $s(\ )$: maximum smoothness corresponds to a linear function, as in the previous section, while minimum smoothness corresponds to exactly fitting each abundance estimate. A convenient way

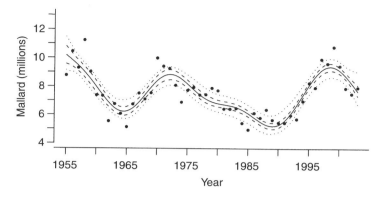

Fig. 5.4. Smoothing spline trend analysis of mallard data, with 8.35 df for the smooth selected using generalized cross-validation. Dashed and dotted lines show 95% bootstrap confidence bands for points on the regression line assuming the true abundances are fixed and random respectively.

to express the amount of smoothing is as the equivalent degrees of freedom (df); this can be loosely interpreted as the number of parameters used in fitting $s(\ )$ (a higher number therefore corresponds to a more wiggly curve). This parameter can be set subjectively, or according to some objective criterion such as minimum Akaike's information criterion (AIC) or generalized cross-validation score. The function gam can estimate the number of degrees of freedom for the smooth using generalized cross-validation—for the mallard data, 8.35 df were selected. The estimated trend (Fig. 5.4) seems to capture very well the major features of the time series that we described in Section 5.3.1, although perhaps underestimating the size of the trough in numbers in 1965 and the peaks in 1970 and around 1999. Residual plots and other graphical diagnostics (Hastie and Tibshirani 1990) were also satisfactory.

Fewster *et al.* (2000) recommend a subjective method of selecting the smoothness: start with a small value for degrees of freedom and increase it until the major features of the data appear to have been identified and further increases 'serve only to roughen the output'. Using this procedure (Fig. 5.5), we conclude that around 15 df are necessary, which is very close to the value that Fewster *et al.* (2000) found worked well for their farmland bird data (they recommend 0.3 times the length of the time series). Relative to the 8.35 df analysis, there is evidence of a small recovery during the 1970–85 decline, and a 'false start' to the 1985–99 recovery we identified earlier.

Given the chosen degrees of freedom, pointwise confidence bands can be easily generated using the mgcv package in R. However, if we used a data-driven method of estimating the degrees of freedom, we may wish to

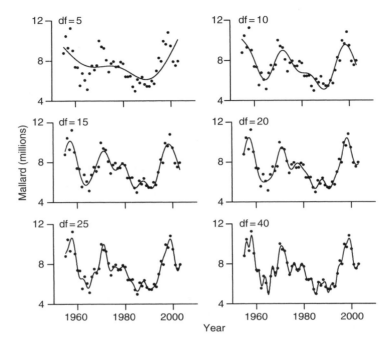

Fig. 5.5. Smoothing spline trend analysis of mallard data, with increasing degrees of freedom for the smooth. Subjectively, the smooth with 15 df captures the major features of the data, while those with greater degrees of freedom start to incorporate short-term 'noise'.

incorporate uncertainty in the estimate of degrees of freedom into the confidence bands. This can be achieved with the following computer-intensive approach. Generate a set of new 'pseudo-abundances' by resampling with replacement from the set of residuals $(\widehat{N}_t - \hat{s}_t)$, $t = 1, \ldots, T$, and adding these resampled values back to the $\hat{s}_t$. Fit the model to these new pseudo-abundances (including re-estimating the degrees of freedom for the smooth), and store the estimated trend values. Repeat this a large number of times (B times). To calculate $100(1 - \alpha)\%$ pointwise confidence bands for the smooth, order the estimated trend values at each time point and select the $(B+1)(\alpha/2)$ and $(B+1)(1-\alpha/2)$ highest values (the bootstrap percentile confidence interval, Buckland 1984). Confidence bands for the mallard data generated using 999 resamples are shown in Fig. 5.4. In this case, they are very similar to the analytic intervals (not shown), because the amount of smoothing varied little between resamples (2.5th and 97.5th quantiles of the distribution of degrees of freedoms were 7.45 and 8.72).

Global confidence bands for the entire smooth are harder to generate. The approach used for linear models is to widen the analytic pointwise bands, to take account of the multiple comparisons implicit in inferring the whole line at once. This is not appropriate for smooths because the global band is not necessarily uniformly wider than pointwise bands (see Hastie and Tibshirani 1990: 60–5). One approach is to display example trajectories from the above residual bootstrap (Hastie and Tibshirani 1990: Fig. 3.8). However, the plot quickly becomes unclear if more than a few trajectories are shown, but displaying only a few trajectories under-represents the variability of the smooth. An alternative is to display the narrowest bands that include the 'middle' $100(1 - \alpha)\%$ of the trajectories. A computer-intensive method of calculating such bands is as follows.

Perform the residual bootstrap as outlined above, and store each bootstrap trajectory. Start with a band that (just) incorporates the whole of every trajectory. We first estimate the upper confidence band. Define the 'upper area' of the band to be the average of the distance between the upper edge of the band and original estimated smooth across all time points times the number of time points. Identify the bootstrap trajectories that contribute to (i.e. touch) the upper edge of the wide starting band. Drop each of these in turn, to identify the one that leads to the biggest loss in upper area. Now drop that trajectory entirely, and again identify all trajectories that contribute to the (new) upper edge. Drop the next largest contributor to the upper area. Continue until $(B + 1)(\alpha/2) - 1$ values have been dropped. Repeat this procedure, this time focussing on the lowest trajectories and using maximum reduction in 'lower area' as the criterion, to estimate the lower confidence band. For this algorithm, 999 resamples may be too few, giving too much Monte Carlo variation. Further research on the properties and performance of the algorithm is needed.

5.3.3.2   *True abundances assumed fixed*

For a given smoothing method and degree of smoothing, the trend estimate will be the same as in the previous subsection. However, because only δ_t is now random, inferences about the variance of the trend estimate will be different—just as with the linear case, the trend will be estimated with greater precision.

In theory, variances and pointwise confidence limits can be calculated using the same principles as with the linear fixed-abundance case (Section 5.3.2.2). To take a relatively simple case, consider estimating smooth trend using a running line smoother (Hastie and Tibshirani 1990: 15–17). Here, the trend estimate in each time period is the predicted value from a linear regression of abundance estimates for times in a fixed size 'window' around that time. Larger window sizes produce smoother trends. The variance of the trend at each time period, assuming fixed N_t, can be calculated using eqn (5.11), but using only the data within the window for that time period. In other words, the variance will vary, depending on the sampling variance of the abundance estimates within the window.

Running line smoothers are conceptually simple, but tend not to perform well in practice (Hastie and Tibshirani 1990), which is why we did not apply them to the mallard data. Estimating analytic variance and confidence limits for more complex, better performing smoothers such as smoothing splines is more difficult. For this reason, in the case of fixed N_t, we propose a computer-intensive method, based on a parametric bootstrap (Davison and Hinkley 1997), which is simple to implement and will work for any smoother. Within each bootstrap replicate, generate a new set of pseudo-abundance estimates using the observed estimates and their associated sampling variance estimates. For example, if we assume for the mallard data that each abundance estimate is independently normally distributed, we can generate a new estimate for each year $t = 1, \ldots, T$ by sampling from normal($\widehat{N}_t, \widehat{\text{var}}(\widehat{N}_t)$). Then fit the smoother to the pseudo-data to obtain a new trend estimate, and store the estimated trend values. Repeat this process many times, and then the stored values can be used to estimate variances and confidence limits as required (see Davison and Hinkley 1997 for details). Figure 5.4 shows 95% pointwise confidence bands for the smooth mallard trend generated by taking the 25th and 975th ordered values at each point from 999 bootstrap resamples (i.e. the percentile method).

5.3.4   Trend estimation when samples covary

5.3.4.1   True abundances assumed random
Here, we relax the assumption that the sampling errors are independent, allowing them to covary. As discussed in Section 5.2.2, we expect positive covariances in the sampling errors from many distance sampling surveys, and we outlined a method for obtaining estimates of the covariances via the bootstrap.

An appropriate framework for estimating trends in this context is *variance components* (also called *random effects*) methods. The approach we outline here is described in more detail in the context of estimating temporal variation in survival rates by Burnham and White (2002) and Franklin *et al.* (2002). We again use the linear trend model

$$\widehat{N}_t = \beta_0 + \beta_1 t + \delta_t + \epsilon_t \qquad (5.22)$$

but now the sampling errors δ_t are no longer assumed to be independent. For the basic method, we assume the process errors ϵ_t are independent normal random variables with mean 0 and variance σ^2. We assume the sampling and process errors are independent of one another.

It is convenient to re-express this model in matrix format:

$$\underline{\widehat{N}} = \mathbf{X}\underline{\beta} + \underline{\delta} + \underline{\epsilon}, \qquad (5.23)$$

where $\mathbf{X}$ is the design matrix, β is the vector of model parameters, and $\underline{\delta}$ and $\underline{\epsilon}$ are the vectors of sampling and process errors. For a linear trend model, $\mathbf{X}$ is a $T \times 2$ matrix with 1s in the first column and $1, \ldots, T$ in the second, and $\underline{\beta}$ is a vector of length 2 (i.e. $(\beta_0, \beta_1)'$). If we denote the covariance matrix of the sampling errors as $\mathbf{W}$, then the covariance matrix of the random part of the model is $\mathbf{D} = \sigma^2 \mathbf{I} + \mathbf{W}$.

We wish to estimate $\underline{\beta}$ and also obtain an estimate of the covariance matrix $C(\hat{\underline{\beta}})$; this will allow us to obtain confidence limits on the trend and perform tests for the significance of the trend. There are potentially several ways to obtain these estimates; the above papers give a method based on the method of moments, which has the main advantage of computational efficiency. In summary, given a value for σ^2, we have:

$$\hat{\underline{\beta}}(\sigma^2) = (\mathbf{X}'\mathbf{D}^{-1}\mathbf{X})^{-1}\mathbf{X}'\mathbf{D}^{-1}\hat{\underline{N}} \qquad (5.24)$$

and

$$C(\hat{\underline{\beta}}) = (\mathbf{X}'\mathbf{D}^{-1}\mathbf{X})^{-1}. \qquad (5.25)$$

Assuming (approximate) normality of the $\hat{\underline{N}}$, the weighted residual sum of squares $[\hat{\underline{N}} - \mathbf{X}\hat{\underline{\beta}}(\sigma^2)]'\mathbf{D}^{-1}[\hat{\underline{N}} - \mathbf{X}\hat{\underline{\beta}}(\sigma^2)]$ has a central χ^2 distribution with $T - r$ degrees of freedom, where r is the number of model parameters, in this case two. Therefore a method of moments estimator of σ^2 is obtained by solving

$$T - r = [\hat{\underline{N}} - \mathbf{X}\hat{\underline{\beta}}(\sigma^2)]'\mathbf{D}^{-1}[\hat{\underline{N}} - \mathbf{X}\hat{\underline{\beta}}(\sigma^2)]. \qquad (5.26)$$

If $\hat{\sigma}^2$ is negative, it is set to zero.

Another potential output from a variance components analysis is shrinkage estimates of $\underline{N}$. These estimates are 'more accurate', on average, than the estimates we already have from the distance sampling analysis, in the sense that they have smaller mean square error if the model is correct. Shrinkage estimates are explained in more detail in the papers referenced above.

For illustration, we applied these methods to the mallard data, although we do not have estimates of sampling covariances, so we have assumed that they are zero. In reality, we expect the true covariances to be close to zero for two reasons. First, the correction factor from ground truthing is estimated independently each year. Second, spatial covariance is likely to be small due to the large number of transects and extensive stratification. Given assumed zero covariances, our estimate of σ^2 should be very similar to the estimate of $\mathrm{var}(N)$ from section 5.2.1, and indeed our estimate from the variance components model was $\hat{\sigma}^2 = 2.41 \times 10^{12}$, identical to three significant figures. The trend estimate was nearly identical to that from the simple linear regression of Section 5.3.2.1: $\hat{\beta}_1 = -1.27 \times 10^4$ with $\widehat{\text{var}}(\hat{\beta}_1) = 2.59 \times 10^8$. Because of the low level of sampling variation in the example, the shrinkage estimates of the N_t were very close to the observed $\widehat{N}_t$, and so are not shown. In summary, for this example, variance components methods add nothing to a simple linear regression analysis. They are only required where the assumptions of the more simple methods are violated, for example, by substantial sampling covariances.

These methods could be extended by allowing for non-independence in the process errors, although more complex fitting methods (such as Markov chain Monte Carlo, MCMC) would then be required. An example using mark-recapture data is given by Johnson and Hoeting (2003). Another extension would be to allow for smooth rather than linear trends. Expanding $\mathbf{X}$ to include polynomial functions of time would be one relatively straightforward way to achieve this. Alternatively, the model could be specified as a generalized additive mixed model (GAMM). GAMMs are an active area of statistical research (Lin and Zhang 1999; Rice and Wu 2001; Mackenzie *et al.* in preparation).

5.3.4.2   *True abundances assumed fixed*

Given an estimate of the sampling variances and covariances, for example, from the bootstrap, the contrasts-based test for trend of Section 5.3.2.2 is readily applied, with the variance of the weighted sum now given by:

$$
\widehat{\text{var}}(\widehat{N}^*) = \sum_{t=1}^{T} c_t^2 \widehat{\text{var}}(\widehat{N}_t \mid \underline{N})
$$

$$
+ 2 \sum_{t_1=2}^{T} \sum_{t_2=1}^{t_1-1} \left(c_{t_1} c_{t_2} \widehat{\text{cov}}(\widehat{N}_{t_1}, \widehat{N}_{t_2} \mid \underline{N}) \right). \tag{5.27}
$$

The other methods given in that section can also be extended in a similar manner.

However, here we propose a more straightforward approach that has the advantage of making only weak distributional assumptions about the sampling errors. The approach involves a nonparametric bootstrap similar to that of Section 5.2.2, but in this case, within each bootstrap replicate, the trend estimate is recalculated using $\widehat{N}_t$ derived from the resampled data. This yields one trend estimate for each bootstrap resample, and inferences can then be made from this collection of estimates about the variance of the trend, confidence limits, and other quantities of interest. The key is to resample according to the survey design (e.g. preserving permanent transects where they appear in a resample); this will preserve the covariance structure in the resamples.

Unfortunately we cannot provide an example of this approach with the mallard data as we do not have the transect-level data. In any case, we expect the results would be very similar to those assuming independence between annual abundance estimates because only weak covariances are expected for the reasons outlined in the previous subsection. An example of a similar analysis is given by Fewster *et al.* (2000) using farmland bird census data taken on permanent plots.

5.4 Spatio-temporal analysis

In the previous section, we focussed on the analysis of estimates of abundance made at the level of the study area. In distance sampling surveys (and many others), abundance estimates can also be made at the level of the transect and these can form the inputs to a rich variety of spatio-temporal analyses. We briefly discuss some of these here.

Because it is rarely possible to estimate the detection function separately for each transect, the transect-level estimates of abundance will usually covary, and this will need to be taken into account in any analysis.

5.4.1 *Transect-level models of trend*

If the survey design includes repeated visits to the same set of transects, we can potentially generalize the models of the previous section to allow for differences in trends and/or other components between transects. For example, the global linear model of eqn (5.6) becomes

$$\widehat{N}_{it} = \beta_{0i} + \beta_{1i}t + \delta_{it} + \epsilon_{it}, \tag{5.28}$$

where i indexes transect, and the log-linear model of eqn (5.7) becomes

$$\log_e(\widehat{N}_{it}) = \beta_{0i} + \beta_{1i}t + \delta_{it} + \epsilon_{it}. \tag{5.29}$$

This kind of model can be fitted using the 'linear route regression' methods commonly used in the analysis of ornithological index data (Geissler and Sauer 1990; Thomas 1996, 1997; Link and Sauer 1997a). ('Route' is synonymous with transect in this context.) In the simpler versions of this method, a linear regression is performed on (log-transformed) abundance estimates independently for each transect. The primary goal of the analysis is to estimate a global linear trend, and this is done by taking a weighted average of the transect-level estimates β_{1i}. Various types of weightings can be used (weighting by estimated average abundance, by area represented by the transect, or by estimated precision); a significant problem is that it is not clear which weighting scheme is best (Thomas and Martin 1996). Variance of the global trend estimate is estimated from between-transect variation in the β_{1i}, so in this sense, the method should produce similar estimates to the bootstrap analysis of Section 5.3.4.2. These methods can be viewed as a type of design-based estimator, where transects provide replicate samples of the global trend.

Route regression methods were developed in the context of index surveys with a large amount of missing data (i.e. transects with missing surveys in many years), and also of unknown transect-level effects in the form of differences between observers. If the primary goal is to estimate global trend and these difficulties do not exist (as will be the case for well-designed distance sampling studies), then the methods presented in previous sections may well be preferable. We note that non-linear route regression methods have been developed (James *et al.* 1996; Link and Sauer 1997a,b).

A simplifying (if unrealistic) assumption is that $\beta_{1i} = \beta_1$, that is all sites follow the same trend. This is the basis of the basic 'sites-by-years' models of ter Braak *et al.* (1994), Pannekoek and van Strien (1996) and Fewster *et al.* (2000); in the last case, a smooth rather than linear global trend model was used. These authors also recommend analyses with trend modelled at intermediate levels (e.g. by region within the study area, or according to covariates such as habitat), and the use of model selection criteria to select the best level to model trend. These methods assume that the $(\epsilon_{it} + \delta_{it})$ are independently and identically distributed, although Fewster *et al.* (2000) used a bootstrap procedure for estimating trend variance similar to that of Section 5.3.4.2.

An intermediate possibility is to treat the β_{1i} as random effects. This gives the extended model:

$$\widehat{N}_{it} = \beta_{0i} + \beta_{1i}t + \epsilon_{it} + \delta_{it}, \tag{5.30}$$

where

$$\beta_{0i} = \beta_0 + \gamma_{0i} \tag{5.31}$$

$$\beta_{1i} = \beta_1 + \gamma_{1i}. \tag{5.32}$$

The γ_{0i} are independently and identically distributed random variables with zero mean, as are the γ_{1i}. Such a model allows estimation of both the global trend (β_1) and transect-level trends (the γ_{1i}). To our knowledge, no one has yet fitted this type of model to wildlife abundance data. As an alternative to allowing γ_{0i} and γ_{1i} to be independent random variables, one could envisage imposing some structure upon them—for example, assuming some spatial dependence. This sort of approach comes close to the kriging-type analyses of spatial statistics. Some covariates could also be included as fixed effects, for example, habitat or other spatial covariates.

These random or mixed effects models potentially could be further extended to allow for smooth trends, using a GAMM framework.

5.4.2 *Spatio-temporal modelling*

The spatial methods of Chapter 4 can be readily extended to include temporal as well as spatial covariates. This allows prediction of spatial density surfaces at successive time points, and by integrating under the spatial surface at each time point, trends in abundance over time can be estimated. Such an approach will have most utility when interest focusses on predicting trend in subsets of the study area, or in comparing trends between regions (perhaps under different management regimes). An example of this kind of approach is Marques (2001: 13–61).

Residual variation in abundance or trend could potentially be modelled as a random effect, providing a link with the mixed-effects methods of the previous subsection (eqn (5.30)).

5.5 Process models

A series of abundance estimates derived from distance sampling surveys can be used to estimate the parameters of a stochastic population dynamics model. This can be useful for a number of reasons. First, with the inclusion of management-related covariates, it allows us to investigate the effect on future abundances of different management regimes. This type of analysis could be used, for example, in a population viability analysis or to set harvest quotas. Second, we can separate the effects of demographic and environmental stochasticity, and investigate the effect of environmental covariates (such as weather) on population vital rates. Third, we can obtain estimates of the true N_t that are consistent with what is biologically possible. This is in contrast with empirical methods, such as shrinkage estimates from variance components models, which can easily suggest, for example, that an increase in abundance has occurred between successive years, the magnitude of which is biologically implausible. A disadvantage of the methods outlined in this section is that they are often more difficult to apply than many empirical methods, and they certainly require more data than simple linear trend methods. They also force us to make explicit assumptions about the population dynamics, although multi-model inference is possible (see below).

A sequence of abundance estimates alone is not very informative for some of the population processes. For example, they cannot alone distinguish between a population with low survival rates and high birth rates and one with high survival rates and low birth rates. Often, there is other information on birth rates and/or survival rates, for example, from ecological studies, that can be used in an integrated analysis together with the abundance estimates.

We outline a framework for specifying biological process models using a simple example, and discuss how the model might be fitted to survey data. This framework and the fitting methods are described in more detail

by Buckland *et al.* (2004) and Newman *et al.* (in preparation), and further
examples are provided by Buckland *et al.* (2004).

5.5.1   *State-space models*

In our example population dynamics model, we follow only two types of
animals: juveniles and adults. Each time-step of the model represents one
year, and each year starts in the autumn with a distance sampling survey of
the population, which yields separate estimates of the number of adults and
juveniles. If we assume that these estimates follow a lognormal distribution,
then we can write down the following *observation process* equations:

$$\widehat{N}_{j,t} \sim \text{lognormal}\left(\log_e(N_{j,t}), \sigma_{j,t}^2\right), \tag{5.33}$$

$$\widehat{N}_{a,t} \sim \text{lognormal}\left(\log_e(N_{a,t}), \sigma_{a,t}^2\right), \tag{5.34}$$

where $\widehat{N}_{j,t}$ and $\widehat{N}_{a,t}$ are the estimated number of juveniles and adults at
time t, $N_{j,t}$ and $N_{a,t}$ are the true numbers, and $\sigma_{j,t}^2$ and $\sigma_{a,t}^2$ are the
sampling variances of the estimates, which we will assume are known.

We now specify the biological part of the model, the *state process*. This
will give us the distribution of $N_{j,t+1}$ and $N_{a,t+1}$ given $N_{j,t}$ and $N_{a,t}$. In
doing this, we have found it convenient to separate out different events that
take place during the year into separate sub-processes, each occurring in
turn. In our example population, we distinguish four sub-processes: a harvest of adults that occurs immediately after the survey, overwinter survival,
age incrementation, and then breeding.

We assume that the number of adults harvested is known, so the harvest
sub-process equations are simple:

$$u_{1,j,t} = N_{j,t}, \tag{5.35}$$

$$u_{1,a,t} = N_{a,t} - c_{a,t}, \tag{5.36}$$

where $u_{1,j,t}$ and $u_{1,a,t}$ are the number of juveniles and adults after harvest
at time t, and $c_{a,t}$ is the number of adults harvested. Extension to the case
where the number of animals harvested is only estimated is straightforward
(by including another observation equation).

We assume that survival is a simple binomial process, with one constant
survival rate for adults and another for juveniles, giving

$$u_{2,j,t} \sim \text{Binomial}\left(u_{1,j,t}, \phi_j\right), \tag{5.37}$$

$$u_{2,a,t} \sim \text{Binomial}\left(u_{1,a,t}, \phi_a\right), \tag{5.38}$$

where $u_{2,j,t}$ and $u_{2,a,t}$ are the numbers of juveniles and adults at time t after
the survival sub-process, and ϕ_j and ϕ_a are the juvenile and adult survival

probabilities. In a more complex model, either or both of these could be density dependent, could depend on covariates (such as winter temperature), could have trends over time, and/or vary randomly over time.

Like the harvest sub-process, the aging sub-process is deterministic. Juveniles become adults, and adults remain as adults:

$$u_{3,j,t} = 0, \tag{5.39}$$

$$u_{3,a,t} = u_{2,j,t} + u_{2,a,t}. \tag{5.40}$$

Lastly, for breeding, we will assume that the number of offspring follows a Poisson distribution, with expected number of offspring per adult a constant, ν:

$$N_{j,t+1} \sim \text{Poisson}\,(u_{3,a,t}\nu)\,, \tag{5.41}$$

$$N_{a,t+1} = u_{3,a,t}. \tag{5.42}$$

The state process model can be conveniently summarized by writing down the expected number of animals using a set of transition matrices:

$$E\left(\underline{N}_{t+1} \mid \underline{N}_t, \underline{c}_t, \underline{\Theta}\right) = \mathbf{BAS}\left(\underline{N}_t - \underline{c}_t\right) \tag{5.43}$$

where

$$\underline{N}_t = \begin{pmatrix} N_{j,t} \\ N_{a,t} \end{pmatrix}, \qquad \underline{c}_t = \begin{pmatrix} 0 \\ c_{a,t} \end{pmatrix}, \qquad \underline{\Theta} = \begin{pmatrix} \phi_a \\ \phi_j \\ \nu \end{pmatrix},$$

$$\mathbf{B} = \begin{pmatrix} 0 & \nu \\ 0 & 1 \end{pmatrix}, \qquad \mathbf{A} = \begin{pmatrix} 0 & 0 \\ 1 & 1 \end{pmatrix}, \qquad \mathbf{S} = \begin{pmatrix} \phi_j & 0 \\ 0 & \phi_a \end{pmatrix}.$$

The expectation is exact in this case, because each sub-process is a linear combination of the states output by the previous sub-process. In more complex models, it can be approximate. Writing the model in this way is nevertheless a useful way to clarify the sequencing of the sub-processes and their average effect (Buckland *et al.* 2004).

The product of the matrices $\mathbf{BAS}$ is a Leslie matrix, a form that is familiar to anyone working with matrix population models (Caswell 2001).

$$\mathbf{BAS} = \mathbf{L} = \begin{pmatrix} \nu\phi_j & \nu\phi_a \\ \phi_j & \phi_a \end{pmatrix}. \tag{5.44}$$

There are three parameters in the model: ϕ_a, ϕ_j, and ν. We wish to estimate these and the unknown states $N_{j,t}$ and $N_{a,t}$, for $t = 1, \ldots, T$.

The above example is relatively simple, but complex models can be constructed using the same approach. The number of states can be much larger, representing animals of different ages, stages, sexes, locations, and even species. There can be many more sub-processes representing movement, transition between stages, inter-species interactions, etc. Processes may be deterministic or stochastic, and may be nonlinear functions of previous states and of covariates. The observation process can include other types of survey data, and some parts of the population may be entirely unobserved.

The general term for this type of model is a *state-space model*. It comprises a set of two linked equations, one describing the *observation process* and the other the *population* or *state process*, together with a description of the starting distribution of the states. Written in generic form, these are:

$$\widehat{\underline{N}}_t \sim F_t(\underline{N}_t, \underline{\Theta}), \tag{5.45}$$

$$\underline{N}_{t+1} \sim G_t(\underline{N}_t, \underline{\Theta}), \tag{5.46}$$

$$\underline{N}_0 \sim G_0(\underline{\Theta}), \tag{5.47}$$

where $F_t(\ )$ is the observation process distribution, $\underline{\Theta}$ are the model parameters, $G_t(\ )$ is the state process distribution, and $G_0(\ )$ is the initial state distribution. Note that eqn (5.45) assumes that the sampling errors are independent, and that eqn (5.46) assumes that the true abundance at $t+1$ depends only on the abundance at t (i.e. is first-order Markov); these assumptions will be relaxed in the next section.

Fitting state-space models to ecological data is often not straightforward. Analytic methods are possible in some restricted cases. For example, if both the observation and state process models are linear and Gaussian, then the Kalman filter (Harvey 1989) can be used. Such models have been applied with good effect to avian census data, combined with separate information from mark-recapture studies, in a series of papers by Besbeas *et al.* (2002, 2003, in press). Approximate analytic methods exist for some nonlinear models (e.g. the extended Kalman filter, Harvey 1989). However, inference usually uses one of two Bayesian Monte Carlo methods: sequential importance sampling (SIS) and MCMC. These two methods are compared in this context by Newman *et al.* (in preparation); examples of SIS applied to ecological data are Trenkel *et al.* (2000), Thomas *et al.* (in press), and Newman and Lindley (in preparation), and examples that use MCMC are Meyer and Millar (1999), Millar and Meyer (2000), and Newman (2000).

5.5.2 *Generalizing state-space models*

An assumption of the standard state-space framework is that the observation process distribution depends only on the current state, not on the states at other time points. This will be violated in distance sampling surveys where there is sampling covariance, as discussed in Section 5.2.2. The framework of the previous section is readily generalized to allow for sampling covariance, for example, by specifying a multivariate log-normal distribution for the observations. In this case, standard SIS algorithms could no longer be used to fit the model because they assume that observations in different years are independent. An alternative approach would be to incorporate explicitly the distance sampling estimation into the model— the input data would now be the perpendicular distances observed on individual lines. This would significantly increase the complexity of the observation model, and require the state model to specify the distribution of animals in space.

Standard state-space models are first-order Markov, that is, the state at time t depends only on the state at the previous time point. For some ecological models, it may be appropriate to allow this dependence to be on states further back in time. An example of this is Newman and Lindley (in preparation). Such an extension poses no problems for either of the Monte Carlo inference procedures. Such generalized state-space models are sometimes referred to as *hidden Markov* or *hidden process* models.

We mentioned in Section 5.2.5 that process modelling can be used to separate out demographic and environmental sources of stochasticity in the true abundances. Demographic stochasticity arises from the stochastic nature of some of the population sub-processes. In the example of the previous section, adult survival is modelled as a binomial process, so the exact number of adults surviving will be a random variable even if the survival probability is fixed. The importance of this source of variation declines as population size increases (the relative variance of a binomial process is inversely proportional to the number of trials). Environmental stochasticity arises from seemingly random variation in environmental conditions that affect the population vital rates such as survival. This source of variation can be equally important for small and large populations. Environmental stochasticity is not accounted for in our example model, but could be added by allowing one or more biological process parameters to be random variables. Such models are called *hierarchical state-space models*. Demographic and environmental stochasticity are discussed in more detail by White (2000) and Caswell (2001).

In many cases we will not be certain about which biological model to use. For example, we may not know which sub-process might be density dependent, or we may not know in advance which covariates are important in governing vital rates. In such cases, we will want to make inferences based on multiple models. This is possible with both Monte Carlo procedures (Newman *et al.* in preparation), although MCMC tools are better developed for the case in which there are numerous potential models.

Another way to relax the strong assumptions about the form of biological relationships in process models would be to use semi-parametric rather than parametric forms for relationships that were not known with certainty. For example, the density-dependent relationship between juvenile survival and adult population size could be specified as a smooth curve, with zero slope at infinite population size. To our knowledge, such semi-parametric models have not yet been developed in the biological literature.

5.6 Other analysis methods

5.6.1 *Time series methods*

Standard time series methods can be divided into two broad categories: spectral analysis and Markov methods. Spectral analysis involves decomposing the time series into a set of simple functions (e.g. sine functions) of different frequencies (Warner 1998). While this may be useful as a data summary technique, it would appear to have little practical relevance to wildlife managers. Markov methods involve models where the count at the current time is a function of counts in previous time periods (e.g. autoregressive integrated moving average, Box *et al.* 1994). Additional features of the data (such as cycles and dependence on covariates) may then be superimposed.

Neither method in standard form takes account of measurement error, which can cause substantial biases; see, for example, the critique by Shenk *et al.* (1998) of the Markov methods of Dennis and Taper (1994). Markov methods are readily extended by including an observation process equation, resulting in an empirical state-space model. These kinds of model were applied to WBPHS data by Jamieson and Brooks (2003) using MCMC.

Empirical state-space models and the biologically-based models of Section 5.5 have similar data requirements and fitting issues. Inferences from biological process models are likely to be more biologically realistic, so they are usually to be preferred. One advantage of empirical methods is that they do not require strong assumptions about the biological population processes.

5.6.2 *Quality control methods*

One aim of many monitoring programmes is to provide an early warning of populations that are in decline. One problem with the trend estimation methods described in previous sections is that it may take several years before a sudden precipitous decline in numbers shows up in the trend estimate. By contrast, statistical quality control methods are designed to give an early warning, by testing each new observation to determine whether it is 'unusual' in the context of the previous observations. Such methods were originally developed for industrial applications, as a way of identifying when a system (e.g. a factory production line) is going 'out of control' so that the system can be stopped and remedial measures taken. Specific methods include cumulative sums (CUSUMs), sequential probability ratio tests (SPRTs), and control charts (Montgomery 1996). For example, CUSUMs involve calculating the sum of deviations of successive observations from their 'target' value (usually the mean of previous values). If this measure exceeds the allowable bounds (usually three standard deviations from the mean, calculated using previous values), then the alarm is raised. Examples of the application of these methods to monitoring biological populations include Schipper *et al.* (1997), Pettersson (1998), and Anderson and Thompson (in press). Rexstad and Debevec (in preparation) developed a metric that explicitly considers the distinction between sampling and population variation.

5.7 Survey design

Long-term monitoring programmes are often initiated without clear objectives. Typically, there might be three stages to the programme: estimation of abundance, using the data from the first survey(s); empirical estimation of trend in abundance, once data from a few surveys are available; and fitting of process models to the time series, perhaps to evaluate the likely effects of different management regimes, when sufficient surveys have been carried out. Often, programmes are developed with only the second objective in mind, in which case it may not be possible to estimate absolute abundance, and process modelling is perhaps conducted some time later as an afterthought. We recommend that all three objectives be addressed if at all possible. Often, surveys are designed to yield indices of abundance, rather than estimates of absolute abundance, on the grounds that fewer parameters need be estimated, and precision is improved. However, in long-term programmes, indices of abundance often show trends generated by changes in detectability rather than changes in abundance (e.g. Pollock *et al.* 2002; Norvell *et al.* 2003). Distance sampling surveys allow detectability and changes in detectability to be modelled, so that bias from such sources is eliminated.

5.7.1 *Repeating transects*

Schemes for sampling through time are often a compromise. If we seek high precision on estimated mean abundance over the time period, then we would select new transects each year, giving good spatial coverage of the region. If however we are only interested in precise estimates of change, we would visit the same set of (permanent) transects each year.

As noted in Section 5.2.2, we expect estimates of abundance over time to be positively correlated when the same transects are surveyed each year. Suppose by chance we chose a design in which transects sampled a higher proportion of high density areas than of low density areas. The abundance estimate is then likely to exceed true abundance. If in the following year, we return to those same transects, and areas of high density remain largely where they were in the previous year, then we again expect an estimate that exceeds true abundance, and positive correlation is induced. The effects of this are most easily seen if there are just two time points, $t = 1, 2$. Then

$$\mathrm{var}(\widehat{N}_1 + \widehat{N}_2) = \mathrm{var}(\widehat{N}_1) + \mathrm{var}(\widehat{N}_2) + 2\,\mathrm{cov}(\widehat{N}_1, \widehat{N}_2) \qquad (5.48)$$

$$\mathrm{var}(\widehat{N}_1 - \widehat{N}_2) = \mathrm{var}(\widehat{N}_1) + \mathrm{var}(\widehat{N}_2) - 2\,\mathrm{cov}(\widehat{N}_1, \widehat{N}_2) \qquad (5.49)$$

and so mean estimated abundance, $(\widehat{N}_1 + \widehat{N}_2)/2$, has higher variance when the same transects are retained than when new transects are used in year two, while estimated change, $\widehat{N}_1 - \widehat{N}_2$, has lower variance.

If the initial sample of transects is chosen according to an appropriate random scheme (e.g. simple random sampling, systematic sampling with a random start, or stratified random sampling), then unbiased estimates of trend with higher precision are obtained by retaining the same transects over time. If however the transects are not chosen randomly, then bias in trend can be anticipated. For example, if sites with particularly good habitat are selected, and permanent transects are located in them, subsequent trend estimates tend to be pessimistic. This is because habitat changes over time will, on average, reduce the suitability of good sites and increase the suitability of poor sites—an example of 'regression to the mean'.

Designs in which some transects are retained and some replaced each year are useful when estimation of both mean abundance and trend in abundance are of interest; they also allow recovery from an initial poor design, relative to a scheme in which all transects are retained (Underwood 2004). Such schemes (called *panel designs*) have been in use since the 1940s in other fields (Jessen 1942), but they have seldom been used for wildlife surveys; a recent example is Atkinson *et al.* (2003). Urquhart and Kincaid (1999) describe designs of ecological surveys that have this property.

There are many types of panel design, based on the idea that sets or panels of units should appear in the same surveys. Common strategies include the serially alternating design, in which panels are sampled in every kth survey, and the rotating panel design, in which one panel leaves the design and one panel enters in each survey. Combinations of these two strategies are possible. Other strategies are described by Urquhart and Kincaid (1999).

Underwood (2004) developed strategies in which a set of permanent transects is selected, using simple random or systematic sampling, and these are supplemented each year by a set of temporary transects. Data from one year are used to estimate animal density as a function of location, and probability proportional to size sampling is used to concentrate the temporary transects for the following year in areas with higher predicted density. This increases the number of detections and improves precision.

5.7.2 *Sample size*

Sample size requirements for distance sampling surveys at a single time point are discussed by Buckland *et al.* (2001: 240–8). Given estimates of the expected encounter rates and variability, one can use their methods to determine the amount of survey effort (transect length or number of points) required to obtain a given coefficient of variation of the abundance estimate. In the context of monitoring for trend, we are interested in how such a coefficient of variation might translate into precision of a trend estimate, or equivalently the power to detect a given level of trend. In this section, we focus on power analysis because the goals of many monitoring programmes are framed in those terms (e.g. 90% power to detect a 50% decline in population size over 25 years, Butcher *et al.* 1993: 199). We assume that the reader is familiar with the basics of statistical power analysis (e.g. Peterman 1990; Taylor and Gerrodette 1993; Thomas and Juanes 1996; Steidl *et al.* 1997; Steidl and Thomas 2001). We focus on simple methods that can be used to provide a quick initial 'ball park' for required power, before briefly discussing more complex power analyses.

5.7.2.1 *Linear trend models—true abundances assumed random*

We return to the linear trend model of eqn (5.6), which could be fitted using simple linear regression. The usual test for the statistical significance of the trend is to calculate

$$t(\hat{\beta}_1) = \frac{\hat{\beta}_1}{\sqrt{\widehat{\mathrm{var}}(\hat{\beta}_1)}}, \tag{5.50}$$

which is assumed to follow a Student's t-distribution with $T - 2$ degrees of freedom, where $\widehat{\mathrm{var}}(\hat{\beta}_1)$ is calculated as

$$\widehat{\mathrm{var}}(\hat{\beta}_1) = \frac{\mathrm{RMS}}{\sum_{t=1}^{T}(t - \bar{t})^2}, \tag{5.51}$$

where RMS is the residual mean square from the regression. When the time steps are equal,

$$\sum_{t=1}^{T}(t-\bar{t})^2 = \frac{T(T-1)(T+1)}{12}.$$ (5.52)

Assuming the test is two-tailed (i.e. we are interested in both significant declines and significant increases), the power $(1-\tau)$ of the t-test is given by

$$1-\tau = 1 - F_t(t_{1-\alpha/2,\nu},\nu,\eta) + F_t(t_{\alpha/2,\nu},\nu,\eta),$$ (5.53)

where $F_t(x,\nu,\eta)$ is the cumulative distribution function of the noncentral t distribution with ν degrees of freedom and noncentrality parameter η, evaluated at x, and $t_{\gamma,\nu}$ is the $100\gamma\%$ quantile from a central t distribution with ν degrees of freedom. The noncentrality parameter η is given by

$$\eta = \frac{\beta_1}{\sqrt{\mathrm{var}(\hat{\beta}_1)}},$$ (5.54)

where β_1 and $\mathrm{var}(\hat{\beta}_1)$ are the true slope and variance of the slope estimator—note the similarity to eqn (5.50). β_1 and $\mathrm{var}(\hat{\beta}_1)$ are not known, but we can calculate the power we would expect for some hypothesized values of these parameters, $\beta_{1,\mathrm{hyp}}$ and $\mathrm{var}_{\mathrm{hyp}}(\hat{\beta}_1)$, where

$$\mathrm{var}_{\mathrm{hyp}}(\hat{\beta}_1) = \frac{\sigma^2_{\mathrm{err,hyp}}}{\sum_{t=1}^{T}(t-\bar{t})^2}$$ (5.55)

and $\sigma^2_{\mathrm{err,hyp}}$ is the hypothesized variance of the error term $(\delta_t + \epsilon_t)$ in eqn (5.6).

If we are investigating the power of a monitoring programme that is already running, we could use the calculated RMS as an estimate of σ^2_{err} and use this value for $\sigma^2_{\mathrm{err,hyp}}$, or we could perhaps calculate an upper confidence limit on σ^2_{err} and use this. If we are planning a new programme, we can obtain an indication of the likely variance of the sampling errors (σ^2_{δ}) from the sample size formulae of Buckland et al. (2001: 240–8). If we can get a defensible hypothesized value for the variance of the process errors (σ^2_{ϵ}), we could obtain $\sigma^2_{\mathrm{err,hyp}}$ by adding the two variances together. One approach would be to perform the power analysis at various values of $\sigma^2_{\mathrm{err,hyp}}$; another would be to estimate values from other similar studies, by subtracting the estimated sampling variance from the RMS. For example, in the mallard study, the RMS from the linear regression was 2.54×10^{12} and the estimated average sampling variance was 1.18×10^{11}, giving an estimate of process error variance of $\hat{\sigma}^2_{\epsilon} = 2.42 \times 10^{12}$.

As we mentioned in Section 5.3.2.1, it is probably more defensible to perform simple linear regression on log-transformed abundance estimates (eqn (5.7)), as the model then corresponds with our usual assumption in distance sampling that the sampling errors are log-normally distributed. It is also easier to justify the assumption of standard linear regression that the errors have the same variance. As we said, $\lambda = \exp(\beta_1)$ is then the average annual rate of change of the population. The required variances to specify $\sigma^2_{\text{err,hyp}}$, which is now on the log scale, can be expressed conveniently as coefficients of variation on the untransformed scale and transformed using

$$\text{var}(\log_e(\cdot)) = \log_e([\text{cv}(\cdot)]^2 + 1). \tag{5.56}$$

For example, the sample size formulae of Buckland *et al.* (2001: 240–8) yield an expected coefficient of variation for the abundance estimate, and this can be readily converted to the required σ^2_δ using the above formula. The hypothesized error variance $\sigma^2_{\text{err,hyp}}$ can be expressed as a coefficient of variation by inverting the above formula.

In the log-linear regression framework, the components of a power analysis are: the hypothesized annual rate of change λ; the error coefficient of variation on the untransformed scale; the number of time periods T; the α-level; and the power $(1 - \tau)$. Given any four of these, eqn (5.53) can be used to calculate the fifth.

As an illustration, we calculated the number of years of monitoring required to achieve a power of $(1 - \tau) = 0.8$ using an α-level of 0.05 over a range of annual rates of change and coefficients of variation (Fig. 5.6(a)). This sort of analysis can be used when designing a monitoring programme

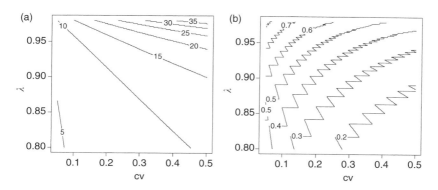

Fig. 5.6. (a) The number of years required so that the power to detect a log-linear population trend is 0.8, for a range of annual rates of population change (λ) and error coefficients of variation (cv), using a two-tailed t-test and assuming $\alpha = 0.05$. (b) The relative size of the population after this number of years.

to help decide what monitoring goals are feasible. For example, if we anticipate an error coefficient of variation of 0.4, it will be 20 years before we
will have a power of 0.8 to detect a change of 0.95 per year (a decline of
5% per year). After 20 years at this rate of change, the population will be
below 40% of its initial size (Fig. 5.6(b); more precisely, $0.95^{20} = 0.36$).

Methods of power analysis for linear and log-linear trend models, including the above models, are discussed by Gerrodette (1987; see also Link
and Hatfield 1990 and Gerrodette 1991). The above analyses can be easily
programmed; alternatively, Gerrodette has written user-friendly software,
available at http://swfsc.nmfs.noaa.gov/PRD/Software/Trends.html.

5.7.2.2 *Linear trend models—true abundances assumed fixed*

Power calculations are very similar to the previous subsection, except that
the standard normal distribution is used in place of the t-distribution, and
only the sampling variance is used in calculating the regression error variance. The statistical significance of the linear regression trend estimate is
therefore determined by calculating

$$z(\hat{\beta}_1) = \frac{\hat{\beta}_1 \mid \underline{N}}{\sqrt{\widehat{\mathrm{var}}(\hat{\beta}_1 \mid \underline{N})}}, \tag{5.57}$$

where $\widehat{\mathrm{var}}(\hat{\beta}_1 \mid \underline{N})$ is calculated using eqn (5.11). The statistic $z(\hat{\beta}_1)$ is
assumed to follow a standardized normal distribution. Power is now
given by

$$1 - \tau = 1 - F_z(z_{1-\alpha/2} - z(\beta_1)) + F_z(z_{\alpha/2} - z(\beta_1)), \tag{5.58}$$

where $F_z(x)$ is the cumulative density function of the standard normal distribution, z_γ is the $100\gamma\%$ quantile from the standard normal distribution,
and $z(\beta_1)$ is

$$z(\beta_1) = \frac{\beta_1 \mid \underline{N}}{\sqrt{\mathrm{var}(\hat{\beta}_1 \mid \underline{N})}}. \tag{5.59}$$

We can now calculate power using hypothesized values for $\beta_1 \mid \underline{N}$ and
$\mathrm{var}(\hat{\beta}_1 \mid \underline{N})$. Note that planning a study when fixed-abundance analyses
are anticipated is easier than the random-abundance case, as we do not
have to provide values for the variance of the process errors.

Preliminary analyses such as those shown in Fig. 5.6 can easily be generated for the fixed-abundance case. The coefficient of variation is now
simply the hypothesized sampling error coefficient of variation. The use of

the z-test rather than the t-test means that the required times are about 1 year less than shown in Fig. 5.6.

5.7.2.3 *More complex power analyses*

In practice, the above methods may be too simplistic for use when planning a major monitoring programme. For example, we may wish to consider the trade-off between transect line length and number of lines, or the effect of permanent transect lines on power—both of which require consideration of the spatial patterns of animal abundance. Another example is power analysis for more complex trend estimation methods. In some cases, more complex analytic power analysis methods are available (van der Meer 1997; Urquhart *et al.* 1998), but a useful general omnibus method is to estimate power via simulation. Data are simulated under the alternative hypothesis, and the proportion of times these data yield a statistically significant result using the analysis method under consideration is an estimate of the power of the test. Examples of simulation studies of power are Link and Hatfield (1990) and Thomas and Juanes (1996).

5.7.3 *Planning long-term studies*

Designing, starting, and then maintaining an effective large-scale, long-term wildlife population monitoring programme is extremely difficult. This is evidenced by the small number of such programmes in existence. We offer the following thoughts, based on our experiences with such programmes.

First priority should be to determine the goals of the programme. What species and areas are under consideration? What sort of population changes are of concern? What action will be taken if the trigger points are exceeded? Under what circumstances should the monitoring programme stop? These goals will need to be reviewed regularly, but any changes should not dilute the main focus of the monitoring programme. Such issues are discussed in detail in Goldsmith (1991).

Statistical design issues are important. These include determining the parameters to estimate (abundance, survival, fecundity, etc.), survey methods, type and layout of samplers, sample size, etc. However, the biggest issues have more to do with funding and maintaining the programme's integrity over the long term. To help with the latter, a technical advisory board for the programme is useful, together with regular publication of results in peer-reviewed journals, and periodic (independent) peer review of the whole programme.

Many changes will occur, expected and unexpected, over the lifetime of a long-term study. These include turnover of important personnel, development of new technologies, field and analysis methods. A project needs to be able to adapt to these changes, and incorporate improvements without compromising the integrity and comparability of the parameter estimates over time. Hence the primary goal should be to obtain unbiased estimates at each time period. As discussed above, indices are not sufficient. In a distance sampling context, assumptions such as $g(0) = 1$ should be tested regularly; for example, double-observer methods may be needed (at least in some years).

If deficiencies are discovered in design or field methods, then they should be corrected. It may be necessary to run old and new protocols in parallel for a few years to maintain comparability, but the old, poor protocol should eventually be dropped. A good example of this is the Common Birds Census, a volunteer-based bird monitoring programme run by the British Trust for Ornithology. It suffered from subjective site placement, and used a mapping census protocol that was labour-intensive, both to collect the data and to prepare them for analysis. After a 6-year overlap, it has now been replaced by the UK Breeding Bird Survey, which uses stratified random placement of samplers and a more efficient sampling scheme (Freeman et al. 2003).

The sampling frame may need to change as the spatial distribution of animals changes. There may be areas of low or zero density not initially sampled that later must be included; conversely permanent extinctions may mean that regular sampling can cease in some areas. A flexible design is needed that can survive such changes. A related problem can occur when the survey region is just the core area of a species' range. Thus the survey estimates rate of change within this core area, which may not be representative at all of what is happening elsewhere: an expanding species will tend to increase faster outside the core area, while a contracting species will decline faster outside the core area. Care is needed in interpreting the results of such studies.

Data quality control, integrity, and management are important. Quality control procedures should involve checking and double-checking the field data. A central repository for data is needed that is accessible to everyone. Issues of data ownership should be resolved. Data should be entered, checked, and corrected as soon as possible (e.g. each day). It is better to do less field work each day and have quality time available for data entry, checking, and preliminary analysis. These data can then be sent to the central repository in a timely manner.

It will be difficult to maintain protocols that give consistently high-quality data because of personnel turnover. A partial solution is to maintain excellent documentation: document the design, the field protocols, data formats, and data codes. Be clear and document reasons for actions and changes. Documentation is also useful in the case of a programme that gets run down due to lack of funding but is subsequently resurrected.

Because of personnel turnover, a continuous training programme is probably necessary for larger programmes. The programme's funding structure should be designed to keep field people and technicians as long as possible (e.g. better to pay well and retain high quality people). High turnover of technicians should be avoided.

An estimate of the parameters monitored is not required for every area in every year if the primary goal is to detect long-term change and the animals are long-lived. Administratively, however, it may be difficult to have some years with no fieldwork (and hence large annual budget variations). One solution is to cycle the sampling over different geographic strata. A programme should start with at least one year of pilot work, after which it may be necessary to have a few years of intensive survey effort to establish a baseline, before falling back to more moderate levels of sampling (and therefore funding).

In conclusion, our intuition, analysis methods, and organizational support structures are mainly suitable for, and based on, short-term, small-scale (often research) studies. These do not carry over well to the large-scale long-term studies that are the focus of this chapter.

6

Methods for incomplete detection at distance zero

J. L. Laake and D. L. Borchers

6.1 Introduction

In conventional distance sampling (CDS) and multiple-covariate distance sampling (MCDS), the detection function is denoted $g(y)$ or $g(y, \underline{z})$ and detection at distance zero ($y = 0$) is certain: $g(0) = 1$ and $g(0, \underline{z}) = 1$. In this chapter, we deal with detection functions in which detection probability at distance zero may be less than one. To distinguish these functions from CDS or MCDS functions, we denote the detection function $p(y, \underline{z})$. When we use $g(\ )$, we mean a detection function that is one at distance $y = 0$. Historically, $g(\ )$ has been used to denote either function, and you will find mention of the '$g(0) < 1$ problem' in the literature. This is a potential source of confusion of which you should be aware. Adding a second detection function symbol, $p(\ )$, is the lesser of two evils—if we used $g(\ )$ for detection functions that may or may not be 1 at $y = 0$, the notation in this chapter would become really confusing. Remember: $p(0, \underline{z}) \leq 1$ but $g(0, \underline{z}) = 1$ always.

Another notation issue we should clarify up front is that we use the term 'observer' (without quotes) to represent a detection occasion. In CDS and MCDS surveys, there is only one occasion; in this chapter, we deal with two occasions. Use of this term should not be viewed as implying that the detection process is limited to visual surveys. For example, an 'observer' could be an acoustic sampler for underwater 'observations' or a sample by radio detection of radio-tagged animals. It represents any method of obtaining a sample of observations from the covered region. Nor should it be taken to imply that there is only one person searching. Each 'observer' could contain multiple persons. Our use of 'observer' is equivalent to 'platform' which is sometimes used when the observers occupy physically separate survey platforms (e.g. Buckland and Turnock 1992).

CDS and MCDS surveys use observed distances to provide a relative measure of detection probability of objects within the covered region (i.e. strips or circular plots). To convert this to an absolute measure, detection of objects at distance zero is assumed to be certain ($p(0) = g(0) = 1$). If it is not ($p(0) < 1$), CDS and MCDS density estimators will be biased because $E(\widehat{D}) = Dp(0)$. It is often reasonable to assume that $p(0) = 1$, but for many surveys, it is likely that $p(0) < 1$.

The bias resulting from assuming $p(0) = 1$ when $p(0) < 1$ can be substantial. For example, Skaug and Schweder (1999) estimated that $p(0)$ was 0.32 on shipboard surveys of minke whales, and Laake *et al.* (1997) estimated that $p(0)$ was 0.29 for experienced personnel and as low as 0.08 for inexperienced personnel on aerial surveys of harbour porpoise. Marine mammals are particularly problematic because they are unavailable to be seen when they are below the water surface for long periods of time. Marsh and Sinclair (1989) coined the terms 'perception bias' and 'availability bias' to describe the reasons animals are not detected by observers. Perception bias occurs when observers miss visible animals because of environmental conditions, fatigue, etc.; availability bias occurs when animals are not available to be detected (e.g. when they are underneath water or hidden by vegetation). While the distinction between the sources can be fuzzy in some circumstances (e.g. an animal only partially hidden), the dichotomy is useful for describing the detection process and estimation techniques. Both may reduce $p(0)$. The term 'visibility bias' is used generically to refer to either or both types.

The uncertainty about $p(0)$ led Pollock and Kendall (1987) to question the utility of distance sampling and to recommend mark-recapture methods to remove visibility bias in aerial surveys. In contrast to distance sampling, mark-recapture techniques attempt to estimate absolute detection probability—using a known 'marked' sample. Two mark-recapture approaches have been used to measure visibility bias: (1) two observers survey the same region and observations of one (or each) observer provides a 'marked' sample for the other, and (2) a sample of animals is marked (e.g. radio-collared) and the proportion detected is measured. The first approach is often called the 'double-count' method (Graham and Bell 1989). If observers count independently, the simplest estimator of abundance from these data is a two-sample mark-recapture (Petersen) estimator (Pollock and Kendall 1987). If the second observer can detect only those missed by the first observer, a removal estimator can be used (Cook and Jacobsen 1979; Pollock and Kendall 1987). As recognized by Graham and Bell (1989) and others, these methods cannot account for animals that are unavailable to both observers; the estimators may remove perception bias but most availability bias may remain. If the observers survey the same region at different times (Buckland and Turnock 1992; Laake *et al.* 1997; Hiby and Lovell 1998), it is possible to remove both sources of bias (although this is not guaranteed). Physical marking of animals and tracking of their positions (the second mark-recapture approach, for example, Samuel *et al.* 1987; Steinhorst and Samuel 1989) can also remove both sources of bias.

Availability bias should not be ignored if it occurs, because it can be a substantial source of bias that may be much larger than perception bias. For example, Marsh and Sinclair (1989) estimated that 83.3% of dugongs were beneath the water surface and unavailable to be detected, while the observer team only missed 2–17% of the visible dugongs within a 200-m strip.

While mark-recapture estimators avoid the assumption that $p(0) = 1$, they are inherently plagued by bias due to unmodelled heterogeneity in capture (detection) probability (Otis *et al.* 1978; Seber 1982; Schweder 1990; Buckland 1992c). This can introduce as much or more bias in mark-recapture estimators as does failure of the assumption that $p(0) = 1$ in conventional distance sampling. Estimators can model heterogeneity by incorporating covariates that affect detection probability (Huggins 1989, 1991; Alho 1990; Pledger 2000). This reduces bias, but eliminates it only if heterogeneity is modelled adequately. To understand the effect of heterogeneity, consider how detection probability is estimated from a survey with two independent observers. Observer 1 detects n_1 animals that are a 'marked' sample for observer 2, and likewise, observer 2 detects n_2 animals that are a 'marked' sample for observer 1. Both observers detect n_3 of the animals. The total number detected is denoted as $n_. = n_1 + n_2 - n_3$. An estimator of detection probability p_j for observer j ($j = 1, 2$) is the fraction of the other observer's detections that (s)he saw: $\hat{p}_1 = n_3/n_2$ and $\hat{p}_2 = n_3/n_1$. Strictly, these are estimators of the conditional probability that observer j detects an animal, given that it was detected by the other observer ($p_{1|2}$ and $p_{2|1}$). These are estimators of p_1 and p_2 only when detections are independent between observers. If a given animal is particularly detectable for both observers (one close to the observers, for example), they will both tend to detect it, and their detections will not be independent. As a result, n_3/n_2 and n_3/n_1 are biased estimators of p_1 and p_2. The bias is caused by 'unmodelled heterogeneity', although 'unmodelled correlation' might be a better name. This bias arises because p_1 and p_2 depend on some variable, the dependence is not modelled, and detections are correlated as a result. See Sections 6.2.3 and 6.2.5 for more on this.

Since the key source of detection probability heterogeneity in distance sampling is distance, a natural development (at least it seems natural after the event) was to combine mark-recapture methods with distance sampling methods. We call these mark-recapture-distance-sampling (MRDS) methods. The early development of these methods occurred in International Whaling Commission (IWC) and is published in their reports (e.g. Butterworth *et al.* 1982; Butterworth and Borchers 1988; Hiby and Hammond 1989). The methods were further developed and generalized, mainly by Buckland and Turnock (1992), Palka (1995), Borchers (1996),

Alpizar-Jara and Pollock (1996), Manly *et al.* (1996), Quang and Becker (1997), Borchers *et al.* (1998a,b), Skaug and Schweder (1999), Laake (1999), and Chen (1999, 2000).

The evolution of distance sampling into the arena of mark-recapture sampling was a useful advance, but it necessarily complicated modelling and data analysis. In CDS, heterogeneity in detection probability is not important because measurements of detection probability are made relative to 'known' certain detection at distance zero. The true detection function is in reality a composite of many different detection functions, but in CDS and MCDS, these need not be modelled because (with appropriate detection function forms) these methods are pooling robust (see Section 11.12). However, pooling robustness does not hold when $p(0) < 1$, so that you need to consider the various covariates that affect detection probability. Thus, methods of incorporating covariates become very important in MRDS.

The composite nature of MRDS likelihood functions (see Section 6.2), the different observation configurations, and the possible incorporation of an availability model expands the range of analyses required for MRDS surveys. To help guide the reader through the maze of possible methods, we have organized the chapter from the general to the specific and we have attempted to provide a framework that accommodates the full range of possible analyses, although we do not provide full details for every method. We describe the general likelihood for MRDS analyses, Horvitz–Thompson-like estimators and estimation approaches in Section 6.2. Following the unifying approach of Borchers *et al.* (2002), we go on to describe the methods in terms of the assumed state and observation models (Section 6.3) and the mark-recapture observation configurations (Section 6.3.3). Using a single example data set for the most part (Section 6.4), we illustrate application of the estimators for the different observation configurations without availability bias (Sections 6.5 to 6.7) and then examine the various estimation approaches that incorporate a state model with an availability process (Section 6.8). All our examples are from line transect surveys. However, the theory developed in this chapter is applicable to point transect surveys. We conclude with sections that discuss some special topics (Section 6.9) and considerations for field methods to implement MRDS (Section 6.10).

6.2 Likelihood and Horvitz–Thompson

In this section, we develop likelihood functions for double-observer distance sampling surveys in which detection at distance $y = 0$ is uncertain. We start with the simple case in which detection probability does not depend on distance at all, and work up to the case in which it depends on distance and a host of other variables as well.

We do not consider likelihood functions that accommodate an availability process here; the likelihoods in this section deal only with animals that are available for the duration of the survey. In Section 6.8, we consider estimators for situations in which animals are not necessarily available to be detected at the time of the survey.

In all cases, we consider only likelihood functions for the number of animals in the covered region (N_c); inferences about abundance or density in the whole survey region can be obtained either using design-based methods, or by modifying the likelihoods we present to include coverage probabilities. We also treat detection units as individuals in this section. If detection units are clusters, the likelihoods apply too, but N_c is then cluster abundance within the covered region. This can be converted into an estimate of individual abundance in the covered region using an estimate of mean cluster size or using a Horvitz–Thompson-like estimator as in eqn (2.16).

Because these surveys involve two independent observers, we have more detection functions to consider than with conventional distance sampling methods. Observer 1 detects an animal that is at distance y and has associated variables $\underline{z}$, with probability $p_1(y, \underline{z}) \le 1$, while observer 2 detects it with probability $p_2(y, \underline{z}) \le 1$. It turns out to be useful to consider two other kinds of detection function as well:

$p_{j \mid (3-j)}(y, \underline{z})$: This is the conditional probability that observer j detects the animal, given that observer $(3 - j)$ detected it, for $j = 1$ or 2.

$p.(y, \underline{z})$: This is the probability that at least one of the observers detects the animal. It is useful to write it as

$$p.(y, \underline{z}) = p_1(y, \underline{z}) + p_2(y, \underline{z})[1 - p_{1 \mid 2}(y, \underline{z})] \qquad (6.1)$$

(If we swap the 1 and 2 subscripts on the RHS, this also equals $p.(y, \underline{z})$, but we will tend to work with the equation as written above.)

6.2.1 Constant detection probability

Suppose that animals were equally detectable at all distances within the covered region (i.e. all $y \le w$). In this case, we drop the arguments of the detection functions for simplicity. We write them as p_1 and p_2. We also assume that detections are independent, so that $p_{j \mid (3-j)} = p_j$ and $p. = p_1 + p_2 - p_2 p_1$.

The likelihood for N_c, p_1, and p_2 given $n.$ is

$$\mathcal{L}_{n.} = \binom{N_c}{n.} p.^{n.} (1 - p.)^{N_c - n.}. \qquad (6.2)$$

We cannot estimate the three unknown parameters (N_c, p_1, and p_2) sensibly from this likelihood because we have only one bit of data, the number of detected objects $n.$. Moreover, the distance data contain no useful information to estimate p_1 and p_2—all they tell us is that they are the same at all distances. This is where ideas from mark-recapture methods are useful.

The basis of the methods is really quite simple. If the animals detected and 'marked' by observer 2 (usually by location rather than physical marking) are typical of the animals in the population, they constitute a representative sub-population of known size. The proportion of this marked sub-population that observer 1 detects is an obvious estimator of the probability that observer 1 detects an animal (p_1). Similarly, observer 1 'marks' animals for observer 2 and we can use these data to estimate p_2. Applying these estimates to the whole population, we can estimate N_c. You can see that the methods rely heavily on the detected sub-population(s) being representative of the whole population, and this can be a problem. In particular, the sub-population(s) might contain a higher proportion of detectable animals than the whole population.

The additional data that come from independent observers are the capture histories. We can implement mark-recapture ideas in the likelihood by adding a component containing the probabilities of observing the three possible capture histories $\underline{\omega} = (1,0)$, $\underline{\omega} = (0,1)$, and $\underline{\omega} = (1,1)$. The new component is

$$\mathcal{L}_\omega = \prod_{i=1}^{n_.} \Pr\{\underline{\omega}_i \,|\, \text{detected}\} = \prod_{i=1}^{n_.} \frac{\Pr\{\underline{\omega}_i\}}{p_.}, \qquad (6.3)$$

where

$$\Pr\{\text{detected by observer 1 only}\} = \Pr\{\underline{\omega}_i = (1,0)\} = p_1(1 - p_2)$$
$$\Pr\{\text{detected by observer 2 only}\} = \Pr\{\underline{\omega}_i = (0,1)\} = (1 - p_1)p_2$$
$$\Pr\{\text{detected by both observers}\} = \Pr\{\underline{\omega}_i = (1,1)\} = p_1 p_2.$$

Combining this likelihood with that of eqn (6.2), we get a likelihood from which we can estimate N_c, p_1, and p_2:

$$\mathcal{L}_{n_.,\omega} = \mathcal{L}_{n_.}\,\mathcal{L}_\omega = \binom{N_c}{n_.} p^{n_.} (1 - p_.)^{N_c - n_.} \prod_{i=1}^{n_.} \frac{\Pr\{\underline{\omega}_i\}}{p_.} \qquad (6.4)$$

6.2.1.1 *Two approaches to estimation*
There are at least two ways this likelihood can be used to estimate N_c. We could obviously estimate it by maximizing the likelihood with respect to N_c, p_1, and p_2. We could, however, also estimate it by maximizing $\mathcal{L}_\omega$ with respect to p_1 and p_2, estimate $p_.$ using the estimates of p_1 and p_2, and then estimate N_c using a Horvitz–Thompson-like estimator. It turns out that

in both cases we get the same estimator and it is one familiar to anyone who has done mark-recapture analysis—the Petersen (or Lincoln–Petersen) estimator (Petersen 1896; Lincoln 1930):

$$\widehat{N}_{\text{Petersen}} = \frac{n_1 n_2}{n_3}. \tag{6.5}$$

Here n_1 is the number of animals detected by observer 1, n_2 is the number detected by observer 2, and n_3 is the number detected by both (the number of 'duplicates').

6.2.2 Detection probability changing with distance

We now consider the case in which detection probabilities depend on distance (y) only. The two detection functions are now written as $p_1(y)$ and $p_2(y)$. These functions have some unknown parameters, which we write as $\underline{\theta}_1$ for observer 1 and $\underline{\theta}_2$ for observer 2. This introduces another component to the likelihood—the familiar distance sampling component:

$$\mathcal{L}_y = \prod_{i=1}^{n.} f.(y_i) = \prod_{i=1}^{n.} \frac{p.(y_i)\pi(y_i)}{E(p.)}, \tag{6.6}$$

where $\pi(y)$ is the probability density function (pdf) of distances from the line in the population (usually $\pi(y) = 1/w$ for line transect surveys), $f.(y)$ is the pdf of distances of animals that are detected by at least one observer, and $E(p.) = \int_0^w p.(y)\pi(y)dy$.

The likelihood component for N_c changes slightly to involve $E(p.)$ rather than $p.$, because $p.$ now depends on y:

$$\mathcal{L}_{n.} = \binom{N_c}{n.} E(p.)^{n.} (1 - E(p.))^{N_c - n.}. \tag{6.7}$$

For brevity, we omit $\underline{\theta}$ from the argument of the likelihood function.

Finally, the capture history component changes slightly, to accommodate y:

$$\mathcal{L}_\omega = \prod_{i=1}^{n.} \Pr\{\underline{\omega}_i \,|\, \text{detected at } y_i\} = \prod_{i=1}^{n.} \frac{\Pr\{\underline{\omega}_i \,|\, y_i\}}{p.(y_i)}, \tag{6.8}$$

where

$$\Pr\{\underline{\omega}_i = (1,0) \,|\, y_i\} = p_1(y_i)[1 - p_{2|1}(y_i)]$$
$$\Pr\{\underline{\omega}_i = (0,1) \,|\, y_i\} = [1 - p_{1|2}(y_i)]p_2(y_i)$$
$$\Pr\{\underline{\omega}_i = (1,1) \,|\, y_i\} = p_{1|2}(y_i)p_2(y_i) = p_{2|1}(y_i)p_1(y_i).$$

Note that we have not assumed independence between the two observers above. If there is independence, $p_{1\,|\,2}(y_i) = p_1(y_i)$ and $p_{2\,|\,1}(y_i) = p_2(y_i)$.

The full likelihood is now:

$$\mathcal{L}_{n_.,y,\omega} = \mathcal{L}_{n_.}\mathcal{L}_y\mathcal{L}_\omega$$

$$= \left[\binom{N_c}{n_.} E(p_.)^{n_.}\, [1 - E(p_.)]^{N_c - n_.} \right]$$

$$\times \left[\prod_{i=1}^{n_.} f_.(y_i) \right] \left[\prod_{i=1}^{n_.} \frac{\Pr\{\omega_i \,|\, y_i\}}{p_.(y_i)} \right]. \tag{6.9}$$

Notice that $\mathcal{L}_{n_.}\mathcal{L}_y$ is the distance sampling likelihood of eqn (2.33), with minor changes of notation.

6.2.2.1 *Approaches to estimation*
As in the case in which detection probability did not depend on y, we can estimate N using either of the approaches used above:

Estimation approach (1): Maximize the full likelihood eqn (6.9) with respect to N_c, $\underline{\theta}_1$, and $\underline{\theta}_2$, or

Estimation approach (2a): Estimate $\underline{\theta}_1$ and $\underline{\theta}_2$ by maximizing $\mathcal{L}_\omega$. Calculate $\hat{p}_.(y_i)$ using these estimates, and then use $\hat{p}_.(y_i)$ in a Horvitz–Thompson-like estimator for N_c:

$$\widetilde{N}_c = \sum_{i=1}^{n_.} \frac{1}{\hat{p}_.(y_i)}. \tag{6.10}$$

Huggins (1989, 1991) and Alho (1990) independently suggested this approach for mark-recapture surveys. Various authors have used it in a distance-sampling context since then.

There are several other possibilities. The first of these is similar to approach (2a) above, but it uses the fact that in distance sampling surveys, we consider $\pi(y)$ to be known, so that once we have an estimate of $p_.(y)$, we also have an estimate of $E(p_.) = \int_0^w p_.(y)\pi(y)dy$. A variant of approach (2a) is therefore:

Estimation approach (2b): Estimate $\underline{\theta}_1$ and $\underline{\theta}_2$ by maximizing $\mathcal{L}_\omega$, as with approach (2a) above. Calculate $\widehat{E}(p_.)$ using these estimates, and then use $\widehat{E}(p_{.})$ in a Horvitz–Thompson-like estimator for N_c:

$$\widehat{N}_c = \sum_{i=1}^{n_{.}} \frac{1}{\widehat{E}(p_{.})}, \tag{6.11}$$

where $\widehat{E}(p_{.}) = \int_0^w \hat{p}_{.}(y)\pi(y)dy$.

Because $\widehat{E}(p_{.})$ is an average of sorts[1], it tends to have lower variance than $\hat{p}_{.}(y)$. In addition, neither $\widetilde{N}_c$ of eqn (6.10) nor $\widehat{N}_c$ of eqn (6.11) is necessarily unbiased, and they can be very biased when their denominators are small. Being an average over all $y \leq w$, $\widehat{E}(p_{.})$ does not involve values as small as the smallest values of $\hat{p}_{.}(y)$; $\widehat{N}_c$ therefore tends to be both less variable and less biased than $\widetilde{N}_c$.

Notice that estimation approaches (2a) and (2b) do not use the likelihood component $\mathcal{L}_y$. This turns out to be convenient for estimation because binary regression methods implemented in many statistical software packages can be used when estimating parameters from the likelihood component $\mathcal{L}_\omega$. It does, however, neglect the information on detection probability that resides in the conventional distance sampling component $\mathcal{L}_y$. In particular, $\mathcal{L}_y$ contains information about the shape of $p_{.}(y)$. With this in mind, a further estimation approach is as follows.

Estimation approach (3): Estimate $\underline{\theta}_1$ and $\underline{\theta}_2$ by maximizing $\mathcal{L}_y\mathcal{L}_\omega$. Calculate $\widehat{E}(p_{.})$ using these estimates, and then use $\widehat{E}(p_{.})$ in a Horvitz–Thompson-like estimator for N_c, as in approach (2b) above. (Note that we could also use $\hat{p}_{.}(y)$ in the Horvitz–Thompson-like estimator for N_c, but as there is no advantage to doing so, we do not pursue this further here.)

N_c can also be estimated using a detection function estimate from a single platform. This can be done for either platform but we consider only estimation for platform 1 here. To this end, it is useful to factorize $\mathcal{L}_\omega$ as follows:

$$\mathcal{L}_\omega = \mathcal{L}_{2|} . \mathcal{L}_{1|2}$$

$$= \left[\prod_{i=1}^{n_{.}} \left[\frac{p_2(y_i)}{p_{.}(y_i)} \right]^{\omega_{i2}} \left[\frac{P_{10}(y_i)}{p_{.}(y_i)} \right]^{1-\omega_{i2}} \right]$$

$$\times \left[\prod_{i=1}^{n_2} p_{1|2}(y_i)^{\omega_{i1}} [1 - p_{1|2}(y_i)]^{1-\omega_{i1}} \right], \tag{6.12}$$

---

[1] You can think of it as a weighted average of $\hat{p}_{.}(y)$ over all y, with weighting $\pi(y)$.

where ω_{ij} is 1 if the ith observation is detected by observer j and 0 otherwise, and $P_{10}(y_i) = p_1(y_i)[1 - p_{2\,|\,1}(y_i)]$ is the probability that observer 1 detects an animal at y_i and observer 2 misses it.

Note that $\mathcal{L}_{1\,|\,2}$ is a binary likelihood so the parameters of $p_{1\,|\,2}(y_i)$ can be estimated using binary regression methods and software.

Estimation approach (4): When detections by observer 1 are independent of those by observer 2, $p_{1\,|\,2}(y) = p_1(y)$, and the parameters of $p_1(y)$ (i.e. $\underline{\theta}_1$) can be estimated by maximizing $\mathcal{L}_{1\,|\,2}$. Calculate $\widehat{E}(p_1)$ using this estimate, and then use $\widehat{E}(p_1)$ in a Horvitz–Thompson-like estimator for N_c:

$$\widehat{N}_c = \sum_{i=1}^{n_1} \frac{1}{\widehat{E}(p_1)}, \tag{6.13}$$

where $\widehat{E}(p_1) = \int_0^w \hat{p}_1(y)\pi(y)dy$. Note that the sum is over all detections by observer 1.

Normally, one would expect estimation approaches (1) to (3) to perform better than approach (4), because they use more of the available data. They use data from each platform to estimate the detection function for the other, while approach (4) does not use data from observer 1 to estimate the detection function for observer 2. This estimation approach is only a sensible option when data from observer 1 cannot reasonably be used to estimate the detection function for observer 2. One such circumstance is when there is responsive movement after observer 2 detects animals. This is discussed in Section 6.6.2 below.

Other approaches to estimation are motivated by consideration of the assumptions made about the independence of detections between observers. We therefore discuss independence issues before considering them.

6.2.3 *Independence issues*

In general, $p.(y) = p_1(y) + p_2(y)[1 - p_{1\,|\,2}(y)]$, but when detections are independent at all distances, $p.(y) = p_1(y) + p_2(y)[1 - p_1(y)]$. An example of detection functions in this case, with $p_1 = p_2$ for simplicity, is shown in Fig. 6.1.

Rather counter-intuitively, the detections by the two observers can be dependent even though the observers are behaving completely independently of one another. This sort of dependence typically arises as a consequence of unmodelled heterogeneity in detection probability.

To see how dependence arises, suppose that detection probability really depends on distance y and cluster size s, but we model it as a function

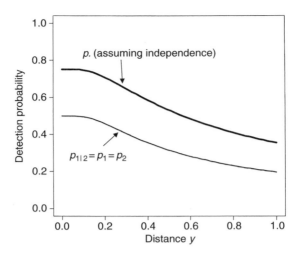

Fig. 6.1. An example of individual and combined detection functions when detections are independent between the two observers. In this case, $p_{1\,|\,2} = p_1$ and $p_{2\,|\,1} = p_2$. For simplicity $p_1 = p_2$ in this example.

of distance only. Suppose also that at distance $y = w$, both observers are able to detect only large clusters, although they can detect all cluster sizes at distance $y = 0$. Consider observer 1 and a cluster that passed by at $y = w$. Before you know whether observer 2 detected this cluster, you have no idea what size of cluster it is and so the probability that observer 1 detects it is the average probability[2] that observer 1 detects a cluster at $y = w$. But after you know that observer 2 detected a cluster, you know it must be a large cluster and therefore that observer 1 was more likely to have detected it: the fact that observer 2 detected it makes it more likely that observer 1 detected it. In other words, the unconditional probability of detection by observer 1 is not equal to the conditional probability of detection, given that observer 2 made a detection: $p_{1\,|\,2}(y) \neq p_1(y)$ (for $y = w$ in this example). Although they were acting independently, the two observers' detections are not independent.

This sort of dependence is illustrated in Fig. 6.2. In this case, detections are independent at distance $y = 0$ ($p_{1\,|\,2}(0) = p_1(0)$) but they are increasingly positively correlated as distance increases ($p_{1\,|\,2}(y) > p_1(y)$ for $y > 0$). Independence only at distance zero in the context of line transect analysis was called 'trackline conditional independence' by Laake (1999). We refer to it here more generally as 'point independence' (independence at the point $y = 0$, but not necessarily elsewhere).

---

[2] Averaged over the distribution of cluster sizes in the population.

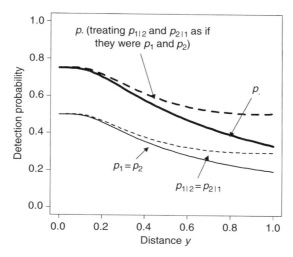

Fig. 6.2. An example of individual detection functions ($p_1 = p_2$) and combined detection functions when detections are independent at distance zero, but increasingly positively correlated at larger distances.

If detections are dependent and this is not modelled, the resulting estimates of detection probability, and of abundance, are biased. For example, the detection function estimate obtained by maximizing likelihood eqn (6.13) for the scenario shown in Fig. 6.2 is an estimate of $p_{1|2}(y)$ (the dashed line in the figure). This is clearly a biased estimator of $p_1(y)$ (the solid line in the figure), because detections are not independent in this case.

Similarly, were we to estimate $p_1(y)$ and $p_2(y)$ for the scenario in Fig. 6.2 by maximizing the likelihood $\mathcal{L}_\omega$ with respect to $\underline{\theta}_1$ and $\underline{\theta}_2$ using the parameterization $\Pr\{\underline{\omega} = (1,1) \mid y\} = p_1(y)p_2(y)$ rather than $\Pr\{\underline{\omega} = (1,1) \mid y\} = p_{1|2}(y)p_2(y)$, our estimates of $p_1(y)$, $p_2(y)$, and $p_.(y)$ would all be positively biased. The estimate of the combined detection function $p_.(y)$ in this case is shown as the dashed line in Fig. 6.2.

Estimates of $p_.(0)$ obtained under the assumption of full independence should not be any more biased than estimates of $p_.(0)$ obtained under the assumption of point independence since the two curves (the solid and dashed lines in Fig. 6.2) coincide at $y = 0$. However, estimates of N_c obtained under the assumption of full independence will tend to be negatively biased compared to estimates obtained under the assumption of point independence because under the former assumption, estimated $p_.(y)$ is higher everywhere except at $y = 0$.

It is of course possible that detections are not independent at any distance, even though observers behave independently. This also typically arises as a consequence of unmodelled heterogeneity, but in this case

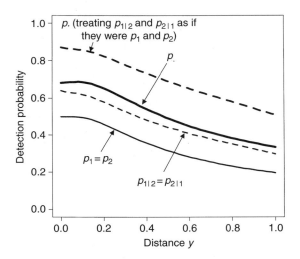

Fig. 6.3. An example of individual detection functions ($p_1 = p_2$) and combined detection functions when detections are positively correlated at all distances.

heterogeneity is such that even at distance $y = 0$ detections are dependent. Fig. 6.3 illustrates this scenario. Notice that the conditional detection functions $p_{1|2}(y)$ and $p_{2|1}(y)$ are higher than the unconditional detection functions $p_1(y)$ and $p_2(y)$. Similarly, the combined detection function under the assumption of full independence $(p_.(y) = p_1(y) + p_2(y)(1 - p_1(y)))$ is higher than the true $p_.(y) = p_1(y) + p_2(y)(1 - p_{1|2}(y))$. Estimates of $p_1(0)$, $p_2(0)$, and $p_.(0)$ obtained under the assumption of either full independence or point independence in this case will be positively biased, and estimates of N_c will be negatively biased. This bias can be substantial.

This kind of bias is well known in mark-recapture studies. It is a consequence of the fact that the marked sub-population is not representative of the whole population—the sub-population is more detectable on average than the whole population.

Finally, we note that in the discussion above, we have assumed dependence takes the form of positive correlation between detection probabilities. Negative correlation can arise when there is unmodelled heterogeneity that causes one observer's detection function to increase while causing the other's detection function to decrease.

An obvious way to deal with dependence induced by unmodelled heterogeneity is to model the effect of the variable(s) causing the heterogeneity. For example, if detection probability depended on y, the Petersen estimator, $\widehat{N}_{\text{Petersen}}$ is negatively biased because heterogeneity due to y is neglected. Similarly, if we use an estimator in which detection probability depends on y only, when in reality it depends on other variables such as animal or cluster size and environmental conditions as well, this estimator will tend to be negatively biased. This can be overcome in principle by modelling the dependence of detection probabilities on these variables (provided they can be observed). We discuss likelihoods for this case in the next section.

6.2.4 *Multiple covariates*

In this section, we extend the likelihoods for the case in which detection probability depends on a vector of covariates $\underline{z}$, in addition to distance y. Detection probabilities are written $p_1(y, \underline{z})$, $p_2(y, \underline{z})$, and $p.(y, \underline{z})$.

In this case, $\mathcal{L}_{n.}$ has the same form, but $E(p.) = \int \int p.(y, \underline{z})\pi(y, \underline{z})dyd\underline{z}$, where $\pi(y, \underline{z})$ is the joint density of y and $\underline{z}$ in the population (of detected and undetected animals). $\mathcal{L}_y$ generalizes to

$$\mathcal{L}_{y,z} = \prod_{i=1}^{n.} f.(y_i, \underline{z}_i) = \prod_{i=1}^{n.} \frac{p.(y_i, \underline{z}_i)\pi(y_i, \underline{z}_i)}{E(p.)} \tag{6.14}$$

and it is convenient to write this as

$$\mathcal{L}_{y,z} = \mathcal{L}_z \mathcal{L}_{y \,|\, z}$$
$$= \left[\prod_{i=1}^{n.} \frac{E(p. \,|\, \underline{z}_i)\pi(\underline{z}_i)}{E(p.)}\right] \left[\prod_{i=1}^{n.} \frac{p.(y_i, \underline{z}_i)\pi(y_i \,|\, \underline{z}_i)}{E(p. \,|\, \underline{z}_i)}\right], \tag{6.15}$$

where $E(p. \,|\, \underline{z}_i) = \int p.(y, \underline{z}_i)\pi(y \,|\, \underline{z}_i)dy$, $\pi(\underline{z}_i)$ is the pdf of $\underline{z}$ in the population, evaluated at $\underline{z}_i$, and $\pi(y_i \,|\, \underline{z}_i)$ is the conditional pdf of y given $\underline{z}_i$, evaluated at y_i.

$\mathcal{L}_\omega$ generalizes to

$$\mathcal{L}_\omega = \prod_{i=1}^{n.} \frac{\Pr\{\underline{\omega}_i \,|\, (y_i, \underline{z}_i)\}}{p.(y_i, \underline{z}_i)} \tag{6.16}$$

and, for estimation of $p_{1|2}(y, \underline{z}_i)$, this can be factorized as

$$\mathcal{L}_\omega = \mathcal{L}_{2|}.\mathcal{L}_{1|2}$$
$$= \left[\prod_{i=1}^{n.} \left[\frac{p_2(y_i, \underline{z}_i)}{p.(y_i, \underline{z}_i)}\right]^{\omega_{i2}} \left[\frac{P_{10}(y_i, \underline{z}_i)}{p.(y_i, \underline{z}_i)}\right]^{1-\omega_{i2}}\right]$$
$$\times \left[\prod_{i=1}^{n_2} p_{1|2}(y_i, \underline{z}_i)^{\omega_{i1}}[1 - p_{1|2}(y_i, \underline{z}_i)]^{1-\omega_{i1}}\right]. \tag{6.17}$$

The full likelihood is

$$\mathcal{L} = \mathcal{L}_n \, \mathcal{L}_z \mathcal{L}_{y \mid z} \mathcal{L}_\omega. \tag{6.18}$$

Because the full likelihood involves the pdf of $\underline{z}$, a plausible model for the density $\pi(\underline{z})$ is required. It can be difficult to specify this reliably, especially if $\underline{z}$ has many components; and the more components it has, the more parameters there are to be estimated using the full likelihood estimation approach (1). In addition, a model for $\pi(y \mid \underline{z})$ is required, where before there was a known pdf $\pi(y)$. In practice, y and $\underline{z}$ are usually assumed to be independent; random line placement ensures this assumption holds.

Estimation approaches (2a) and (2b) can be applied directly in the presence of additional variables $\underline{z}$, the only modification being that detection probability must be parameterized in terms of y and $\underline{z}$.

Estimation approach (3), however, requires specification of a form for $\pi(\underline{z})$ and because this may be difficult to do reliably, a modified version of approach (3) can be used. The modification uses the assumption that y and $\underline{z}$ are independent, but it does not involve $\pi(\underline{z})$.

Estimation approach (5): Estimate $\underline{\theta}_1$ and $\underline{\theta}_2$ by maximizing $\mathcal{L}_{y \mid z} \mathcal{L}_\omega$. Calculate $\widehat{E}(p \mid \underline{z}_i)$ using these estimates, and then use $\widehat{E}(p \mid \underline{z}_i)$ in a Horvitz–Thompson-like estimator for N_c, as in approach (3) above.

Estimation approach (4) can be applied directly, using $\mathcal{L}_{1 \mid 2}$ in eqn (6.17).

6.2.5 Unobserved heterogeneity

It is often not possible to observe all the variables that strongly influence detection probability. A diagnostic to indicate unmodelled heterogeneity, together with some means of dealing with unobserved heterogeneity, would therefore be useful.

If heterogeneity is such that detections between the two observers are not independent anywhere, diagnosing this is difficult or impossible (Link 2003). If, however, point independence applies and detections are independent at $y = 0$, there is some prospect of diagnosing this. The key to diagnosis is the difference between the solid and dotted lines in Fig. 6.2. Consider detections by observer 1 on a line transect survey. The shape of the distribution of perpendicular distances of this observer's detections gives an estimator of the shape of $p_1(y)$—the lower solid curve in the figure. An estimator of the shape of $p_{1 \mid 2}(y)$—the lower dashed curve in the figure—is given by the shape of a histogram comprising the proportion of detections by observer 2, which observer 1 also saw, tallied by distance interval. If the shape of the latter is different from that of the former, this is an indication that there is dependence. (If the latter is flatter than the former, this indicates positive correlation, as in Fig. 6.2.) This suggests that variables other than y should be included in the model. If, once all observed variables have been included, there remains an indication of dependence, a method that accommodates some unmodelled heterogeneity is required. Because it is not always possible to diagnose unmodelled heterogeneity, it is as well to try methods that accommodate unmodelled heterogeneity and use a model selection criterion to select the best model.

Laake (1999) developed methods for point independence (with independence at distance $y = 0$) for line transect surveys, and Borchers *et al.* (in preparation *a*) generalized these to include variables other than y. The methods use $\mathcal{L}_\omega$ and $\mathcal{L}_{y\,|\,z}$ (or $\mathcal{L}_y$) and a parameterization of p that incorporates point independence. This is discussed in more detail in Section 6.3.2.

Pledger (2000) developed a very general method for mark-recapture studies which incorporates both observed and unobserved sources of heterogeneity. The methods are powerful and in principle applicable to independent observer distance surveys, although to our knowledge they have not been applied in this context yet. The disadvantage of Pledger's method in a distance sampling context is that, unlike the methods of Laake (1999) and Borchers *et al.* (in preparation *a*), they do not use the density $\pi(y)$, which is conventionally assumed known in distance sampling. Nor do they easily accommodate the detection function forms commonly used in distance sampling, although this is not a major obstacle.

6.3 State and observation models

In general, detection probability depends on many variables other than distance. Depending on circumstances, it may change with observer, environmental conditions, habitat, size or behaviour of the animals, season, etc. Provided that detection at distance $y = 0$ is certain, estimates need not be biased if all variables other than y are neglected (Section 11.12). However, when detection at distance $y = 0$ is uncertain, and double platform survey methods are used to estimate abundance, bias is introduced by neglecting variables that affect detection probability—and this bias can be large. It is therefore important to include all variables that have a significant impact on detection probability when using double-platform methods. Thus the construction of likelihood functions and estimators is more complicated than when $p(0) = 1$.

It is useful to think of models for distance surveys in terms of a model for the state of the population (a 'state model') and a model for the process of observing the population (an 'observation model'). The state model describes things like how many animals are present, how they are distributed in the survey region, how they cluster into groups, pods, flocks, etc., and so on. The observation model describes how animals are observed, given their state.

The state and observation models contained in the conventional distance sampling likelihood function $\mathcal{L}_n \mathcal{L}_y$ of Section 6.2.2 are:

State model: Animals are equally likely to be anywhere in the covered region (uniformly distributed) and they are distributed independently of one another. This uniform distribution model is implicit in the form of $\pi(y)$, which is $1/w$ for line transects and $2y/w^2$ for point transects (where w is the truncation distance). Independence is implicit in the fact that $\pi(y_1, \ldots, y_{N_c}) = \prod_{i=1}^{N_c} \pi(y_i)$. (This fact is not immediately apparent from $\mathcal{L}_n \mathcal{L}_y$, but it is used to derive this likelihood.)

Observation model: An animal at distance y is detected with probability $p(y)$ and animals are detected independently of one another.

More general state and observation models are needed for independent observation likelihoods. We outline some of these briefly below.

6.3.1 *State models*

6.3.1.1 *Spatial location*

The most common state model for spatial location is given above (independent uniform distribution of animals), although there are situations in which this is not appropriate. The spatial modelling methods of Chapter 4 use non-uniform spatial state models.

If the lines or points systematically follow or avoid some geographic feature that affects animal density, the assumption of uniform distribution is likely to be violated. An example of this is when transect lines follow tracks—animal density on the line will tend to differ from that away from it, due to habitat differences, disturbance, etc. In this case, conventional line transect methods run into real difficulty because they are unable to separate the effect of changing animal density with distance, from the effect of changing detection probability with distance. Independent observation methods, however, are able to estimate detection probability from capture histories alone, without any assumptions about animal distribution. The proportion of the sub-population 'marked' by observer 2 that observer 1 detects provides an estimate of the detection function of observer 1 (provided the observers independently detect the animals), no matter how many animals there were at each distance. More formally, notice that we can estimate detection probabilities from likelihood component $\mathcal{L}_\omega$ in eqns (6.8) and (6.16) and that these likelihoods do not involve $\pi(y)$.

Unlike CDS and MCDS methods, estimation approach (2) above ((2a) and (2b)) does not require uniform animal distribution and can therefore be used to estimate density and abundance within covered regions that are located along geographic features. Note, however, that in this case density within the covered region may be quite different from density outside the covered region, and treating the covered region as representative of the whole region may introduce substantial bias. This is not a concern if the whole region is covered; this situation is rare, but occurs, for example, in a migration count, if all animals pass within the visible range of the observer (Section 11.8).

6.3.1.2 *Animal movement*

Non-uniform animal distribution can also arise as a result of responsive animal movement. Here too, estimation approach (2) above can give valid estimates of density and abundance within the covered region at the time of the survey, and here too drawing inferences about density in the whole region from density in the covered region can be very misleading. For example, if animals are attracted into the covered region from an area much larger than that of the covered region, the area effectively surveyed is much larger than the covered region. Estimating abundance in the whole region without accounting for this will result in large positive bias. In order to estimate abundance reliably using data gathered after responsive movement, you need to know (or estimate) the extent of the responsive movement. A state model for animal movement is required to do this using maximum likelihood methods.

Most distance sampling surveys are considered to be instantaneous so that only movement before detection is relevant. In the case of independent observation surveys, however, some time may elapse between detection by observer 1 and by observer 2; in this case, movement after first detection becomes more of an issue. Moreover, detections of the same animal at different times provide data from which movement can be estimated. This is an area that might well see development in future, with improved technology allowing animal positions to be measured more accurately. Increasingly, point counts of scarce species are conducted by attracting animals to a point using calls, song or bait; if for example, a sample of animals can be radio-tagged, these provide the data for fitting a movement model.

In Section 6.6.2 below, we discuss a method that implicitly deals with responsive movement, without an explicit movement model.

6.3.1.3 *Cluster size*

Many animal populations occur in clusters, but most distance sampling methods do not include a state model for the cluster size distribution. This is because most are not based on a full likelihood—an estimate of cluster abundance is just multiplied by an estimate of mean cluster size. If inference for a clustered population is based on the full likelihood, a model for the cluster size distribution needs to be included: $\pi(s)$, where

s is cluster size. Poisson, negative binomial, and gamma distribution models are candidates for $\pi(s)$, the last two being more flexible than the former. Examples of applications with state models for cluster size include Borchers *et al.* (1998b), who used a negative binomial to model Antarctic minke whale cluster size, and Chen and Cowling (2001), who used a gamma distribution to model southern bluefin tuna school size.

6.3.1.4   *Availability*

The likelihoods and estimation approaches outlined above apply only to available animals. N_c above does not include animals that are unavailable for detection (because they are under water, under cover, etc.). If some fraction of the population in the covered region is unavailable, the estimates of N_c obtained using methods above will be estimates of the available component of the population, not the whole population. This is true for CDS, MCDS, and the methods of this chapter. However, when detection at distance zero is less than certain, there is an additional possible source of bias from the availability process. This is the bias that arises from unmodelled heterogeneity. If the probability that an animal is detected depends on when and where it was available while in detection range, availability can be considered a source of heterogeneity in detection probability, and neglecting it will therefore introduce bias (see Sections 6.2.3 and 6.2.5 above). If an animal's availability history makes it more detectable to one observer and less detectable to the other, neglecting the availability introduces negative bias in detection probability and positive bias in $\widehat{N}_c$. An example would be visual surveys combined with acoustic surveys of whales, if whales can only be heard under water where they cannot be seen. If an animal's availability history that makes it more detectable to one observer also makes it more detectable to the other, neglecting the availability introduces positive bias in detection probability and negative bias in $\widehat{N}_c$ (Schweder *et al.* 1991). This positive dependence is the more usual case, such as in aerial visual surveys of whales with independent observers.

In this section, we outline some availability models that describe the availability/unavailability process probabilistically.

6.3.1.5   *Static availability*

The simplest kind of availability model is one in which animals are either available or unavailable for the duration of the period that they are in detectable range. This would be the case for aerial line transect surveys of pack-ice seals, for example. On these surveys, seals are either hauled out for the duration of the period they are visible from the aircraft, or they are off the ice and unavailable for the duration. In such cases, an appropriate availability model for a single animal would be $\pi(a) = P_v^a(1 - P_v)^{1-a}$, where $a = 1$ if the animal is available at the time of the survey, and $a = 0$ otherwise, and P_v is the probability that the animal is available. If animals become available independently of one another, then the availability model for all N_c animals is $\prod_{i=1}^{N_c} P_v^{a_i}(1-P_v)^{1-a_i}$. If there is dependence in availability between animals, if different animals have different probabilities of being available, or if availability depends on time or environmental conditions, a more complicated model is required.

6.3.1.6 *Discrete availability*

If animals become available for an instant only, the availability model needs to model the frequency with which an animal is available while in detectable range, and the times at which it becomes available. The simplest scenario is when the probability that an animal becomes available in the next instant, remains constant, and is independent of the animal's availability history and of other animals. In this case, an independent Poisson point process model is appropriate. More complicated models are required to accommodate inter-animal differences, dependence between animals, and dependence on time or environmental conditions. In general, we denote the availability history of an animal $\mathcal{H}$, and we denote the availability state model $\pi(\mathcal{H})$ (Section 6.8.3).

Line transect surveys of North Atlantic minke whales (which are visible only for a very short time when surfacing to breathe) are a case in which discrete availability models are appropriate. A simple Poisson point process model was used by Skaug and Schweder (1999) to model whale availability, while Schweder *et al.* (1996) used a Poisson distribution with a random (gamma-distributed) rate parameter. Schweder *et al.* (1999) used simulation to generate an availability model that is consistent with independent radio-tag data from the whales. Barlow (1999) also used simulation to accommodate discrete availability due to diving behaviour for a variety of large whales.

If animals switch between being available and being unavailable while in detectable range, but are available for some time when they are detectable, discrete availability models may be inadequate. They can still be used if one defines an observable instantaneous event associated with becoming available (the instant of first availability, for example) to be the availability event. However, to apply discrete availability methods in this case, all detections of animals other than at the instantaneous event must be ignored. An alternative is to develop what we call an intermittent-availability state model.

6.3.1.7 *Intermittent availability*

When animals are available for more than an instant and their availability can change when they are within detectable range, we call this 'intermittent availability'. One example is line transect surveys of marine mammals in which the animals are at the surface for more than a brief moment when they surface. Another is point transect surveys of songbirds in which the duration of a songburst is significant. In these cases, a state model for availability must include components for the process of becoming available and for the duration of the availability.

McLaren (1961) used a non-probabilistic model for intermittent availability in which animals were alternately available and unavailable for constant amounts of time. Eberhardt (1978) suggested using McLaren's model to correct for availability bias in line transect studies. Barlow *et al.* (1988) used a modification of McLaren's model to correct for availability bias in aerial surveys of harbour porpoise. Laake *et al.* (1997) suggested a probabilistic model based on an alternating renewal process for intermittent availability of harbour porpoise from aerial surveys and compared it to the models described by McLaren (1961) and Barlow *et al.* (1988). Hiby and Lovell (1998) suggested a similar probabilistic model for intermittent availability of harbour porpoise.

6.3.1.8   *Other state models*

There may be many variables affecting detection probability other than those mentioned above. Some examples of independent-observer surveys in which a host of such variables have been investigated are Borchers *et al.* (1998a), Schweder *et al.* (1999) and Southwell *et al.* (in preparation). In the likelihood functions developed above, these are referred to collectively as z, and the state model for z, which models the distribution of z in the population, is denoted $\pi(z)$. For most such variables, it is reasonable to assume that they are independent of y, but if not, the joint density $\pi(y, z)$ must be specified.

6.3.2   *Observation models*

Distance sampling methods are founded on the idea that relative frequencies of detection at different distances contain information about how detection probability depends on distance. Mark-recapture methods are founded on the idea that the proportion of recaptures contains information about detection probability. MRDS methods combine both sources of information and the observation model must accommodate both methods.

Mark-recapture sampling requires two or more samples from the same survey region (we will only consider two samples), and for MRDS, at least one of the samples must contain data adequate for distance sampling.

As in CDS and MCDS, the MRDS observation model is represented by detection functions together with the assumption that detections are independent to some degree (full independence or point independence as described below). To incorporate heterogeneity in detection probability from a variety of sources, the detection function can include an explanatory covariate vector $\underline{z}$ in addition to distance y. One way (but not the only way) of formulating an MRDS detection function is as a scaled version of an MCDS detection function:

$$p(y, \underline{z}) = p(0, \underline{z})g(y, \underline{z}). \tag{6.19}$$

Here $p(0, \underline{z})$ could have logistic or any other suitable form. CDS and MCDS methods alone cannot estimate the intercept $p(0, \underline{z})$: distance sampling data contain information on the shape ($g(y, \underline{z})$) but not on the intercept of the detection function. This is apparent if you consider the distance sampling component of the likelihood $\mathcal{L}_{y \mid z}$ of eqn (6.15) and replace $p.(y, \underline{z})$ with the representation in eqn (6.19). In this case, $p.(0, \underline{z})$ cancels and the likelihood component $\mathcal{L}_{y \mid z}$ only includes $g.(y, \underline{z})$:

$$\mathcal{L}_{y \mid z} = \prod_{i=1}^{n.} \frac{g.(y_i, \underline{z}_i)\pi(y_i \mid \underline{z}_i)}{\int_0^w g.(y, \underline{z}_i)\pi(y \mid \underline{z}_i)\,dy}. \tag{6.20}$$

The mark-recapture component is needed to estimate $p.(0, \underline{z})$. There are various ways that the mark-recapture samples can be collected to estimate $p.(0, \underline{z})$, and there are several configurations for the observers that we describe below (Section 6.3.3). The appropriate configuration will depend partially on the assumed state model. In describing observation models, we will consider the most general configuration with two independent observers and we will use the notation from Section 6.2. In this discussion, we are assuming that there is no uncertainty in determining whether an observation is a duplicate or not. Section 6.9.1 addresses uncertainty in duplicate identification. Also, we assume no measurement error in distance and the other covariates.

The two observers have detection functions denoted $p_1(y, \underline{z})$ and $p_2(y, \underline{z})$: $p_1(y, \underline{z}) = \Pr\{\underline{\omega} = (1, 0)\} + \Pr\{\underline{\omega} = (1, 1)\}$ and $p_2(y, \underline{z}) = \Pr\{\underline{\omega} = (0, 1)\} + \Pr\{\underline{\omega} = (1, 1)\}$. We use the dot subscript to represent the probability that at least one of the observers makes the detection, $p.(y, \underline{z}) = 1 - \Pr\{\underline{\omega} = (0, 0)\}$.

If observers' detections are independent, then the probabilities for the three observable capture histories are as follows:

$$\Pr\{\underline{\omega} = (1, 0)\} = p_1(y, \underline{z})[1 - p_2(y, \underline{z})], \tag{6.21}$$

$$\Pr\{\underline{\omega} = (0, 1)\} = [1 - p_1(y, \underline{z})]p_2(y, \underline{z}),$$

$$\Pr\{\underline{\omega} = (1, 1)\} = p_1(y, \underline{z})p_2(y, \underline{z}),$$

where

$$p.(y, \underline{z}) = p_1(y, \underline{z}) + p_2(y, \underline{z}) - p_1(y, \underline{z})p_2(y, \underline{z}). \qquad (6.22)$$

The probability distribution for $\mathcal{L}_\omega$ is computed by dividing the probabilities by $p.(y, \underline{z})$ (eqn (6.16)). The independence assumption was made by Alpizar-Jara and Pollock (1996), Manly et $al.$ (1996), Quang and Becker (1997), Borchers et $al.$ (1998a,b), Chen (1999, 2000), and Chen and Lloyd (2000). However, if there is heterogeneity in detection probability beyond that caused by y and the covariate vector $\underline{z}$, this can induce dependence between detections (Section 6.2.3).

Borchers et $al.$ (in preparation a) weakened the independence assumption by parameterizing $\mathcal{L}_\omega$ in terms of conditional detection functions $p_{1\,|\,2}(y, \underline{z})$ and $p_{2\,|\,1}(y, \underline{z})$. The conditional detection functions are related to the unconditional detection functions in the following manner: $p_j(y, \underline{z}) = p_{j\,|\,3-j}(y, \underline{z})\delta(y, \underline{z})$. The function $\delta(y, \underline{z})$ quantifies the dependence from unmodelled heterogeneity and can be expressed as:

$$\delta(y, \underline{z}) = \frac{\mathrm{cov}[p_1(y, \underline{z}), p_2(y, \underline{z})]}{p_1(y, \underline{z})p_2(y, \underline{z})} + 1, \qquad (6.23)$$

(Borchers 1996) where the expectation for the covariance is over unspecified variables that affect detection probability other than y and $\underline{z}$. The independence assumption is equivalent to assuming that $\delta(y, \underline{z}) = 1$. We refer to this as 'full independence'. Chen (1999, 2000) and Chen and Lloyd (2000) used a heterogeneity index α in a similar fashion to modify probabilities. However, their function α is fundamentally different because it represents heterogeneity that is included in the model covariates, whereas $\delta(y, \underline{z})$ represents heterogeneity from unspecified variables not included in the model.

Allowing for possible dependence, we get the following alternative representation for the capture history probabilities:

$$\mathrm{Pr}\{\underline{\omega} = (1, 0)\} = \frac{p_{1\,|\,2}(y, \underline{z})[1 - p_{2\,|\,1}(y, \underline{z})]}{\delta(y, \underline{z})}, \qquad (6.24)$$

$$\mathrm{Pr}\{\underline{\omega} = (0, 1)\} = \frac{[1 - p_{1\,|\,2}(y, \underline{z})]p_{2\,|\,1}(y, \underline{z})}{\delta(y, \underline{z})},$$

$$\mathrm{Pr}\{\underline{\omega} = (1, 1)\} = \frac{p_{1\,|\,2}(y, \underline{z})p_{2\,|\,1}(y, \underline{z})}{\delta(y, \underline{z})},$$

which in turn leads to the following expressions for the conditional capture history probabilities (given detection), which are required to evaluate $\mathcal{L}_\omega$:

$$\frac{\Pr\{\underline{\omega} = (1,0)\}}{p.(y,\underline{z})} = \frac{p_{1|2}(y,\underline{z})[1 - p_{2|1}(y,\underline{z})]}{p_.^c(y,\underline{z})}, \tag{6.25}$$

$$\frac{\Pr\{\underline{\omega} = (0,1)\}}{p.(y,\underline{z})} = \frac{[1 - p_{1|2}(y,\underline{z})]p_{2|1}(y,\underline{z})}{p_.^c(y,\underline{z})},$$

$$\frac{\Pr\{\underline{\omega} = (1,1)\}}{p.(y,\underline{z})} = \frac{p_{1|2}(y,\underline{z})p_{2|1}(y,\underline{z})}{p_.^c(y,\underline{z})},$$

where

$$p_.^c(y,\underline{z}) = p_{1|2}(y,\underline{z}) + p_{2|1}(y,\underline{z}) - p_{1|2}(y,\underline{z})p_{2|1}(y,\underline{z}). \tag{6.26}$$

From eqn (6.25), it is clear that we can only estimate $p_{1|2}(y,\underline{z})$ and $p_{2|1}(y,\underline{z})$ from $\mathcal{L}_\omega$, and using them, we can derive an estimate of $p_.^c(y,\underline{z})$. Also, even though we formulated the probabilities in terms of $\delta(y,\underline{z})$, it is not contained in $\mathcal{L}_\omega$.

So where does that lead us in regards to estimation of $p.(y,\underline{z})$ and N_c? From $\mathcal{L}_\omega$, we can only estimate $p_.^c(y,\underline{z})$, but we need $p.(y,\underline{z})$. However, from $\mathcal{L}_{y,z}$ we can estimate the shape of $p.(y,\underline{z})$ ($g.(y,\underline{z})$), but not its intercept, $p.(0,\underline{z})$. Even though we do not use it for estimation, $\delta(y,\underline{z})$ provides the conceptual link between $p_.^c(y,\underline{z})$ and $p.(y,\underline{z})$:

$$\delta(y,\underline{z}) = \frac{p_.^c(y,\underline{z})}{p.(y,\underline{z})}. \tag{6.27}$$

To estimate $p.(0,\underline{z})$, we must assume that $\delta(0,\underline{z}) = 1$. Based on the assumption of 'point independence' at $y = 0$, our estimator for $p.(y,\underline{z})$ is:

$$\hat{p}.(y,\underline{z}) = \hat{p}^c(0,\underline{z})\hat{g}.(y,\underline{z}), \tag{6.28}$$

where $\hat{p}^c(0,\underline{z})$ is derived from $\mathcal{L}_\omega$ and $\hat{g}.(y,\underline{z})$ from $\mathcal{L}_{y,z}$. In comparison, the estimator for $p.(y,\underline{z})$ based on 'full independence' ($\delta(y,\underline{z}) = 1$) can be derived from eqn (6.22) solely using $\mathcal{L}_\omega$. In the next two sections, we discuss point independence and full independence in more detail.

6.3.2.1 *Point independence*
The function $p^c(y,\underline{z})$ is the combined detection function you would get were you to treat the conditional detection functions as unconditional and assume independent detections. The function p^c is estimated from $\mathcal{L}_\omega$ and the shape of $p.$ is estimated from $\mathcal{L}_{y,z}$. The function $\delta(y,\underline{z})$ is their ratio. To see the significance of eqn (6.27), take another look at Fig. 6.2 and recall

that $\delta(y, \underline{z})$ measures the degree of dependence in detection probability between platforms. Eqn (6.27) says that the ratio of the top dashed and top solid curves in the figure quantifies the degree of dependence in detection probability between observers.

This is of very limited use on its own, because when there is dependence, we cannot estimate the solid curve without assumptions about the dependence. However, using the conventional state models for y ($\pi(y) = 1/w$ for line transects and $\pi(y) = 2y/w^2$ for point transects), we can estimate the shape of the combined detection probability function $p.()$ from $\mathcal{L}_{y \mid z}$. All that is then required to quantify the dependence is an assumption about the relative height of $p.(y, \underline{z})$ and $p^c_.(y, \underline{z})$ at some point. This is the basis of methods based on a point independence assumption: when detections are independent at a point, $p.(y, \underline{z})$ and $p^c_.(y, \underline{z})$ are the same height at that point: $\delta(0, \underline{z}) = 1$ if the point is $y = 0$. Knowing the heights of $p.(y, \underline{z})$ and $p^c_.(y, \underline{z})$ at a point allows us to scale their shapes appropriately, and from their ratio at other points we can quantify $\delta(y, \underline{z})$. This in turn allows us to estimate $p.(y, \underline{z})$ (the quantity of interest) from $p^c(y, \underline{z})$ (the quantity estimated from the mark-recapture likelihood component $\mathcal{L}_\omega$). In practice, estimation does not proceed with these explicit steps but they are useful for understanding the ideas on which point independence estimation is based.

As noted above, implementation of point independence in estimation requires specification of an additional functional form. While we could specify a functional form for $\delta(y, \underline{z})$ such that $\delta(0, \underline{z}) = 1$, it is easier and adequate to specify a functional form for the shape of $p.(y, \underline{z})$. When $p.(y, \underline{z})$ is parameterized as $p.(0, \underline{z})g.(y, \underline{z})$, the MCDS component $g.(y, \underline{z})$ is its shape.

Note that inference under an assumption of point independence is only possible because we treat $\pi(y \mid \underline{z})$ as known and this makes the shape $g.(y, \underline{z})$ estimable from $\mathcal{L}_{y \mid z}$. Without known $\pi(y \mid \underline{z})$ or additional information from which to estimate $\pi(y \mid \underline{z})$, full independence must be assumed. The cost of relaxing the full independence assumption to one of point independence is that $\pi(y \mid \underline{z})$ must be treated as known. In a distance sampling context this is hardly a cost, since $\pi(y \mid \underline{z})$ is conventionally treated as known anyway.

By contrast, Chen and Lloyd (2000) estimate three probability density functions, $p_1()$, $p_2()$, and $p_3()$ (defined below), and use an assumption of full independence to allow estimation of the shape of $\pi(y \mid \underline{z})$. This has important implications for use in distance sampling if there is unmodelled heterogeneity. The approach of Chen and Lloyd (2000) could be modified to assume known $\pi(y \mid \underline{z})$ and point independence. In effect, this would be equivalent to the Palka (1995) estimator using kernel density estimation for the detection functions. Equation (12) of Chen and Lloyd (2000) would then be equivalent to eqn (12) of Laake (1999) for the Palka estimator. (Note: the latter equation is missing w in the numerator of the right-hand side.)

While the material in the remainder of this section is not needed to understand point independence, we provide it to document the development of the idea and different formulations that were less optimal. The remainder of this section can be skipped if you are not interested in these details.

The idea of point independence is due to Laake (1999), who developed it for univariate data (y only). His development was based on the estimator of Palka (1995) that included detection functions for each observer, $p_1(y, z)$ and $p_2(y, z)$, and a detection function for the duplicates, $p_3(y, z)$. Her estimator was not explicitly based on a likelihood and was effectively a Petersen estimator using abundance (or density) estimates rather than numbers seen:

$$\widehat{N} = \frac{\widehat{N}_1 \widehat{N}_2}{\widehat{N}_3}, \tag{6.29}$$

where each $\widehat{N}$ is an abundance estimate from CDS using the appropriate set of observations and estimated detection function. The key component here is that her estimator effectively did not in general impose $p_3(y) = p_1(y)p_2(y)$, but it did implicitly assume that $p_3(0) = p_1(0)p_2(0)$. Laake (1999) cast these assumptions into a likelihood framework and illustrated the advantages of the weaker point independence assumption with real survey data from a known population.

Laake (1999) used the following representation for the capture history probabilities with point independence:

$$\Pr\{\underline{\omega} = (1, 0)\} = p_1(y, z) - p_3(y, z), \tag{6.30}$$
$$\Pr\{\underline{\omega} = (0, 1)\} = p_2(y, z) - p_3(y, z),$$
$$\Pr\{\underline{\omega} = (1, 1)\} = p_3(y, z),$$

where $p_3(y, z)$ is a separate detection function that is fitted to the duplicate detections. With this representation, $p.(y, z) = p_1(y, z) + p_2(y, z) - p_3(y, z)$. An alternative formulation that allows dependence is (Skaug and Schweder 1999):

$$\Pr\{\underline{\omega} = (1, 0)\} = p.(y, z) - p_2(y, z), \tag{6.31}$$
$$\Pr\{\underline{\omega} = (0, 1)\} = p.(y, z) - p_1(y, z),$$
$$\Pr\{\underline{\omega} = (1, 1)\} = p_1(y, z) + p_2(y, z) - p.(y, z),$$

where $p.(y, z)$ is a separate detection function that is not constrained to satisfy eqn (6.22). Both of these formulations have three separate detection functions that are not constrained relative to each other except that $p_3(0, z) = p_1(0, z)p_2(0, z)$ or $p.(0, z) = p_1(0, z) + p_2(0, z) - p_1(0, z)p_2(0, z)$.

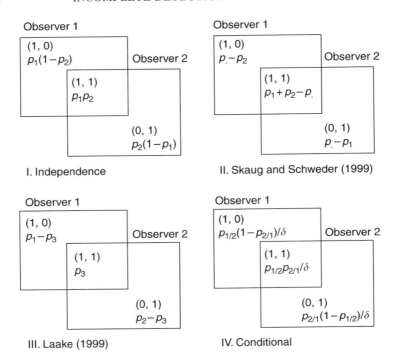

Fig. 6.4. Venn diagrams illustrating different representations for the probability structure of the observable capture histories for two independent observers. Each observer is represented by a box and the three sections of the boxes represent the capture histories: $(1, 0)$—observer 1 only; $(1, 1)$—seen by both (duplicate); and $(0, 1)$—observer 2 only. In each case, the sum of the probabilities within the box is p_1 for observer 1 and p_2 for observer 2, or $p_{1|2}$ and $p_{2|1}$ for the conditional structure. The total probability is $p.$ in each case. Structures II–IV allow dependence and subsequently have three different quantities to represent the probabilities.

For comparison, the four probability structures described (including the full independence case) are shown as Venn diagrams in Fig. 6.4.

With the formulation of Laake (1999) and Skaug and Schweder (1999), the functions $p_1(y, \underline{z})$, $p_2(y, \underline{z})$, and either $p_3(y, \underline{z})$ or $p.(y, \underline{z})$, must be used in both $\mathcal{L}_\omega$ and $\mathcal{L}_{y,z}$. Because the functions are not constrained relative to one another except at $y = 0$, it is possible to get negative probabilities for the capture histories in the process of numerically maximizing the joint likelihood. While it is possible to constrain the probabilities in the optimization, achieving convergence can become a practical problem with some data sets in which one of the observers is poor relative to the other so that $p_j(y, \underline{z})$ is quite close to $p.(y, \underline{z})$. This situation can also be represented as $p_{j|3-j}(y, \underline{z})$ being very close to unity. This illustrates the advantage of the formulation used by Borchers *et al.* (in preparation *a*). By using conditional detection functions, no constraints are needed other than constraining $p_.(y, \underline{z})$ and $p_{j\,|\,3-j}(y, \underline{z})$ to be probabilities, which is done through specification of their functional form. An additional advantage of the formulation of Borchers *et al.* (in preparation *a*) is that $\mathcal{L}_{\omega}$ and $\mathcal{L}_{y,z}$ can be maximized separately rather than jointly because they do not share parameters.

6.3.2.2 *Full independence*

Under full independence, $p_1(y, \underline{z}) = p_{1\,|\,2}(y, \underline{z})$ and $p_2(y, \underline{z}) = p_{2\,|\,1}(y, \underline{z})$, so that we need only specify forms for the conditional detection functions in order to estimate the unconditional probabilities of detection $p_1(\ )$, $p_2(\ )$, and $p_.(\ )$, and hence to estimate abundance.

While various authors have assumed full independence, they have not used the same likelihoods for estimation. For example, Alpizar-Jara and Pollock (1996) only used $\mathcal{L}_{\omega}$ whereas Chen (1999) used both $\mathcal{L}_{\omega}$ and $\mathcal{L}_y$, and Borchers *et al.* (1998b) used both $\mathcal{L}_{\omega}$ and the full likelihood $\mathcal{L}_n\,\mathcal{L}_z\mathcal{L}_{y\,|\,z}\mathcal{L}_{\omega}$.

Alpizar-Jara and Pollock (1996) and Chen (1999) analysed the same set of data from a known population (Laake 1978), using distance as the only covariate. The estimate of Chen (1999) was considerably higher and closer to the true answer. Chen (1999) attributed the discrepancy to heterogeneity introduced by the grouping of distances into bins by Alpizar-Jara and Pollock (1996). While this may explain part of the difference, the most likely reason is the differences in their likelihoods. Borchers *et al.* (1998b) suggested this was a possibility in explaining the differences in their conditional and full likelihood results.

The likelihood component $\mathcal{L}_{y\,|\,z}$ contains information on the shape of $p_.(y, \underline{z})$, while the mark-recapture component $\mathcal{L}_{\omega}$ contains information about the shape and intercept of the conditional detection function $p^c(y, \underline{z})$. If full independence holds, $\mathcal{L}_{y\,|\,z}$ should provide little additional information on detection probability over and above that contained in the mark-recapture component. When full independence is assumed but does not hold, the shape of the unconditional detection function $p_.(y, \underline{z})$ will not correspond to the shape of $p_.^c(y, \underline{z})$ (as illustrated in Section 6.2.3), and inference using both will result in an estimate of detection function shape somewhere between $p_.(y, \underline{z})$ and $p_.^c(y, \underline{z})$.

6.3.2.3 *Functional forms for detection probability models*

An observation model comprises representations for the detection functions and the assumption of independence (full or point). The advantage of point independence is that $\mathcal{L}_y$ (or $\mathcal{L}_{y,z}$) and $\mathcal{L}_{\omega}$ do not share parameters. If full independence is assumed, then those likelihood components share parameters and must be maximized jointly, if both are used. For both levels of independence, a model for the conditional detection functions $p_{j\,|\,3-j}(y, \underline{z})$

must be specified. Additionally, for point independence a model for the unconditional detection function $g_.(y, \underline{z})$ must be specified.

The logistic is a convenient form for the conditional detection functions $p_{j\,|\,3-j}(y, \underline{z})$, to incorporate covariates and allow $p_{j\,|\,3-j}(0, \underline{z})$ to be constrained to the unit interval:

$$p_{j\,|\,3-j}(y, \underline{z}) = \frac{\exp\left\{\mathbf{Z}\underline{\beta}\right\}}{1 + \exp\left\{\mathbf{Z}\underline{\beta}\right\}},\qquad(6.32)$$

where $\mathbf{Z}$ is the design matrix for y and $\underline{z}$, and $\underline{\beta}$ is the parameter vector.

Alternatively, $p_{j\,|\,3-j}(y, \underline{z})$ can be modelled as a composite of a logistic for $p_{j\,|\,3-j}(0, \underline{z})$ and an MCDS form for $g_{j\,|\,3-j}(y, \underline{z})$ as in Laake (1999):

$$p_{j\,|\,3-j}(y, \underline{z}) = \frac{\exp\left\{\mathbf{Z}\underline{\beta}\right\}}{1 + \exp\left\{\mathbf{Z}\underline{\beta}\right\}} \times g_{j\,|\,3-j}(y, \underline{z}).\qquad(6.33)$$

The advantage of this alternative form is the consistency of applying similar MCDS forms for $g_.(y, \underline{z})$ and $g_{j\,|\,3-j}(y, \underline{z})$ with point independence. The disadvantage of the composite form is an increase in the complexity of model specification and possibly an increase in the number of estimated parameters. Laake (1999) was forced to use the composite form to enforce point independence with the restrictive formulation of eqn (6.30).

For point independence, an MCDS model can be used for $g_.(y, \underline{z})$, or any other functional form that is restricted to the unit interval such that $g_.(0, \underline{z}) = 1$. One alternative model for $g_.(y, \underline{z})$ is a logistic equation scaled such that $g_.(0, \underline{z}) = 1$:

$$g_.(y, \underline{z}) = h(y, \underline{z})/h(0, \underline{z}),\qquad(6.34)$$

where

$$h(y, \underline{z}) = \frac{\exp\left\{\mathbf{Z}\underline{\theta}\right\}}{1 + \exp\left\{\mathbf{Z}\underline{\theta}\right\}}.\qquad(6.35)$$

$\mathbf{Z}$ is the design matrix for y and $\underline{z}$, and $\underline{\theta}$ is the parameter vector. In principle, any strictly positive function for $h(y, \underline{z})$ could be used, but typical assumptions of monotonicity with distance should be considered. Note that we differentiate between $\underline{\beta}$ and $\underline{\theta}$ because they are different parameters in different detection functions.

Table 6.1 provides a summary of some possible observation models. Model 1 is equivalent to those used by Manly *et al.* (1996), Borchers *et al.* (1998b), and Quang and Becker (1997). Point independence model 4 is the approach suggested by Laake (1999) and was used by Evans-Mack *et al.* (2002) and Innes *et al.* (2002). Models 3 and 5 are more convenient

Table 6.1. Summary of observation models including assumption about independence (full or point) and functional forms for detection probability. Numbers in brackets are references to equations in the text. For full independence, a separate $g_.(y, \underline{z})$ is not needed because it is determined by $p_{j\,|\,3-j}(y, \underline{z})$

Model	Independence	$p_{j\,|\,3-j}(y, \underline{z})$	$g_.(y, \underline{z})$
1	Full	(6.32)	
2	Full	(6.33)	
3	Point	(6.32)	MCDS form
4	Point	(6.33)	MCDS form
5	Point	(6.32)	(6.34)
6	Point	(6.33)	(6.34)

than Model 4 because the conditional detection functions can be estimated efficiently using generalized linear model (GLM) software and the detection function shape can be estimated using Distance software, due to the independence of the likelihood components.

6.3.3 *Observation configurations*

There are various ways that the mark-recapture samples can be collected to estimate $p_.(0, \underline{z})$, but regardless of method (e.g. visual, acoustic, tagged), the important differences relate to the symmetry of the observation roles and any dependence between the samples, which we have denoted as the 'observation configuration'. We have identified three different configurations that have been used or could be used: (1) independent, (2) trial, and (3) removal. The configuration defines both the way the data are collected and the manner in which they can be analysed, and in some cases they can overlap (i.e. data collected based on one configuration can be analysed as a different configuration).

6.3.3.1 *Independent configuration*

If the observers survey independently, their roles are symmetric. Each observer provides trials for the other and the detection functions for both observers can be estimated. The observers can be on the same or different survey platforms, and can be visual, acoustic, or any combination of methods to detect and locate observations. The observers can survey simultaneously or at different times (Hiby and Lovell 1998) as long as duplicate detections can be ascertained or modelled. To provide for independence but allow real-time determination of duplicates, a separate coordinating recorder is useful. This avoids the need for post-survey analysis to identify duplicates. Independent observation is the most flexible configuration in terms of analysis options. All three of the observable capture histories are possible and the probability distribution is given by eqn (6.25). Unlike the other configurations, the data can also be analysed as if they were collected in either the trial or the removal configuration. Analysis with the independent configuration uses more data than the trial configuration and the estimator should be more precise. Also, it does not require the limiting equality assumption of the removal configuration. Many applications of MRDS have been based on the independent configuration (Palka 1995; Alpizar-Jara and Pollock 1996; Manly *et al.* 1996; Quang and Becker 1997; Laake 1999; Skaug and Schweder 1999; Chen 2000). The use of the Petersen estimator to estimate visibility bias outside the context of distance sampling (Pollock and Kendall 1987; Graham and Bell 1989; Carretta *et al.* 1998) is also an example of the independent configuration.

6.3.3.2   *Trial configuration*

In this configuration, the roles of the observers are asymmetric. Observer 2 collects a sample of observations that are used as trials to estimate the detection function for observer 1: $p_{1|2}(y, \underline{z})$ and $p_1(y, \underline{z})$. There is no attempt to estimate the detection functions $p_{2|1}(y, \underline{z})$ and $p_2(y, \underline{z})$ corresponding to the observer that is generating the trials. There are only two values for the capture history: $\underline{\omega} = (0, 1)$ and $\underline{\omega} = (1, 1)$, a trial missed by observer 1 and a trial seen by observer 1, respectively. The probability distribution of the observable capture histories is:

$$\frac{\Pr\{\underline{\omega} = (0, 1)\}}{p_2(y, \underline{z})} = 1 - p_{1|2}(y, \underline{z}) \qquad (6.36)$$

$$\frac{\Pr\{\underline{\omega} = (1, 1)\}}{p_2(y, \underline{z})} = p_{1|2}(y, \underline{z})$$

An obvious example of a trial configuration is to use radio-tagged animals (Samuel *et al.* 1987; Steinhorst and Samuel 1989) as the trial generator (observer 2). The locations of radio-tagged animals relative to the observer provide detection trials at different distances. Because radiolocation does not depend on detection, there is no need for $p_2(y, \underline{z})$. For visual surveys, the detections of observer 2 are treated as trials for observer 1. Observer 2 can be aware of detections by observer 1 and this level of awareness can be useful to help ascertain whether the trial observation was detected or not by the other observer in real time, rather than depending on post-survey analysis of duplicate detections. However, the selection of the trials should be made independent of the detections by observer 1. If there is visual or auditory communication between the observers, the trial observations should be made prior to possible detection by observer 1. Observer 2 should locate and track an animal and then determine whether observer 1 detects the animal. The method proposed by Buckland and Turnock (1992) with separate survey platforms for observers (e.g. helicopter searching forward of a ship's path) was designed and analysed as a trial configuration. A similar approach can be implemented with two observers on the same survey platform if one of the observers searches ahead of the other (Palka 1995; Borchers *et al.* 1998a; Evans-Mack *et al.* 2002). The trial configuration is advantageous in the following situations: (1) the animals are only available at discrete times, and within the observers' field of view, the animal could become available more than once, or (2) the animals respond and move prior to detection (Section 6.6.2). In addition, duplicate identification may be more reliable because there can be one-way communication between observers.

6.3.3.3   *Removal configuration*

In some circumstances, it is not possible for either observer to survey independently of the other (e.g. observers in close proximity that cannot be isolated). In those cases, a removal configuration can be used to implement a limited MRDS analysis. In the removal configuration, the first observer makes detections of which the second observer is fully aware and the second observer detects observations that are missed by the first observer. As in the trial configuration, the roles of the observers are asymmetric. The conditional detection function $p_{2\,|\,1}(y, \underline{z})$ is unity because observer 2 always 'detects' what observer 1 detected because they are aware of the observations. The removal configuration is slightly different from the other configurations in that the 'marked' sample for observer 1 is composed of those observer 1 detected, together with those that (s)he missed but that were seen by observer 2. If you consider all of these observations to have been seen by the second observer, then they can be viewed as a set of trials for the first observer, even though they were not set up in that fashion. As the name suggests, the removal configuration is equivalent to removal mark-recapture models (Otis *et al.* 1978), and with only two samples, it is not possible to estimate sample-specific probabilities because there are only two possible capture histories ($\underline{\omega} = (0, 1)$ and $\underline{\omega} = (1, 1)$) and thus only one estimable parameter. We cannot estimate the probability that observer 2 detects what observer 1 misses. Thus, we must make some assumption. The most common assumption is that it equals $p_{1\,|\,2}(y, \underline{z})$. In this case, the conditional probability distribution of the capture histories is:

$$\frac{\Pr\{\underline{\omega} = (0, 1)\}}{p_2(y, \underline{z})} = \frac{p_{1\,|\,2}(y, \underline{z})[1 - p_{1\,|\,2}(y, \underline{z})]}{2p_{1\,|\,2}(y, \underline{z}) - [p_{1\,|\,2}(y, \underline{z})]^2},$$

$$\frac{\Pr\{\underline{\omega} = (1, 1)\}}{p_2(y, \underline{z})} = \frac{p_{1\,|\,2}(y, \underline{z})}{2p_{1\,|\,2}(y, \underline{z}) - [p_{1\,|\,2}(y, \underline{z})]^2}.$$

(6.37)

To satisfy the equality assumption, there should not be any observer differences, which is rather limiting; however, one can partly allow for innate differences in the abilities of visual observers by rotating personnel through the observer roles. This latter approach was used by Cook and Jacobsen (1979) in a mark-recapture setting. In a visual survey, there should not be any inherent differences in the conditions for each observer (e.g. different sized windows in an aircraft). Also, the equality assumption is unlikely to be satisfied if two different types of detection, such as visual and acoustic, are used.

6.4   Example data

To illustrate many of the MRDS analyses, we have used data from surveys of a known population of golf tees, which was used throughout Borchers *et al.* (2002). The data were derived from independent distance sampling surveys by eight different observers of a population of 250 clusters (760 individuals) of golf tees. The tees, of two colours (green = 0 and yellow = 1), were placed in clusters of between one and eight in a survey region of 1680 m^2, either exposed above the surrounding grass (exposure = 1), or at least partly hidden by it (exposure = 0). They were surveyed by the 1999 statistics honours class at the University of St Andrews[3], so while golf tees are clearly not animals (or plants), the survey was real, not simulated. The entire survey region was covered by eleven 8-m wide strips ($2w = 8$ m) that totalled 210 m in length. We have treated each cluster of golf tees as a single 'animal', with size equal to the number of tees in the cluster. For most of the example analyses, we pooled the eight independent surveys into two teams of four surveys (1–4 and 5–8) to increase the sample size. The results from the two observer teams were then treated as the two mark-recapture samples, with each sample generated by distance sampling. Duplicate detections were determined based on the (known) locations of the tees. We used the distance from the known location to the line for our example analyses. We have used colour, size, exposure, and platform (observer team) as potential covariates in some of the analyses.

We briefly describe some summary statistics (Table 6.2) and initial analyses from the example data to illustrate some of the ideas we have introduced so far, and to help develop the results of the analyses given in the following sections. Team 1 detected 124 and team 2 detected 142 of the 250 tee clusters, and 104 of their detections were duplicates. A CDS analysis with a half-normal detection function yielded estimates of 212 and 215 clusters for the teams respectively. The estimates are 85% and 86% of the true N, which approximately matches the proportion detected within 0.5 m of the line (Table 6.2). Even after pooling the results of four different

---

[3] We are grateful to Miguel Bernal for making these data available to us. They were collected by him as part of a Masters project at the University of St Andrews.

Table 6.2. Summary statistics by perpendicular distance interval for golf tee data with two observer teams; n_1 is the number seen by team 1, n_2 is the number seen by team 2, n_3 is the number of duplicates, and the total number seen $n. = n_1 + n_2 - n_3$. The true number N in each bin is known, so the true proportions detected can be calculated as $p_1 = n_1/N$ and $p_2 = n_2/N$. The true proportion detected was always less than the conditional probabilities $p_{1\,|\,2} = n_3/n_2$ and $p_{2\,|\,1} = n_3/n_1$

Distance (m)	n_1	n_2	n_3	$n.$	N	p_1	$p_{1\,|\,2}$	p_2	$p_{2\,|\,1}$
0–0.5	29	27	25	31	32	0.91	0.93	0.84	0.86
0.5–1	19	20	18	21	27	0.70	0.90	0.74	0.95
1–1.5	27	32	26	33	34	0.79	0.81	0.94	0.96
1.5–2	13	17	11	19	27	0.48	0.65	0.63	0.85
2–3	26	33	21	38	66	0.39	0.64	0.50	0.81
3–4	10	13	3	20	64	0.16	0.23	0.20	0.30
Total	124	142	104	162	250	0.50	0.73	0.57	0.84

observers, it does not appear that $g(0) = 1$ for either team. However, if we pool the observations of both teams, only one of the 32 tee clusters within 0.5 m of the line was missed.

A simple Petersen estimator applied to the two samples yielded an estimate of 169, and the same estimator applied to the data stratified by each of the six distance intervals (Table 6.2) reduced the heterogeneity due to distance and increased the estimate to 190. However, for each distance interval, the conditional probabilities $p_{1\,|\,2}$ and $p_{2\,|\,1}$ are greater than the true probabilities, suggesting that additional heterogeneity beyond distance remains, as demonstrated in Section 6.2.3. The remaining heterogeneity could possibly be explained by the effects of size, colour, and exposure on detection probability.

Notice that the difference between the conditional probabilities and true probabilities is much less for the distance intervals close to the line (Table 6.2), which suggests that the remaining heterogeneity is greatest at larger distances. Using the conditional probabilities, we can predict the expected values for each of the capture histories ($\underline{\omega} = (1, 0)$, $\underline{\omega} = (0, 1)$, $\underline{\omega} = (1, 1)$, and $\underline{\omega} = (0, 0)$) assuming independence. Because we know the true N within each distance interval for this example, we can use a chi-square test of independence. For the distance interval 0–0.5 m, the expectations are: 25.5, 4.1, 2.0, and 0.3 respectively, with observed values of 25, 4, 2, and 1. A chi-square test does not reject the hypothesis of independence ($\chi^2 = 1.4$, $df = 1$, $P = 0.24$). However, for the distance interval 3–4 m, with expectations 4.4, 10.3, 14.8, and 34.5 and observed values 3, 7, 10, and 44, the independence assumption is not justified ($\chi^2 = 5.72$, $df = 1$, $P = 0.02$). These results demonstrate that point independence is a reasonable assumption for these data without additional covariates, but full independence is not reasonable without additional covariates that could explain the remaining heterogeneity.

Using these example data with a known N, in the following three sections we will illustrate analyses and estimation with different configurations and assumptions about the state model. We begin with the most widely used, the independent observer configuration, and then consider situations in which the trial and removal configurations are useful. In each, we have assumed that the population is always available to be detected. In all of these example analyses, we have used custom software written in the statistical language R (R Development Core Team 2003). These routines are currently being incorporated into the Distance software package. For the most part in the examples, we have not included standard errors to avoid clutter, which might detract from the main points of the examples. Also these data are rather unusual in that the covered region and survey region are the same, so there is no spatial sampling variance, which is typically the major component of variance. Estimates of variance can be constructed using methods described by Borchers *et al.* (1998a) and Innes *et al.* (2002).

6.5   Estimation for independent configuration

With the independent configuration, we will illustrate three different analyses of the golf tee data using different portions of the likelihood and independence assumptions to demonstrate the impact of unmodelled heterogeneity. At first we will use distance as the only covariate, and then we will expand the analysis to consider size, colour, and exposure covariates that were collected with these data.

6.5.1   *Distance only*

We initially consider some analyses that use only distance, because they are simpler, and serve to highlight the potential problems with unmodelled heterogeneity. We go into these initial examples in more detail than later examples in the hope that this will aid understanding of the most important concepts. We will also ignore the clustering of tees and estimate the number of tee clusters (which we know to be 250)—but we refer to them as individual tees.

6.5.1.1   *Example 1*
For the first example, we will consider a situation in which we do not assume that the distribution of tees is uniform within the covered region. We will treat the two observation teams as an independent configuration, and we will assume full independence. More formally, the state and observation

models are:

State model: All tees are always available to be detected and $\pi(y)$ is unspecified and not necessarily uniform.

Observation model: A tee at distance y is detected with probability $p.(y)$ and tees are detected independently of one another by two independent observers with probabilities $p_{1|2}(y) = p_1(y) = p_{2|1}(y) = p_2(y)$. Detection probability is specified by Model 1 from Table 6.1.

Using eqns (6.8) and (6.32), we can construct the specific likelihood for this example:

$$\mathcal{L}_\omega(\beta_0, \beta_1) = \left[\prod_{\underline{\omega}_i=(1,0)} \frac{p_1(y_i)[1 - p_2(y_i)]}{p.(y_i)} \right]$$
$$\times \left[\prod_{\underline{\omega}_i=(0,1)} \frac{p_2(y_i)[1 - p_1(y_i)]}{p.(y_i)} \right] \left[\prod_{\underline{\omega}_i=(1,1)} \frac{p_1(y_i)p_2(y_i)}{p.(y_i)} \right], \tag{6.38}$$

where

$$p_1(y_i) = p_2(y_i) = \frac{\exp\{\beta_0 + \beta_1 y_i\}}{1 + \exp\{\beta_0 + \beta_1 y_i\}} \tag{6.39}$$

and

$$p.(y_i) = 2\left[\frac{\exp\{\beta_0 + \beta_1 y_i\}}{1 + \exp\{\beta_0 + \beta_1 y_i\}} \right] - \left[\frac{\exp\{\beta_0 + \beta_1 y_i\}}{1 + \exp\{\beta_0 + \beta_1 y_i\}} \right]^2. \tag{6.40}$$

In this example, we have assumed that there are no differences between the observation teams. The parameter β_0 is the common intercept for both observation teams, and the probability of detecting tees at $y = 0$ is

$$p_1(0) = p_2(0) = \frac{\exp\{\beta_0\}}{1 + \exp\{\beta_0\}} \tag{6.41}$$

$$p.(0) = \frac{2\exp\{\beta_0\}}{1 + \exp\{\beta_0\}} - \frac{\exp\{2\beta_0\}}{[1 + \exp\{\beta_0\}]^2}.$$

The parameter β_1 is the common slope for distance y. The maximum likelihood estimates (MLE) for the parameters can be obtained using numerical methods to find the values that maximize the natural logarithm of eqn (6.38), or by using standard logistic regression software (GLM of a binary random variable and logit link) and implementing an iterative offset as described by Buckland *et al.* (1993a) and in Section 11.8.3. However, it

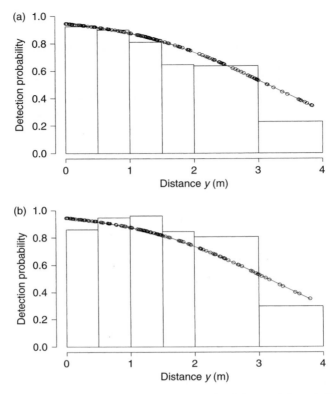

Fig. 6.5. Fitted detection probability models for (a) $p_{1\,|\,2}(y)$ and (b) $p_{2\,|\,1}(y)$ overlaid on histograms of conditional detection probability for the example 1 analysis of tee data.

is important to note that, while the correct MLEs are obtained using the iterative-offset GLM, the reported log-likelihood value from the standard software is not the value of the logarithm of eqn (6.38) at the MLEs. This value is easily obtained by substituting the MLEs into eqn (6.38) and taking the logarithm. This becomes important in model selection using AIC, when the model set includes models that are necessarily solved by numerically maximizing the log-likelihood.

Using the $n_{.} = 162$ observed distances and capture histories, the MLEs are $\hat{\beta}_0 = 2.881$ and $\hat{\beta}_1 = -0.918$. The estimated detection probability at $y = 0$ for a single observation team is 0.947 and the estimated probability that at least one team sees a tee at $y = 0$ is 0.997. The estimated detection probability at $y = w = 4$ m for a single team is 0.312, and for both teams combined, 0.526. Only one of 32 tees was missed by both teams (97% detected) within 0.5 m, so the correspondence between the predicted and true values for small distances is reasonably close. However, of the 64 tees at distances between 3 and 4 m, only 20 were detected (31%) by at least one team, and that is considerably less than the predicted detection probability of 52.6% for $y = 4$ m.

We can explore the model fit by comparing the predicted detection function against equivalent histogram estimators. First we will examine the conditional detection functions $p_{1|2}(y)$ and $p_{2|1}(y)$. These are the same as the unconditional detection functions in this analysis because we have assumed full independence. We can construct histogram estimators of the respective conditional detection functions by computing the ratios n_3/n_2 and n_3/n_1 within each distance interval from Table 6.2. The fitted conditional detection functions are reasonable representations of the data (Fig. 6.5), but there is some suggestion of lack of fit associated with differences between observation teams, or with some underlying covariate.

Next we can explore the fit of the unconditional detection function $p.(y)$ against a histogram estimator of the same function. The histogram estimator is simply:

$$\hat{p}.(y) = \frac{n_{.k}}{\widehat{N}_k}, \quad c_{k-1} \le y < c_k, \tag{6.42}$$

where c_k are cut points for the distance intervals, $n_{.k}$ is the number of observations in interval k, and $\widehat{N}_k$ is the estimated abundance in interval k. We can obtain the total abundance estimate $\widehat{N} = 180.2$ and the set of $\widehat{N}_k$ using estimation approach (2a) (eqn (6.10)). The fitted $p.(y)$ matches the equivalent histogram estimator perfectly (Fig. 6.6), but this should not be surprising because the $\widehat{N}_k$ are created from the $p.(y)$. However, clearly there is a problem because the estimated total abundance is substantially less than the true N and is less than either CDS estimate from the two individual observation teams. If we compare $p.(y)$ to a histogram of the true detection probabilities (not typically known), we see that $p.(y)$ is substantially higher, especially for larger distances (Fig. 6.7).

In this example, we have assumed nothing about $\pi(y)$, but we can derive a histogram estimate of it as:

$$\hat{\pi}(y) = \frac{\widehat{N}_k}{\widehat{N}(c_k - c_{k-1})}, \quad c_{k-1} \le y < c_k. \tag{6.43}$$

A plot of this estimated histogram (Fig. 6.8) suggests a non-uniform function with greater density close to the line. If the expected distribution for $\pi(y)$ was uniform, this would indicate a lack of fit, which is the case in this example. We could use the uniformity assumption with estimation approach (2b) (eqn (6.11)) to estimate N, but there is very little benefit with $\widehat{N} = 186.1$ and no change in the estimated detection functions.

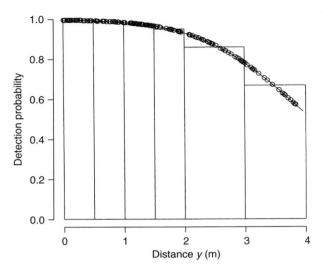

Fig. 6.6. Fitted detection probability model $p_{\cdot}(y)$ overlaid on a histogram estimate of unconditional detection probability for the example 1 analysis of tee data.

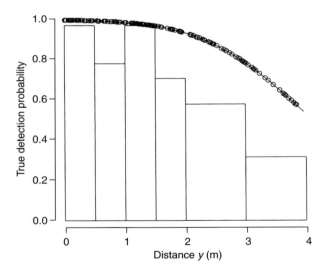

Fig. 6.7. Fitted detection probability model $p_{\cdot}(y)$ for the example 1 analysis of tee data, overlaid on a histogram of the true detection probability.

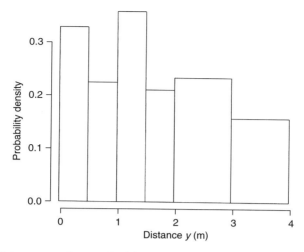

Fig. 6.8. Histogram estimate of $\pi(y)$ for the example 1 analysis of tee data.

This example should perhaps not be considered MRDS—it is actually just a mark-recapture analysis using distance as an explanatory covariate. The estimation scheme has not included one of the fundamental assumptions of distance sampling, namely that $\pi(y)$ is known. In the next example, we use the same data and conduct the analysis using the uniformity assumption.

6.5.1.2 *Example 2*

In this example, we make the additional assumption in the state model that tees are independently distributed within the covered region and follow a uniform distribution. The observation model is the same as in the first example.

State model: All tees are always available to be detected and tees are independently distributed within the covered region with $\pi(y) = 1/w$.
Observation model: A cluster of tees at distance y is detected with probability $p.(y)$ and tee clusters are detected independently of one another by two independent observers with probabilities $p_{1\,|\,2}(y) = p_1(y) = p_{2\,|\,1}(y) = p_2(y)$. Detection probability is specified by Model 1 from Table 6.1.

The likelihood for this analysis is $\mathcal{L}_y\mathcal{L}_\omega$, where $\mathcal{L}_\omega$ is given by eqn (6.38) and $\mathcal{L}_y$ is given by eqn (6.6), with $p.(y)$ specified by eqn (6.40). For this combined likelihood, we can no longer use standard logistic regression software and solve for the MLEs; $\hat{\beta}_0$ and $\hat{\beta}_1$ require numerical maximization of the likelihood. This likelihood is an example of estimation approach (3), and $\widehat{N}$ is constructed using eqn (6.11).

Using the $n_{.} = 162$ observed distances and capture histories, the MLEs are $\hat{\beta}_0 = 3.140$ and $\hat{\beta}_1 = -1.120$. The estimated detection probability at $y = 0$ for a single observation team is 0.959, and the estimated probability that at least one team sees a tee at $y = 0$ is 0.998. Neither is much different from the results from example 1. However, the estimated detection probabilities at $y = w = 4$ m have changed substantially. For a single team, the estimated probability is 0.208, and the probability is 0.372 that at least one team sees the tee. While these latter estimated probabilities are closer to the true probabilities, they still remain high.

As in the previous example, we can explore graphically the model goodness of fit. The histograms for the conditional detection functions are estimated as in example 1, but the histogram estimator for $p_{.}(y)$ is modified slightly to accommodate the uniformity assumption. We estimate $\widehat{N}_k$ in eqn (6.42) by partitioning total estimated abundance based on uniformity:

$$\widehat{N}_k = \widehat{N} \left[\frac{c_k - c_{k-1}}{w} \right]. \tag{6.44}$$

The fitted conditional detection functions (Fig. 6.9) appear visually to be somewhat better, but the graphs are deceiving because we are visually evaluating the entire line, and we should evaluate based on the number of data points. We can evaluate the goodness of fit for the conditional detection functions by comparing the deviances ($-2\times$ log-likelihood) of the two fitted models for $\mathcal{L}_\omega$. The deviance increased by 2.4 for this fit. The models for the conditional detection functions are identical, so the deviance could not improve. The increase in the deviance has occurred because of the inconsistency between $\mathcal{L}_\omega$ and $\mathcal{L}_y$, due to the failure of the full independence assumption. The estimates we get in this analysis maximize the combined likelihood, and are a compromise between fitting the conditional detection functions and the unconditional detection function (Fig. 6.10). If we compute $\mathcal{L}_y$ using the estimates from example 1 and compare the deviances for the combined likelihood, we find that the estimates from example 1 would increase the combined deviance by 2.3. Not suprisingly, if $\pi(y)$ is uniform, we do better by using $\mathcal{L}_\omega \, \mathcal{L}_y$.

What is surprising is $\widehat{N} = 196.1$! The estimated abundance is higher, but well short of the CDS estimates and the true abundance. If we compare $p_{.}(y)$ and the true detection probabilities (not typically known), we see that the $p_{.}(y)$ are closer to the true values but remain substantially higher, especially for larger distances (Fig. 6.11). Clearly, there is still some remaining unmodelled heterogeneity.

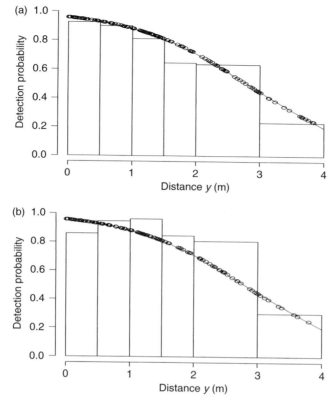

Fig. 6.9. Fitted detection probability models for (a) $p_{1\,|\,2}(y)$ and (b) $p_{2\,|\,1}(y)$ overlaid on histograms of conditional detection probability for the example 2 analysis of tee data.

6.5.1.3   *Example 3*

In both of the examples so far, we have assumed full independence. Example 1 is equivalent to the approach of Alpizar-Jara and Pollock (1996) in analysing the stake data of Laake (1978), and example 2 was the approach taken by Chen (2000) in analysing the same data. In analyses of both the stake and tee data, inclusion of $\mathcal{L}_y$ improved the estimate, but in each case the estimate was still lower than an estimate from conventional distance sampling. With an appropriate distance sampling design, if we remove the assumption of complete detection at distance zero, we should not expect to lessen the estimate (attractive movement is a possible exception: Section 6.6.2) because $p_{\cdot}(0) \leq 1$. With the full independence assumption, any remaining unmodelled heterogeneity will bias the estimator. To avoid this problem, we must weaken this strong assumption and

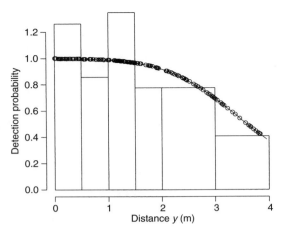

Fig. 6.10. Fitted detection probability model $p_.(y)$ overlaid on a histogram estimate of unconditional detection probability for the example 2 analysis of tee data.

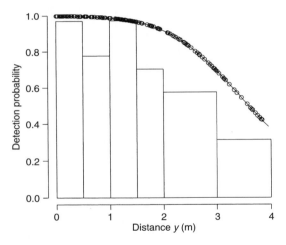

Fig. 6.11. Fitted detection probability model $p_.(y)$ for the example 2 analysis of tee data, overlaid on a histogram of the true detection probability.

only assume independence at distance zero (point independence). In this example, we modify the observation model from example 2 to assume point independence.

State model: All tees are always available to be detected and tees are independently distributed within the covered region with $\pi(y) = 1/w$.

Observation model: A tee at distance y is detected with probability $p.(y)$ and tees are detected independently of one another by two independent observers with probabilities $p_1(y)$ and $p_2(y)$, with the only restriction that point independence at $y = 0$ holds (i.e. $p_1(0) = p_{1\,|\,2}(0) = p_2(0) = p_{2\,|\,1}(0)$). Detection probability is specified by Model 3 from Table 6.1.

As in example 2, the likelihood for this analysis is $\mathcal{L}_\omega \mathcal{L}_y$. $\mathcal{L}_\omega$ is given by eqns (6.38)–(6.40), except that p_1 and p_2 are denoted specifically as conditional detection functions, $p_{1\,|\,2}$ and $p_{2\,|\,1}$, respectively. $\mathcal{L}_y$ is given by eqn (6.6) except that $p.(y)$ is specified as:

$$p.(y) = p.(0)g.(y) = p.(0)\exp\left\{-\frac{y^2}{2\sigma^2}\right\}, \qquad (6.45)$$

where $p.(0)$ is computed from the conditional detection functions (eqn (6.41)) and $\sigma = \exp(\theta_0)$. We have chosen a simple half-normal form for $g.(y)$, but we could have used other functional forms such as the hazard-rate or logistic. The parameter θ is related to σ through a log-link, so that θ can be unconstrained and $\sigma > 0$.

It is important to recognize the difference between the conditional and unconditional detection functions for each observer when we assume point independence. To prevent possible confusion, we give the explicit forms of each for this example. The conditional detection functions are:

$$p_{1\,|\,2}(y) = p_{2\,|\,1}(y) = \frac{\exp\{\beta_0 + \beta_1 y\}}{1 + \exp\{\beta_0 + \beta_1 y\}} \qquad (6.46)$$

and after some algebraic simplifications with eqns (6.27), (6.28), (6.40) and (6.41), the unconditional detection functions are:

$$
\begin{aligned}
p_1(y) = p_2(y) = & \frac{p_{1\,|\,2}(y)}{\delta(y)} = \frac{p_{2\,|\,1}(y)}{\delta(y)} = p_{1\,|\,2}(0)g.(y)\left[\frac{2 - p_{1\,|\,2}(0)}{2 - p_{1\,|\,2}(y)}\right] \\
= & \left[\frac{\exp\{\beta_0\}}{1 + \exp\{\beta_0\}}\right]\exp\left\{-\frac{y^2}{2\exp\{2\theta_0\}}\right\} \\
& \times \left[\frac{2 - \exp\{\beta_0\}/(1 + \exp\{\beta_0\})}{2 - \exp\{\beta_0 + \beta_1 y\}/(1 + \exp\{\beta_0 + \beta_1 y\})}\right] \qquad (6.47)
\end{aligned}
$$

Even though the likelihood is a product of $\mathcal{L}_\omega$ and $\mathcal{L}_y$, as in example 2, each component likelihood can be maximized separately because $\mathcal{L}_\omega$ is solely a function of β_0 and β_1, and $\mathcal{L}_y$ is solely a function of θ_0. As in example 1, we can use standard logistic regression software with the iterative offset approach to solve for the MLEs, $\hat{\beta}_0$ and $\hat{\beta}_1$. Separately, we can numerically maximize $\mathcal{L}_y$ to find the MLE $\hat{\theta}_0$. The total log-likelihood value is simply the sum of the component log-likelihoods. $\widehat{N}$ is constructed using eqn (6.11) with representation of $p.(y)$ as given above in eqn (6.45).

Using the $n. = 162$ observed distances and capture histories, the MLEs are $\hat{\beta}_0 = 2.881$, $\hat{\beta}_1 = -0.918$ and $\hat{\theta}_0 = 0.923$. The values of $\hat{\beta}_0$ and $\hat{\beta}_1$ are identical to the results of example 1. This is expected because they are derived from the same likelihood with the same data. The value of $\hat{\theta}_0$, or equivalently $\hat{\sigma} = \exp(\hat{\theta}_0) = 2.517$, is the same value that would be obtained from a CDS analysis by Distance in fitting a half-normal detection function to the $n. = 162$ observed distances. Thus, an MRDS analysis assuming point independence can be viewed as two separate analyses:

(1) a CDS or MCDS analysis of the unique observations; and

(2) a mark-recapture analysis of the capture history data (using distance as a covariate) to estimate $p.(0)$.

The estimated detection probabilities at $y = 0$ are the same as in example 1 because we assume independence at $y = 0$. However, the estimated detection probabilities at $y > 0$ are quite different. At $y = w = 4$ m, for a single team the estimated probability computed with eqn (6.47) is 0.167, and using eqn (6.45) the probability is 0.282 that at least one team sees the tee at the same distance. These values are much closer to the true probabilities than in either of the first two examples.

As in the previous examples, we can explore graphically the model goodness of fit. The conditional detection functions are identical to those in example 1 as shown in Fig. 6.5. The histogram estimator for $p.(y)$ is constructed as in example 2, and the fitted model is a clear improvement in comparison to example 2 (Fig. 6.12). The most notable improvement is the comparison to the true detection probabilities (Fig. 6.13), which we know only because this is an artificial example.

The three examples represent three different models for the same data and we can use AIC for model selection. To do this, we need to compute the log-likelihood value of $\mathcal{L}_y\mathcal{L}_\omega$ for all three models, even though $\mathcal{L}_y$ was not used in example 1. The deviance values ($-2\times$ log-likelihood) are 697.4, 695.1, and 692.0 respectively. The models have two, two, and three parameters, so the AIC values are 701.4, 699.1, and 698.0, respectively. Although the evidence is not overwhelming, there is support for assuming point independence and fitting the required extra parameter.

By assuming point independence, the MRDS analysis for an independent configuration will as expected always produce an estimate that is at least as large as an equivalent CDS or MCDS analysis because $p.(0) \leq 1$. From a CDS analysis of the unique observations, assuming a simple half-normal detection function, we get $\widehat{N} = 231.3$, and from this MRDS analysis, we get $\widehat{N} = 232.0$ because $\hat{p}.(0) = 0.997$. Even though the estimate is approaching the true $N = 250$ and $\hat{p}.(0)$ is nearly unity, this does

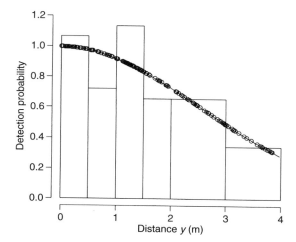

Fig. 6.12. Fitted detection probability model $p_.(y)$ overlaid on a histogram estimate of unconditional detection probability for the example 3 analysis of tee data.

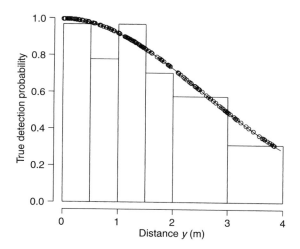

Fig. 6.13. Fitted detection probability model $p_.(y)$ for the example 3 analysis of tee data, overlaid on a histogram of the true detection probability.

not mean that the true $p_.(0) = 1$. Point independence may be a better assumption for these data, but that does not mean that there is not additional unmodelled heterogeneity at $y = 0$ that would inflate $\hat{p}_.(0)$. We consider additional models for these data below. These model heterogeneity in detection probability due to differences in colour, exposure, and size.

6.5.2 *Distance and covariates*

Detection probability will typically depend on distance and many other variables as well. Here we expand on the previous examples and consider how tee colour, exposure, and size (number of tees) affect detection probability and how they can be incorporated into an MRDS analysis with the independent configuration of observers. We consider the same analysis approaches as in the above three examples, but they are expanded to include covariates: (1) to use $\mathcal{L}_\omega$ with full independence assumption; (2) to use $\mathcal{L}_{y\,|\,z}\mathcal{L}_\omega$ with full independence; and (3) to use $\mathcal{L}_{y\,|\,z}\mathcal{L}_\omega$ with point independence. However, we will focus primarily on a comparison of approaches 1 and 3.

We approach the analysis as we would with any real data set for which we did not know the answer. We propose a sequence of plausible models for consideration and select the model with minimum AIC. While we consider many different models containing different combinations of covariates, we will only consider a logistic (eqn (6.32)) for $p_{j\,|\,3-j}(y,\underline{z})$ and a half-normal (eqn (6.45)) for $g_{.}(y,\underline{z})$. We will not give the specific functions for each model we consider, but instead we will give the functions for a sufficiently complex model as an example.

We use the fairly standard model notation $D*P = D + P + D : P$, where $D*P$ means a model that includes the main effects D, P and their interaction $D : P$. We will consider models for $p_{j\,|\,3-j}(y,\underline{z})$ that all contain distance ($D \equiv y$) and possibly the covariates platform (observation team), tee colour, tee exposure, and tee cluster size. We will use the abbreviations D, P, C, E, and S respectively for these variables. P has two values (0 for observation team 1 and 1 for observation team 2), C has two values (0 for green and 1 for yellow), E also has two values (0 for non-exposed and 1 for exposed), and S can range from 1 to 8, and is treated as continuous. The model $D + C + S + C : S$ for $p_{j\,|\,3-j}(y,\underline{z})$ is parameterized as:

$$
p_{j\,|\,3-j}(y,\underline{z}) = \frac{\exp\left\{\beta_0 + \beta_1 y + \beta_2 C + \beta_3 S + \beta_4(C \times S)\right\}}{1 + \exp\left\{\beta_0 + \beta_1 y + \beta_2 C + \beta_3 S + \beta_4(C \times S)\right\}}. \quad (6.48)
$$

The above parameterization for the factor variable C is often called a treatment contrast. While C has only two levels in this case, factors with more than two levels can be treated in a similar fashion. In the above model, β_0 is the intercept ($y = 0$) for green tees of size 0 and $\beta_0 + \beta_2$ is the intercept for yellow tees of size 0. Yellow is an additive treatment effect relative to a baseline of green. Those intercepts are rather meaningless because $S \geq 1$. The intercept for a single green tee is $\beta_0 + \beta_3$ and the intercept for a cluster of three yellow tees is $\beta_0 + \beta_2 + 3(\beta_3 + \beta_4)$.

For point independence models, we also need to specify $g_{.}(y,\underline{z})$ to describe the effect of distance on the probability that at least one observer

Table 6.3. Model specifications and model selection results for point independence (PI) analysis and full independence (FI) analysis (using $\mathcal{L}_\omega$) of tee data with independent configuration of observer teams. Abundance estimates are also given for PI models. (True $N = 250$.) The model for $g.(y, \underline{z})$ is only relevant for PI models

		PI			FI
	$p_{j\,\vert\,3-j}(y, \underline{z})$	$g.(y, \underline{z})$	ΔAIC	$\widehat{N}$	ΔAIC
1	D	1	64.2	232.0	67.6
2	D	C	59.6	236.9	
3	D	S	66.2	232.0	
4	D	E	66.2	232.0	
5	D	$C + E$	60.7	237.0	
6	D	$C + S$	60.5	236.0	
7	$D + P$	C	55.9	236.8	64.7
8	$D + C$	C	54.4	237.0	55.4
9	$D + S$	C	59.2	236.9	66.6
10	$D + E$	C	25.4	237.5	37.3
11	$D + C + P$	C	50.7	236.9	52.6
12	$D + S + P$	C	55.5	236.9	63.7
13	$D + E + P$	C	21.8	237.4	33.3
14	$D + C + P + E$	C	6.7	238.6	14.5
15	$D + S + P + E$	C	19.9	237.5	31.1
16	$D + S + P + C$	C	49.8	237.0	50.5
17	$D + P + C + S + E$	C	4.3	239.0	11.5
18	$D + P + C + S + E + C : E$	C	2.8	250.8	12.9
19	$D + P + C + S + E + S : E$	C	6.2	239.1	13.4
20	$D + P + C + S + E + S : C$	C	1.7	239.0	14.9
21	$D + P + C + S + E + S : C + C : E$	C	0.0	252.0	16.1
22	$D + P + C + S + E + S : C + S : E$	C	2.7	240.7	17.8
23	$D + P + C + S + E + E : C + S : E$	C	4.7	251.2	14.8
24	$D + P + C + S + E + E : C + S : E + C : S$	C	0.7	271.8	18.2

detects an object. Covariates can be incorporated into the scale parameter to 'modify' the effect of distance on detection probability. For example, presume that half of the single green tees are seen at $y = 2$ and half of the single yellow tees at $y = 4$ are seen. The detection probability of yellow tees is the same as green tees at a distance (scale) twice that of green tees. In the half-normal model, the scale parameter model can also be conceptualized as the standard deviation. For a scale parameter model of $C + S$, the detection function is:

$$g.(y, \underline{z}) = \exp\left\{-\frac{y^2}{2\left[\exp\left\{\theta_0 + \theta_1 C + \theta_2 S\right\}\right]^2}\right\}. \qquad (6.49)$$

The scale parameter for a single green tee is $\exp\{\theta_0 + \theta_2\}$ and for a single yellow tee, it is $\exp\{\theta_0 + \theta_1 + \theta_2\}$. If $\theta_1 = 0.694$, the scale parameter of single yellow tees is twice $(\exp\{0.694\} = 2)$ as large as single green tees.

For full independence, we need only to select a model for $p_{j\,|\,3-j}(y, \underline{z})$, and for point independence, we can select the model for $g_.(y, \underline{z})$ separately from the model selection for $p_{j\,|\,3-j}(y, \underline{z})$. As stated previously, we can view these as two separate analyses: a mark-recapture (MR) analysis for $p_{j\,|\,3-j}(y, \underline{z})$ and in this case, an MCDS analysis for $g_.(y, \underline{z})$. However, total AIC is computed using $\mathcal{L}_\omega \mathcal{L}_{y\,|\,z}$.

In analysing the data, we considered 24 possible models (Table 6.3) with six different models for $g_.(y, \underline{z})$ and 19 different models for $p_{j\,|\,3-j}(y, \underline{z})$. Model 1 is the same as in examples 1 and 3 above. All of the point independence models had smaller AIC values than the equivalent full independence models. We also fitted models 17–21 assuming full independence, and maximized $\mathcal{L}_\omega \mathcal{L}_{y\,|\,z}$ as in example 2 above. In each case, the fitted model provided a smaller AIC than the full independence model based solely on $\mathcal{L}_\omega$, but the resulting AIC was always greater than for the equivalent point independence model.

Model 21 with point independence had the smallest AIC and is the best of the fitted models. Yellow tees had a larger scale (they could be seen further away) than green tees (Table 6.4) but no other covariates were supported by the data for $g_.(y, \underline{z})$. A fairly complex model was selected for the conditional detection functions, which included distance, platform, colour, size, exposure, and interactions of colour with size and exposure. The direction of the estimated effects matched expectations. Yellow tees were more likely to be detected than green tees. Exposure was very important for green tees and less important for yellow tees. Yellow tees were primarily seen because of their colour contrast with the grass whereas green tees were primarily seen because of their profile above the grass. Larger clusters of yellow tees were easier to detect than smaller clusters of yellow tees, but size made very little difference for green tees.

As in the previous examples, we can examine graphical representations of the fitted functions to evaluate the model fit. Unlike the previous examples, we no longer have a single detection function line to compare to the histogram estimators. However, we can compute an average estimated detection probability by weighting the covariate-specific detection probabilities by the estimated proportion of the population with those covariate values:

$$\hat{p}(y) = \frac{\sum_{i=1}^{n} \hat{p}(y, \underline{z}_i)/\widehat{E}(\hat{p}(\underline{z}_i))}{\sum_{j=1}^{n} 1/\widehat{E}(\hat{p}(\underline{z}_j))}. \tag{6.50}$$

Both of the conditional detection functions coincide well with the histogram estimators (Fig. 6.14). As expected, there is much more variation in conditional detection probability for a single team than in the probability

Table 6.4. Parameter estimates and standard errors for selected $p_{j\,|\,3-j}(y,\underline{z})$ and $g.(y,\underline{z})$ models in analysis of tee data with independent configuration of observer teams

Function	Parameter	Effect	Estimate	Standard error
$p_{j\,|\,3-j}(y,\underline{z})$	β_0	Intercept	-0.341	0.694
	β_1	Distance	-1.641	0.191
	β_2	Platform	0.642	0.339
	β_3	Colour	1.678	0.848
	β_4	Size	-0.149	0.177
	β_5	Exposure	5.132	0.582
	β_6	Colour:Size	0.718	0.232
	β_7	Colour:Exposure	-2.297	0.710
$g.(y,\underline{z})$	θ_0	Intercept	0.563	0.134
	θ_1	Colour	0.613	0.276

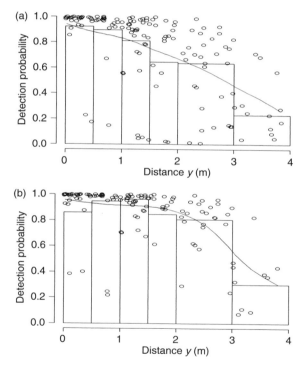

Fig. 6.14. Fitted detection probability models for (a) $p_{1\,|\,2}(y)$ and (b) $p_{2\,|\,1}(y)$ overlaid on histograms of conditional detection probability for the analysis of tee data with covariates.

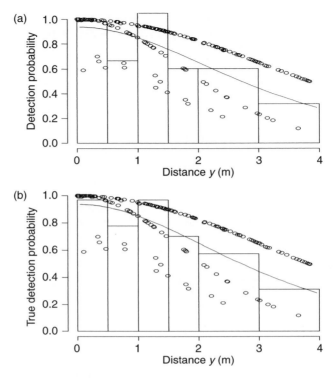

Fig. 6.15. Fitted detection probability model $p_.(y)$ for the analysis of tee data with covariates, overlaid on (a) a histogram estimate of unconditional detection probability and (b) a histogram of the true detection probability.

that an object is detected by at least one team (Fig. 6.15). It is also apparent from Fig. 6.15 that $p_.(0, \underline{z})$ is nearly unity for most tee clusters, which is the reason that most of the points align on two curves, one for yellow tees and one for green tees. The exceptions were the 14 unexposed green tee clusters that were detected.

This example is also useful to demonstrate the potential pitfalls of over-fitting and the hazard of obtaining small estimated probabilities with the Horvitz–Thompson-like approach. Although we did not include it in Table 6.3, we also fitted the model $D + P + C^*S^*E$ for $p_{j\,|\,3-j}(y, \underline{z})$, which contains all two-way interactions of C, S, and E, and their three-way interaction. The model fitting failed to converge with sufficient accuracy regardless of the number of iterations, and the resulting estimate of abundance was ridiculously high (>1 million). The problem was that the $n_. = 162$ observations were effectively partitioned so finely, that for one of the subsets, none of the observations seen by team 2 (the better team) were seen by team 1, resulting in estimates of detection probability that were essentially zero. Using simulation, Borchers (1996) demonstrated the potential problem of small estimated detection probabilities with the Horvitz–Thompson estimator. In distance sampling applications, he recommended truncating larger distances and using $E(p.)$ rather than $p.(y)$. If we express $E(p. \mid z_i)$ as $p.(0, z_i)E_y(g.(y, z_i))$, then it is clear that while truncation can help by increasing $E_y(g.(y, z_i))$, little can be done if $p.(0, z_i)$ is very small. Over-fitting and overly complex models should be avoided to prevent the problem.

The estimated abundance of 252 tee clusters for the selected model is almost too good to be believed. However, we will illustrate a weakness of the analysis to demonstrate the realism of this example. While we derived a close estimate of total abundance, if we consider abundance estimates of tee classes (exposed yellow, exposed green, unexposed yellow, and unexposed green), we do not do so well. The estimates were 78, 74, 58, and 42, compared with the true values of 66, 47, 76, and 61. We have overestimated the number of easily detected exposed tees, and underestimated the number of unexposed tees, which are much harder to detect. The discrepancy may be due to sampling error or an additional unmodelled source of heterogeneity like grass height or variation in degree of exposure, or model mis-specification. It may also be the result of conditioning on the set of observed covariates in using $\mathcal{L}_{y \mid z}$. It may be that inclusion of $\pi(z)$ in the likelihood may provide an improvement if we can assume that $\pi(z)$ is independent of $\pi(y)$.

6.6 Estimation for trial configuration

In this section, we provide a couple of examples for the trial configuration for observers. In the first example, we analyse the golf tee data using the same functions and models of Section 6.5.2, but with the likelihoods for the trial configuration. In the second example, we demonstrate the usefulness of the trial configuration in coping with responsive movement for a line transect survey for common dolphins. In both examples, we provide less detail on the specific likelihood functions than in the previous section; however, the specific likelihoods can be surmised from the generic likelihoods in Section 6.2.

6.6.1 *Distance and covariates*

To illustrate the trial configuration with the golf tee data, we use the $n_2 = 142$ observations from team 2 as trials for team 1; $n_3 = 104$ of these were detected by team 1. We are only interested in estimating $p_{1 \mid 2}$ and p_1. We demonstrate the results under full independence using only $L_{1 \mid 2}$, and under point independence using $L_{1 \mid 2}L_{y \mid z}$, where $L_{y \mid z}$ is constructed using the $n_1 = 124$ observations from team 1. We use the same models

Table 6.5. Model specifications and model selection results for PI and FI analysis (using L_ω) of tee data with trial configuration in which team 2 creates trials for team 1. Abundance estimates are also given for PI models. (True $N = 250$.) The model for $g_1(y, \underline{z})$ is only relevant for PI models

	$p_{1\,|\,2}(y, \underline{z})$			PI		FI
			$g_1(y, \underline{z})$	ΔAIC	$\widehat{N}$	ΔAIC
1	D		1	52.8	223.9	53.6
2	D		C	46.0	233.4	
3	D		S	54.6	224.0	
4	D		E	54.7	224.0	
5	D		$C + E$	47.0	233.7	
6	D		$C + S$	47.8	233.6	
7	$D + C$		C	36.4	231.8	35.2
8	$D + S$		C	47.9	233.5	55.3
9	$D + E$		C	20.7	229.4	34.5
10	$D + C + E$		C	0.0	228.1	6.4
11	$D + S + E$		C	22.4	229.5	36.1
12	$D + S + C$		C	38.1	232.0	36.8
13	$D + C + S + E$		C	1.8	228.6	8.1
14	$D + C + S + E + C : E$		C	2.2	235.7	6.5
15	$D + C + S + E + S : E$		C	3.8	228.6	10.1
16	$D + C + S + E + S : C$		C	2.7	228.3	9.0
17	$D + C + S + E + S : C + C : E$		C	2.9	235.1	7.2
18	$D + C + S + E + S : C + S : E$		C	4.4	228.1	11.3
19	$D + C + S + E + E : C + S : E$		C	4.2	235.5	8.6
20	$D + C + S + E + E : C + S : E + C : S$		C	4.2	237.8	9.6

as in Section 6.5.2, except that we need not consider models with a platform effect, because we are estimating detection probability for a single platform only.

In the trial configuration, fewer data are used and we would expect to select a simpler model than from the analysis of these data with the independent configuration. A point independence model with only the main effects of distance, colour, sex, and exposure for the conditional detection function and colour for the unconditional detection function had the minimum AIC= 399.3 (Table 6.5). The AIC for the best full independence model was considerably larger at 405.7. A plot of the unconditional detection function (Fig. 6.16) does not suggest any lack of model of fit, but the estimated detection probabilities are higher than the true values (Fig. 6.16) suggesting that there is remaining unmodelled heterogeneity.

The trial configuration is very useful in some situations (e.g. responsive movement), but it does have a couple of disadvantages. First, the trial configuration is more susceptible to heterogeneity. The number of trials will

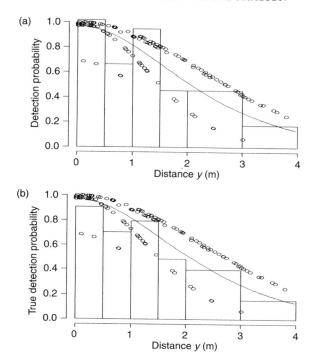

Fig. 6.16. Fitted detection probability model $p_1(y)$ for the trial configuration analysis of tee data with covariates, overlaid on (a) a histogram estimate of unconditional detection probability and (b) a histogram of the true detection probability.

often be much less than for an independent configuration. This will cause AIC to select a simpler model, possibly with important effects excluded. In addition, the abundance estimate is based on the detection probability from a single team, rather than two teams as in the independent configuration. In general, $p_1(0, \underline{z}) \leq p_.(0, \underline{z})$, so the possible amount of remaining heterogeneity is greater with the trial configuration. Also, there are some modelling limitations because the abundance estimate is constructed solely from the observations from a single platform. The conditional detection function $p_{1\,|\,2}(y, \underline{z})$ is derived from the n_2 trials, but the abundance estimate is computed from the n_1 observations. If a covariate used in $p_{1\,|\,2}(y, \underline{z})$ does not represent the full range of the covariate values in the data, there is no basis for prediction. For example, consider a factor covariate like exposure in the golf tee data. If the n_2 trials only contained exposed tees, but some of the $n_1 - n_3$ observations with history $\underline{\omega} = (1, 0)$ (seen by team 1 only) were of unexposed tees, there would be no reasonable basis to estimate $p_1(0, \underline{z}) = p_{1\,|\,2}(0, \underline{z})$ for unexposed tees. When the factor has only two levels and one is missing, clearly it must be excluded, but if it has three or more levels, the levels can be collapsed so there are no missing levels. Although this problem is less severe with continuously varying covariates, it can occur if the prediction range is outside of the range contained in the trials.

These problems can be minimized by using the better observers to generate trials. This will create a larger number of trials with the broadest range of possible covariate values. However, the one downside is that n_1 may then be reduced, resulting in less precise results. Although the trial configuration does have some disadvantages relative to the independent configuration, it is the best alternative in certain circumstances, as illustrated in the following example.

6.6.2 *Distance, covariates, and responsive movement*

One of the main reasons to use a trial/observation configuration is to accommodate responsive animal movement. We cannot illustrate this with the St Andrews golf tee data set because tees do not move. Instead, we use a data set from the Faroese component of the 1995 North Atlantic Sighting Survey (NAMMCO 1997: 135–7). The data were gathered on a shipboard double-platform survey of common dolphins in the northeast Atlantic. The example is taken from Cañadas *et al.* (in preparation). The observation configuration involved observer 2 (called the 'tracker' by Cañadas *et al.*) searching with binoculars far ahead of the vessel, and observer 1 (called the 'primary') searching much closer to the vessel with naked eye. With the help of a third observer, observer 2 tracked detected animals (hence the term 'tracker') until they were detected by observer 1 or passed abeam of the vessel undetected. Tracking was done mainly to improve the reliability of duplicate identification. Primary detected a total of 76 schools, tracker detected 77 schools, of which 52 were also detected by primary. The distribution of perpendicular distances of primary detections, truncated at 0.07 nautical miles[4] (nm), is shown in Fig. 6.17.

The following explanatory variables were available in addition to perpendicular distance y: observer, sea state, swell height and angle, glare width and strength, horizontal and vertical angle of the sun, wind direction, weather type, cue type, animal behaviour, aspect (orientation of the animals relative to the observer), and cluster size. These variables constitute the vector z.

The primary detection function and animal abundance were estimated under the assumption of full independence as well as under the assumption of point independence.

---

[4] This truncation distance is small compared to other studies on common dolphins. We suspect that the scarcity of sightings at greater perpendicular distances is mainly a consequence of a large proportion of effort in Beaufort 2 and above, small group sizes, and a tendency to concentrate search effort ahead of the vessel.

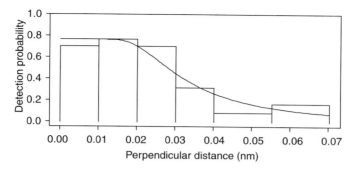

Fig. 6.17. Common dolphin survey: the perpendicular distances distribution of primary detections, together with the primary detection function estimated assuming point independence (see text).

6.6.2.1 *Full independence*

The estimates of abundance under the assumption of full independence of primary detections from tracker detections (i.e. $p_{1|2}(y, \underline{z}) = p_1(y, \underline{z})$) was obtained using estimation approach (4). This involved estimating $p_{1|2}(y, \underline{z})$ by maximizing the likelihood component $\mathcal{L}_{1|2}$ of eqn (6.13), using binary regression software and a logistic form for $p_{1|2}(y, \underline{z})$. AIC selected only cluster size and sea state for inclusion in the final model. The estimated detection function in the perpendicular distance dimension only is shown in Fig. 6.18, overlaid on the proportions of tracker sightings that were also seen by primary. Abundance was estimated to be 351,000 animals, with cv = 0.24.

The detection function intercept averaged over all $\underline{z}$, $p_1(0)$, was estimated by

$$\hat{p}_1(0) = \frac{\sum_{i=1}^{n_1} \hat{p}_1(0, \underline{z}_i)/\widehat{E}(\hat{p}_1(\underline{z}_i))}{\sum_{j=1}^{n_1} 1/\widehat{E}(\hat{p}_1(\underline{z}_j))}, \tag{6.51}$$

where $\widehat{E}(\hat{p}_1(\underline{z}_j)) = \int_0^w \hat{p}_1(y, \underline{z}_j)\pi(y)dy$. The resulting estimate of $p_1(0)$ was 0.80, with cv = 0.14.

6.6.2.2 *Point independence*

Cañadas *et al.* (in preparation) also estimated abundance under an assumption of point independence of primary detections from tracker detections. More specifically, the detection function was assumed to have the form $p_1(y) = p_1(0)g_1(y)$. The intercept $p_1(0)$ was estimated as above; the shape of the primary detection function $g_1(y)$ was estimated by maximizing the likelihood component $\mathcal{L}_y$ of eqn (6.6) using Distance 4.0. This is the shape shown in Fig. 6.17. Truncation was necessary for reliable estimation of the detection function using CDS methods. The resulting estimate of abundance was 1,012,000 (cv = 0.40)—an estimate of N that

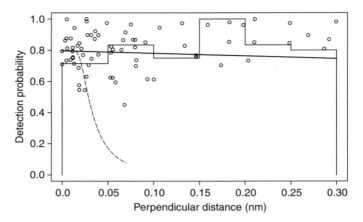

Fig. 6.18. Common dolphin survey: the perpendicular distances distribution of the proportion of tracker detections that were detected by the primary observer (the histogram), together with the fitted logistic detection function (solid line). The distances are those at the time the tracker detected the animals. Dots are the estimated detection probabilities for individual sightings. The detection function estimated under the assumption of point independence (with truncation at 0.07 nm) is shown as a dashed line.

is nearly three times that obtained under the full independence assumption! In addition, the shape of the detection function estimated under the point independence assumption (from primary data only—shown in Fig. 6.17) is very different from that estimated under the full independence assumption (from the mark-recapture data—shown in Fig. 6.18). The former (dashed line) is shown on the same scale as the latter (solid line) in Fig. 6.18. These differences require explanation.

6.6.2.3 *Discussion*
Common dolphins can exhibit strong attraction to survey vessels. Examination of the perpendicular distances of duplicates at the time they were detected by tracker and by primary suggest that this was the case here—see Fig. 6.19. Moreover, it is difficult to believe that such extreme difference between the estimated detection function shapes could come about as a consequence of unmodelled heterogeneity (failure of the full independence assumption). The very large difference in the estimated detection functions and between the two point estimates of abundance is almost certainly due to dolphins being attracted to the survey vessel.

At the time that the tracker (who is searching far ahead with binoculars) detects animals, they have not responded (or have not responded much) to the presence of the vessel. By the time they are within the primary's detection range, the animals have moved in towards the trackline, making

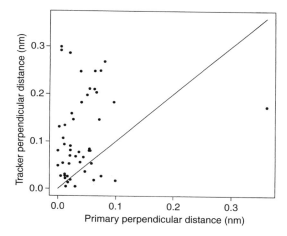

Fig. 6.19. Common dolphin survey: the perpendicular distances of duplicates at the times they were detected by the primary observer (x-axis) and by the tracker (y-axis). The diagonal line corresponds to no movement. Points below the line correspond to movement away from the transect line, while those above the line correspond to movement towards it.

the perpendicular distance distribution of primary detections much more 'spiked' than it would be had the animals not responded. In this case, $\pi(y)$ at the time primary detects animals is not uniform, and the fall-off in detection frequency due to a declining detection probability cannot be separated from that due to the presence of fewer animals at larger distances. The estimate of $g_1(y)$ obtained from program Distance assumes that all the fall-off is due to a decline in detection probability, and increasingly underestimates detection probability as perpendicular distance increases. This in turn results in a huge positive bias in estimated abundance.

To be able to accommodate responsive movement using estimation approach (4), it is necessary to assume that $p_1(y \mid \underline{z}) = p_{1\mid 2}(y \mid \underline{z})$—that is, primary detections are independent of tracker detections. If there is unmodelled heterogeneity, this assumption will fail to some extent. With a trial-observer configuration and responsive movement, a point independence assumption is not tenable,[5] because we cannot separate the effects of responsive movement and unmodelled heterogeneity on the shapes of histograms like Figs 6.17 and 6.18.

In the case of this common dolphin survey, there is additional evidence that substantial attractive movement occurred (a priori knowledge of animal behaviour and the evidence in Fig. 6.19). In addition, exceptionally severe unmodelled heterogeneity would be required to generate a difference

---

[5] Conversely, with a trial-observer configuration and a point independence assumption, the ability to deal with responsive movement breaks down.

in shape as dramatic as that between Figs 6.17 and 6.18. Given these facts, it is unwise to use an estimation method that does not correct for responsive movement, and we are much more comfortable with an assumption of full independence than with an assumption of no responsive movement.

A similar example of the trial configuration to cope with responsive movement, but with point transects, is provided by Klavitter (2000) and Klavitter *et al.* (2003). They describe an example in which surveys were conducted of the Hawaiian hawk (*Buteo solitarius*) at a sample of points, for 10 min at each point, using playback recordings designed to elicit a response. Analysis of the survey data with CDS methods yielded an unbelievably high estimate of abundance for the endangered species because of attractive movement prior to detection by the observer. Using marked hawks to generate trials, they were able to estimate the amount of attractive movement and obtain a more appropriate estimate of detection probability (or effective area surveyed) using the Buckland and Turnock (1992) method.

Buckland and Turnock (1992) proposed this method for dealing with responsive movement, although there have been some developments in estimation methods since then. The central idea of the method is that if the y values at the time the tracker detected animals (i.e. before responsive movement) are used in estimating $p_1(y, \underline{z})$, the resulting estimate of $p_1(y, \underline{z})$ includes a component due to responsive movement (attraction after animals pass through the tracker search area but before they pass through the primary search area raises the probability, avoidance lowers it) as well as a component due to reduced ability to detect animals at distance.

We should not leave this example without pointing out an important design issue. If animals can move into the region surveyed by primary from outside the region surveyed by tracker, the method will yield positively biased estimates of abundance. Before responsive movement (when animals are uniformly distributed with respect to distance from the line), these animals fall outside the surveyed region and should therefore not be included in estimates of N_c. By contrast, animals detected by tracker which start at, or move to, a distance beyond the range at which primary can detect them, do not cause bias—they start in the covered region, and being outside the range of detection by primary just makes the primary detection probability for these animals tend to zero.

6.7　Estimation for removal configuration

The removal configuration is not strictly mark-recapture because all animals 'marked' by observer 1 are detected by observer 2 with certainty. Thus, marked animals can be considered either as removed or as 'recaptured'. Either way, the capture history (1,0) is not observable and you cannot estimate the capture/detection probability for observer 2. However, there are 'unmarked' animals missed by observer 1 and seen by observer 2, so the probability of capture can be estimated if we assume no differences between platforms (observers). We can formulate the analysis in terms of capture histories, and the likelihood requires a minor modification for the removal configuration. Here we analyse the same golf tee data with the removal configuration that we analysed with the independent and trial configurations in Sections 6.5 and 6.6.

For the removal configuration, $n_2 = n.$ and $n_3 = n_1 \leq n.$. The $n.$ trials for observer 2 are all 'seen', but only n_1 of the $n.$ trials for observer 1 are seen. We can estimate $p_{1|2}$ by maximizing L_ω using the probability distribution given by eqn (6.37). For a point independence model, we can separately estimate the parameters of $g.(y, \underline{z})$ using the $n.$ observations in $L_{y|\underline{z}}$. The unconditional detection function for the $n.$ observations is:

$$p.(y, \underline{z}) = [2p_{1|2}(0, \underline{z}) - p_{1|2}^2(0, \underline{z})]g.(y, \underline{z}). \tag{6.52}$$

Using the $n. = 162$ trials with $n_1 = 124$ detected by team 1, we assumed point independence and fitted the same sequence of models as with the trial configuration (Table 6.6). As in the trial configuration example, the

Table 6.6. Model specifications and model selection results for point independence analysis of tee data with removal configuration in which team 2 detects tees missed by team 1

	$p_{1|2}(y, \underline{z})$	$g.(y, \underline{z})$	ΔAIC	$\widehat{N}$
1	D	1	36.1	231.8
2	D	C	31.5	236.7
3	D	S	38.0	231.8
4	D	E	38.1	231.8
5	D	$C + E$	32.6	236.8
6	D	$C + S$	32.3	235.8
7	$D + C$	C	21.9	237.3
8	$D + S$	C	33.4	236.7
9	$D + E$	C	14.1	237.3
10	$D + C + E$	C	0.0	242.0
11	$D + S + E$	C	16.1	237.3
12	$D + S + C$	C	23.9	237.3
13	$D + C + S + E$	C	2.0	242.3
14	$D + C + S + E + C : E$	C	4.0	242.2
15	$D + C + S + E + S : E$	C	4.0	242.6
16	$D + C + S + E + S : C$	C	2.9	243.8
17	$D + C + S + E + S : C + C : E$	C	4.8	240.8
18	$D + C + S + E + S : C + S : E$	C	4.1	268.6
19	$D + C + S + E + E : C + S : E$	C	6.0	242.5
20	$D + C + S + E + E : C + S : E + C : S$	C	6.1	268.3

best model contained main effects of distance, colour, and exposure for the conditional detection function and colour for the unconditional detection function. The abundance estimate of $\widehat{N} = 242$ was closer to the true value than the trial configuration estimate because it was based on the $n_.$ observations rather than just the n_1 observations in the trial configuration. Because $n_.$ contained 31 of the 32 tee clusters less than 0.5 m from the line, $p_.(0)$ was close to unity. In this particular example, there were small differences between observation teams, especially at $y = 0$, so it was reasonable to use the removal configuration. That will not always be the case and the removal configuration should be used as a last resort.

6.8   Dealing with availability bias

Three kinds of availability were described in Section 6.3.1 above. In this section, we structure our discussion of estimation methods that use state models under the same headings: static availability, and the two forms of dynamic availability, discrete and intermittent. We also consider design-based estimation which avoids explicitly incorporating an availability process in the estimation.

Before we do that, we should note that in some applications, availability is not well defined. For example, it is difficult to say exactly when an animal in water becomes available for detection by observers in an aircraft. An animal on the surface is definitely available and at depth it is definitely not. In between there may be a continuum of increasing availability but no well-defined switch between unavailability and availability. This difficulty might be surmountable by imposing a strict definition of availability on the continuum (e.g. it is only considered to be available if it is actually at the surface), but in order for this to be really useful, one needs to be able to estimate the proportion of time the animal is available according to this definition. To illustrate with a silly example: if the availability definition were something like 'animals are only available when you can see the whites of their eyes', how could you estimate the proportion of time you can see the whites of their eyes? The definition of availability must be clear and unambiguous, and such that the availability process can be estimated unambiguously.

6.8.1   *Static availability*

Because statically available animals are either continuously available or continuously unavailable for detection while within detectable range, we can correct availability bias by using methods developed for continuously available animals together with an estimate of the proportion of the population that is available.

6.8.1.1 *Horvitz–Thompson-like estimator*

Static availability bias can easily be corrected using a Horvitz–Thompson-like estimator, assuming that an estimate of the proportion of the population that is available for detection can be obtained (independently of the distance survey). Suppose that the probability that an animal is available for detection on the survey is P_v. With uniform coverage probability P_c, the probability that it is in the covered region and is available for detection is $P_v P_c$. The probability that animal i with characteristics $\underline{z}_i$ is included in the sample is $P_v P_c P_a(\underline{z}_i)$, where $P_a(\underline{z}_i) = \int_0^w p(y, \underline{z}_i)\pi(y \mid \underline{z}_i)dy$. If we knew this probability, we could use it in a Horvitz–Thompson estimator to estimate abundance without bias. Because we do not usually know P_v and $P_a(\underline{z}_i)$, we use the following Horvitz–Thompson-like estimator:

$$\widehat{N} = \sum_{i=1}^{n} \frac{1}{\widehat{P}_v P_c \widehat{P}_a(\underline{z}_i)}, \tag{6.53}$$

where $\widehat{P}_v$ is an estimate of the probability that an animal is available for detection—which we assume to be the same for all animals. This estimator can readily be generalized for cases in which probability of being available depends on location $\underline{u}_i$ or animal characteristics $\underline{z}_i$; the difficulty in this case is likely to be in obtaining reliable estimates of the probability of being available, $P_v(\underline{u}_i, \underline{z}_i)$.

6.8.1.2 *Maximum likelihood estimation*

In the case of MCDS models, the probability that animal i at distance y_i with characteristics $\underline{z}_i$ is available and is detected is $g(y_i, \underline{z}_i)P_v$. Under the assumption that animals become available independently of each other, a full MCDS likelihood function that includes the availability process is obtained just by replacing $g(y_i, \underline{z}_i)$ with $g(y_i, \underline{z}_i)P_v$ in eqns (2.33) and (2.34). Similarly, a full CDS likelihood function that includes the availability process is obtained by replacing $g(y_i)$ with $g(y_i)P_v$ in eqn (2.43).

In the case of MRDS models, the full likelihood is obtained by replacing $p(y_i, \underline{z}_i)$ with $p(y_i, \underline{z}_i)P_v$ in eqn (6.18). Note that P_v does not appear in the likelihood component $\mathcal{L}_\omega(\underline{\theta})$, which is conditional on animals having been detected.

More generally, the probability of being available might depend on location $\underline{u}_i$ or animal characteristics $\underline{z}_i$, so that P_v becomes $P_v(\underline{u}_i, \underline{z}_i)$ in the above.

In all these cases, the distance survey itself is insufficient for estimation of P_v; an additional source of data such as satellite or radio-transmitter tagging is needed.

6.8.2   *Hazard-rate models for dynamic availability*

Dynamic availability is more difficult to accommodate than static availability because the probability of detecting an animal in the covered region depends not only on the probability that the animal becomes available within detectable range, but *where* it becomes available relative to the observer, and *for how long*. If it becomes available close to the observer, the probability of detecting it will be higher than if it only becomes available far from the observer; if it becomes available frequently within detectable range, it is more likely to be detected than if it becomes available rarely, and the longer it is available, the more likely it is to be detected.

To deal with this more complex scenario, we need to take account of the fact that the detection process actually takes place in two spatial dimensions—although conventional distance sampling methods reduce it to one dimension, y. To deal with detection probability in two dimensions, it is useful to work with a 'hazard-rate function' (Section 11.4).

We should note at this point that virtually all the methodological development in this area has taken place in the context of line transect or cue counting surveys. As a result, our discussion of the problem is for the most part related to line transect or cue counting surveys. Point transect surveys with stationary animals are in principle simpler to deal with because the animals are always at the same distance from the observer, whereas in line transect surveys they move relative to the observer. Point transect surveys with moving animals may be more difficult to deal with.

Many one-dimensional detection functions are purely empirical (they are convenient forms that work well), but the 'hazard-rate' model of Hayes and Buckland (1983) was developed by considering a line transect detection process in two dimensions and integrating out the along-trackline dimension. This is relatively straightforward if animals are continuously available for detection, but when they are discretely available, integration can be done algebraically only in special cases. We outline the derivation of the continuous-availability model here by way of introduction to the models we cover below for dynamic availability.

Consider a line transect survey on which there is an undetected animal at location (x, y) relative to the observer, where y is perpendicular distance to the animal and x is its along-trackline distance. The position of the animal relative to the observer will change with time; if the animal is stationary, it will trace out a path parallel to the trackline so that y will stay constant but x will depend on time, t. It is therefore convenient to write it as a function of time: $x(t)$.

Suppose that the probability that an undetected animal at $(x(t), y)$ at time t is detected in the next short time interval of length dt is $Q(x(t), y)dt$. The function $Q(x, y)$ is called the hazard function; it gives the instantaneous probability of detecting an as yet undetected animal at (x, y).

Using a result from survival analysis,[6] one can show that the probability of detecting an animal that is continuously available for detection from time $t = 0$ to time T, from a platform moving at speed $dx(t)/dt$ at time t, is

$$g(y) = 1 - \exp\left\{-\int_0^T Q(x(t), y)\frac{dx(t)}{dt}\,dt\right\}. \qquad (6.54)$$

Hayes and Buckland (1983) showed that with constant observer speed, a variety of plausible hazard functions $Q(x(t), y)$ lead to the same detection function form

$$g(y) = 1 - \exp\left\{-\left(\frac{y}{\sigma}\right)^{-b}\right\}, \qquad (6.55)$$

where σ and b are parameters to be estimated and $g(0) = 1$. This has become one of the standard key function forms for analysis of conventional distance survey data. However, modelling the detection function in terms of a hazard function has much wider applicability than this. In particular, it is useful for dealing with situations in which animals are not continuously available for detection.

6.8.3 *Discrete availability: animal-based*

Schweder (1974) was the first to develop a hazard probability model for line transect surveys with discrete availability. The approach he and others have since developed is sometimes technically challenging, although the basic ideas are fairly straightforward. We outline the basic ideas here and leave the reader to follow up on references given below for the mathematical details of the more complicated discrete availability hazard probability models.

Suppose that a stationary animal becomes available within detection range for an instant at times $t_1, \ldots, t_k$ and locations $(x_{t_1}, y), \ldots, (x_{t_k}, y)$ relative to the observer. For brevity, the set of times and locations that the animal becomes available is called its availability 'history',[7] $\mathcal{H}$, where

$$\mathcal{H} = \begin{pmatrix} t_1 & x_{t_1} & y \\ t_2 & x_{t_2} & y \\ \vdots & \vdots & \vdots \\ t_k & x_{t_k} & y \end{pmatrix} \qquad (6.56)$$

---

[6] If you think of detection as 'death', the analogy with survival analysis is more apparent.

[7] On a point transect survey with stationary animals, all the locations in the history are the same single point at radial distance r.

The hazard function $Q(x, y, \underline{z})$ is the probability of detecting an as yet undetected animal with variables $\underline{z}$ that becomes available at (x, y) at time t. (We assume throughout that this probability does not depend on t.) The probability that the animal is detected, given that it has availability history $\mathcal{H}$, is

$$p(\mathcal{H}, y, \underline{z}) = 1 - \prod_{i=1}^{k} (1 - Q(x_{t_i}, y, \underline{z})). \qquad (6.57)$$

In general, we cannot evaluate this probability even if we know the hazard function, because we do not observe the complete availability history $\mathcal{H}$. However, if we have a state model for the availability process $\pi(\mathcal{H})$, we can evaluate the expectation over the availability process: $E(p(\mathcal{H}, y, \underline{z}))$. This is the probability of detecting an animal with variables $\underline{z}$ whose availability process is given by $\pi(\mathcal{H})$ (i.e. an animal from the population of interest) and this probability can be used in likelihood functions and estimators.

More specifically, if given $\mathcal{H}$, detections are assumed to be independent between observers,

$$\Pr\{\underline{\omega} = (1, 0) \,|\, (y, \underline{z})\} = E\left\{p_1(\mathcal{H}, y, \underline{z}) \left[1 - p_2(\mathcal{H}, y, \underline{z})\right]\right\}$$
$$\Pr\{\underline{\omega} = (0, 1) \,|\, (y, \underline{z})\} = E\left\{\left[1 - p_1(\mathcal{H}, y, \underline{z})\right] p_2(\mathcal{H}, y, \underline{z})\right\}$$
$$\Pr\{\underline{\omega} = (1, 1) \,|\, (y, \underline{z})\} = E\left\{p_1(\mathcal{H}, y, \underline{z}) p_2(\mathcal{H}, y, \underline{z})\right\} \qquad (6.58)$$

and

$$p_1(y, \underline{z}) = \Pr\{\underline{\omega} = (1, 0) \,|\, (y, \underline{z})\} + \Pr\{\underline{\omega} = (1, 1) \,|\, (y, \underline{z})\}$$
$$p_2(y, \underline{z}) = \Pr\{\underline{\omega} = (0, 1) \,|\, (y, \underline{z})\} + \Pr\{\underline{\omega} = (1, 1) \,|\, (y, \underline{z})\}$$
$$p_.(y, \underline{z}) = \Pr\{\underline{\omega} = (1, 0) \,|\, (y, \underline{z})\} + \Pr\{\underline{\omega} = (0, 1) \,|\, (y, \underline{z})\}$$
$$+ \Pr\{\underline{\omega} = (1, 1) \,|\, (y, \underline{z})\}. \qquad (6.59)$$

These expressions can be used in the likelihood functions developed in Section 6.2.4. The difficulty in evaluating the likelihoods, and in estimation in general, is evaluation of the expectations over the availability process. To do so, we need first to know the availability state model $\pi(\mathcal{H})$ and second, be able to evaluate the expectation algebraically or by some other means. Moreover, additional data on availability may be required to estimate the availability state process model parameters. These data often come from tagged animals from whom a history of availabilities can be obtained.

The simplest discrete availability state model is a Poisson point process. Skaug and Schweder (1999) obtained analytic expressions for the expectations in this case and used supplementary data obtained from tagged animals to estimate the parameter of the point process.

6.8.3.1 *Using observed along-trackline distances*

Notice that eqns (6.58) do not use the along-trackline distances of detections. Likelihood functions constructed with the probabilities in eqns (6.58) neglect the along-trackline distance, x. Likelihoods that include x have been developed but they are rather more complicated. We describe them in outline below. For ease of reference, we call models that include both along-trackline distance data and perpendicular distance data '(x, y)-models' and models that include only perpendicular distance data 'y-models'.

One would expect (x, y)-models to have some advantages over y-models because the former include additional relevant data. Cooke (unpublished) demonstrated that there is information on the availability process in the (x, y) data. One can get an intuitive feel of why this is so by considering the expected proportions of 'simultaneous duplicates' (detections of an animal at the same instant that it becomes available, by both observers) and 'delayed duplicates' (detection of the same animal by both observers, but with one detecting the animal at a later availability event than the other). If animals became available so rarely that they could be available for detection only once while in detectable range, all duplicates would be simultaneous duplicates. If, on the other hand, animals became available many times while within detectable range, a relatively high proportion of duplicates would be delayed duplicates.

Thus the proportion of delayed duplicates tells us something about the availability process, and the rate at which animals become available can be estimated using data on simultaneous and delayed duplicates. This gives (x, y)-models an advantage over y-models for discretely available animals, because it can be difficult and expensive to gather independent data on the availability rate.

Okamura *et al.* (2003) developed (x, y)-models, using a Poisson point process availability model. They also developed models that use data on whether duplicates were simultaneous or delayed, but not the data on the along-trackline distance. We refer to these as 'y_{dd}-models' (subscript 'dd' for 'delayed duplicate'). They investigated the models' properties by simulation and found that (x, y)-models performed better than y_{dd}-models, as might have been expected. They also found that both kinds of model performed better (sometimes very much better) when given the true availability rate than when left to estimate it. This suggests that there is sometimes little information on availability rate in the simultaneous- and delayed-duplicate data. While (x, y)-models and y_{dd}-models represent a substantial advance over y-models for animals with a discrete availability process, they do not remove the need for independent data on the availability process entirely.

6.8.3.2 *Using resightings data*

Cooke (1997) and Schweder *et al.* (1997, 1999) developed hazard probability models for surveys in which detected animals are tracked after initial detection. When one observer detects an animal, (s)he follows it, observing all subsequent availability events if possible, until it is seen by the other observer or it passes abeam unseen by the other observer. Such surveys provide additional data over and above those used in the methods described above. These are the binary data generated from tracking animals: 'success' on each resighting corresponds to detection by the other platform of the availability event, 'failure' to the other observer missing the event.

The models developed by these authors also included components to deal with uncertain duplicate identification and measurement error in locating detections. They used independent data (from tagged animals) on the availability process and did not estimate parameters of the availability process from survey data. However, it seems possible that the models could estimate these parameters if there was no measurement error and no duplicate mis-identification.

Schweder *et al.* (1997) proposed a compound Poisson point process model (with a rate parameter distributed as a gamma random variable) for the availability process, and evaluated the associated likelihood by simulation. Schweder *et al.* (1999) did not use an explicit availability state model; instead they resampled availability histories from the sample of tagged animals. Cooke (1997) used both these approaches.

6.8.4 *Discrete availability: cue-based*

A cue is an instantaneous availability event, such as a whale blow. Cue-based estimation methods for discretely available animals are analytically much simpler than animal-based methods, and could be applied to the line transect survey data used by the (x, y)-methods above.

Cue counting methods were developed by Hiby (1985) and are described in Buckland *et al.* (2001: 191–7). These methods estimate the mean number of cues produced per unit area per unit time. This is achieved by using point transect methods to estimate the number of cues within radial distance w of the observer(s) over the duration of the survey, and dividing this estimate by the product of survey duration and area of the searched sector of radius w. The result is then divided by an estimate of the mean number of cues per animal per unit time (the cue rate), to give an estimate of the number of animals in the survey region. The cue rate is usually estimated from external data.

If the probability of detecting a cue at distance $y = 0$ is unity, MRDS methods are unnecessary. If cues at distance $y = 0$ can be missed (as may be the case on aerial surveys), MRDS methods are useful. The point transect component of double-observer cue-counting surveys is amenable to analysis using the likelihood functions and estimation methods developed in Sections 6.2 and 6.3, but using cues rather than animals as the detection unit. Hiby *et al.* (1989) developed a double-observer maximum likelihood method based on grouped distances rather than exact distances. (Sections 6.2 and 6.3 assume exact distances.) Borchers *et al.* (in preparation *b*) developed likelihoods and estimators for exact distance data. The likelihoods of Hiby *et al.* (1989) and Borchers *et al.* (in preparation *b*) include allowance for measurement errors, which on point transect and cue counting surveys, can cause substantial bias (Buckland *et al.* 2001: 195–6).

6.8.5 *Intermittent availability*

At least for marine mammals, a state model based on intermittent availability is more appropriate for aerial surveys than ship surveys because the animal is typically visible for some distance beneath the water surface, and often the time that the animal is at or near the surface is longer than the time it is within detectable range. By contrast, discrete availability is more appropriate for ship surveys because the animals are only available to be seen at the surface, and the duration of the surfacing is short relative to the time that they are within detectable range. Thus, the examples of intermittent availability that we describe here (Laake *et al.* 1997; Hiby and Lovell 1998; Hiby 1999; Hain *et al.* 1999) are of aerial surveys of marine mammals.

The interplay of the availability and observation models is also different for aerial surveys. While the two-dimensional model for detection is the most realistic, it is most appropriate for ship surveys at relatively slow speeds, where the observer focuses forward, whereas aircraft surveys are conducted at much faster speeds with the observers focusing primarily downward rather than forward. An additional minor point is that the two-dimensional models would be difficult to implement for aerial surveys, because they require measurement of the forward distance x as well as y, which is typically done in shipboard surveys by measuring a radial distance and angle from the line. Measuring forward detection distance x for aerial surveys would require electronic measurement of the time between initial sighting and when it was abeam.

For aerial survey applications with intermittent availability (Laake *et al.* 1997; Hiby and Lovell 1998; Hiby 1999; Hain *et al.* 1999), the observation model has been reduced to the y dimension, but in two slightly different ways. Hiby and Lovell (1998) and Hiby (1999) presumed that porpoise at perpendicular distance y were only detected abeam of the aircraft ($x = 0$) with probability $g(y)$, whereas Laake *et al.* (1997) assumed that animals at perpendicular distance y could be detected within an interval of forward distance $0 \leq x \leq \xi(y)$, that could depend on y. In relation to the two-dimensional model for discrete availability as described by eqn (6.57), this is equivalent to limiting $k = 1$ and $Q(x,y) = g(y)$ for $0 \leq x \leq \xi(y)$ and $Q(x,y) = 0$ for $x > \xi(y)$ or $x < 0$. Thus, if an animal is available or becomes available within the detectable range $0 \leq x \leq \xi(y)$, it has probability $g(y)$ of being detected, but the forward distance when it becomes available does not matter. The limitation that $x > 0$ (forward of being abeam) is used here as a convenience and can be easily removed. The forward distance $\xi(y)$ will depend on the assumed observation model. For example, for a single observer on one side of the aircraft, McLaren (1961) and Laake *et al.* (1997) assumed a quarter circle detection range (i.e. fixed radius), Barlow *et al.* (1988) assumed a rectangular range (i.e. constant $\xi(y)$), Hain *et al.* (1999) assumed a sector of a circle both forward ($x > 0$) and behind ($x < 0$), and Hiby (1999) assumed $\xi(y) = 0$. The speed of the aircraft and the size of the detectable range $\xi(y)$ determines the length of time, $T(y)$, in which a previously unavailable animal may become available within detectable range. (Note that Laake *et al.* (1997) used x for perpendicular distance and their $w(x)$ is equivalent to $T(y)$ here.)

The same availability process model was used by Laake *et al.* (1997), Hiby and Lovell (1998), and Hiby (1999). Each assumed that animal availability, the surfacing-diving process, is an alternating Poisson process (Cox 1962), which means that both the length of time at the surface (available) and the length of the dive (unavailable) are independent exponential random variables. As described by Laake *et al.* (1997) (a modification of their eqn (4)), the probability that an animal will be available within detectable range is:

$$P_v(y) = \Pr\{\text{animal at } y \text{ is at surface}\} \qquad (6.60)$$

$$= \frac{\lambda}{\lambda + \mu} + \frac{\mu[1 - \exp\{-\lambda T(y)\}]}{\lambda + \mu} \qquad (6.61)$$

$$= 1 - \frac{\mu \exp\{-\lambda T(y)\}}{\lambda + \mu},$$

where λ is the rate parameter of the dive process and μ is the rate parameter of the surfacing process. The average length of a dive (d) is $E(d) = 1/\lambda$ and the average length of a surfacing (s) is $E(s) = 1/\mu$; the sum of these is the average length of a surface-dive cycle.

Laake *et al.* (1997) used the model to describe availability of harbour porpoise for a calibration survey to measure aerial detection of porpoise that were tracked by shore-based observers (a trial configuration). By contrast, Hiby and Lovell (1998), and Hiby (1999) used the model to describe availability of harbour porpoise for an independent configuration of tandem aircraft or of a single aircraft making repeat runs along the transect. In each case there was an intervening time t between the possible detections of the same porpoise. For Laake et $al.$ (1997), the aerial survey was independent of the shore-based survey, so t varied substantially between encounters, whereas there was less variability in t for the tandem aircraft (Hiby and Lovell 1998), which were flown one following the other at a preset but somewhat varying distance. In the single aircraft circle-back method of Hiby (1989), the circle back distance (time) was preset but variable. However, in all of these applications, the relevant availability function is the probability that a porpoise at the surface is still at the surface or back at the surface after an interval of time t. We express that availability function as (a modification of eqn (8) from Laake et $al.$ 1997):

$$P_v(y \,|\, t) = 1 - \frac{\mu}{\lambda + \mu} \left[1 - \exp\{-(\lambda + \mu)t\}\right] \exp\{-\lambda T(y)\}. \qquad (6.62)$$

If we denote $S = E(s)$ and $C = E(d) + E(s)$, this availability function is equivalent to $U(t)$ from Hiby (1999), if $T(y) = 0$. The length of time that animals are in detectable range relative to C will determine the importance of the assumption about $T(y)$.

Note that $P_v(y \,|\, 0) = 1$, as it should. With a range of $t \geq 0$ values, Laake et $al.$ (1997) demonstrated that perception bias (probability of detecting an available animal) and availability bias (due to the availability process) can be estimated separately. Perception bias is obtained directly from the results of detection trials when $P_v(y \,|\, 0) = 1$. In the methods of Hiby and Lovell (1998) and Hiby (1999), $t > 0$ and this makes it less feasible to separate the two sources of bias. They used $g(y)$ such that $g(0) = 1$ and Laake et $al.$ (1997) effectively used $p(y)$ with $p(0) \leq 1$ even though they described it as $g(y)$. The tandem aircraft and circle-back methods could use $p(y)$ most easily by using independent observers within one or each aircraft.

As t gets large, eqn (6.62) converges to (6.61). This is the basis for one of the methods of design-based estimation without a specified availability process as described in the next section. If t is sufficiently large, the probability of being available (at the surface) is independent for the two platforms and the estimated detection probability from the trials incorporates both availability and perception.

Before we leave this topic, we will discuss a few more aspects of these applications. Because t is often greater than zero, animals are likely to move during the intervening time, which complicates matching observations between platforms. In each of the applications we have described above, a movement model was used. Laake *et al.* (1997) used the movement model to develop a deterministic criterion for matching detections that allowed the disparity in distance to increase with t. Hiby and Lovell (1998) and Hiby (1999) used a more satisfying method that quantified the uncertainty as described in Section 6.9 below. While both applications used the same model for intermittent availability, it is certainly not the only model; Laake *et al.* (1997) suggest that an exponential distribution may not be a good approximation for length of dives, and discuss other approaches.

6.8.6 *Design-based availability estimation*

Suppose that a subset of the population of interest had been marked on a previous occasion far enough in the past that their locations now were independent of their locations at marking. Suppose in addition that you were certain to detect every animal in the covered region. The fraction of the marked sub-population that you observe on the survey would be an estimate of $P_c \times P_v$, the probability that an animal in the marked sub-population is in the covered region and is available for detection. If, in addition, animals did not move out of the covered region between marking and your survey, $P_c = 1$ for the marked sub-population. In this case, the fraction of the marked sub-population that you observe on the survey is an estimate of the probability that an animal is available (P_v), irrespective of whether the availability process is static, discrete, or intermittent.

With an appropriate survey design, you can estimate the probability that an animal is available without an availability state model, and you can do this for any kind of availability process.

To estimate without bias, it is necessary that animals are as likely to be available to you whether or not they were available to be marked. In other words, it is necessary that there is no correlation between their availability at the time of marking and their availability at the time of your survey. Positive correlation in availability will result in positive bias in $\widehat{P}_v$: the proportion of the marked sub-population that is available would be higher than the proportion of the whole population that is available. Negative correlation will result in negative bias in $\widehat{P}_v$: the proportion of the marked sub-population that is available would be lower than the proportion of the whole population that is available.

The key is therefore to have a design that ensures no correlation between availability for the two observers in a double-platform distance survey. There are two considerations here:

Timing: The time gap between when animals are within detectable range of one observer and within detectable range of the other should be suffi-ciently large. The required period depends very much on the nature of the availability process. (Hence it is not quite true to say these design-based methods require no availability state model—you need to know or estimate enough about the availability process to know what is an adequate time period.) If animals' availability cycle is short (e.g. harbour porpoise, whose cycle is of the order of 2 min), the period can be shorter than if it is long (e.g. hauled-out pack-ice seals, whose cycle is some hours long).

Observation method: It may be possible for observer 1 and observer 2 to use different kinds of observation method. The ideal is that availability to the method used by observer 1 is completely unrelated to availability to the method used by observer 2. This is difficult or impossible to achieve when both observers use visual cues to detect animals, so consideration should be given to use of another method by one observer. Live traps for observer 1 and visual detection for observer 2 represent one possibility in terrestrial studies, although even in this case there may be correlation in availability because more active animals might be more detectable by both methods. A combination of visual and acoustic detection methods might be useful in surveys of vocalizing marine mammals (Borchers 1999). Here too there may be correlation if there is correlation between surfacing and vocalizing.

In the interests of keeping things simple, our discussion thus far in this section has had no distance component to it. We introduce it in general terms now and deal with it in a bit more detail below.

Suppose for the moment that animals do not move around the study area between detection occasions, and that detection probability depends on y alone. The fraction of animals detected (and 'marked') by observer 1 at distance y that is also detected by observer 2 provides an estimate of $P_v p_{2\,|\,1}(y)$ (which is equal to $P_v p_2(y)$ if detections by observers 1 and 2 are independent). If we let $p^*_{2\,|\,1}(y) = P_v p_{2\,|\,1}(y)$ and $p^*_{1\,|\,2}(y) = P_v p_{1\,|\,2}(y)$, we can proceed as we did for continuously available animals and use the likelihoods and estimation methods of Section 6.2, but with $p^*_{2\,|\,1}(y)$ and $p^*_{1\,|\,2}(y)$ in place of $p_{2\,|\,1}(y)$ and $p_{1\,|\,2}(y)$. Doing so implicitly incorporates estimation of the probability of animals being available for detection. Providing that there is no correlation in availability for observer 1 and observer 2, this method corrects for availability bias without the need for an explicit availability state model. If availability depends on y, the approach is also valid even though the shapes of $p^*_{2\,|\,1}(y)$ and $p^*_{1\,|\,2}(y)$ will be different from the shapes of $p_{2\,|\,1}(y)$ and $p_{1\,|\,2}(y)$.

If detection probability depends on additional covariates $\underline{z}$ (which it almost invariably will do), we can use $p^*_{2\,|\,1}(y, \underline{z})$ and $p^*_{1\,|\,2}(y, \underline{z})$ in place of $p_{2\,|\,1}(y, \underline{z})$ and $p_{1\,|\,2}(y, \underline{z})$, and use the likelihoods and estimation methods of Section 6.2.

If animals do move between observation occasions, this needs to be considered. If movement is such that the location of animals in the survey region when being 'marked' is completely independent of their location in the survey region on the subsequent observation occasion, and of the observer, the above estimation method holds, but in this case $p^*_{2|1}(y, \underline{z})$ includes the coverage probability P_c. That is, $p^*_{2|1}(y, \underline{z}) = P_c P_v p_{2|1}(y, \underline{z})$ (and similarly for $p^*_{1|2}(y, \underline{z})$).

If movement is such that animal locations between detection occasions are not independent, care needs to be taken so that there are no animals beyond the detectable range of the observer 'marking' animals, but within detectable range of the observer surveying the population with marked animals in it. See Section 6.6.2 for more discussion and an example.

6.9   Special topics

Cooke and Leaper (unpublished) developed a very general framework for line transect survey methods that encompasses almost all the methods described here. It also allows some features we have not discussed (random observer effect models, for example). The interested reader is referred to this paper for further details.

6.9.1   Uncertain duplicate identification

In our discussion of mark-recapture methods for distance surveys above, we have treated recaptures (duplicates in distance sampling terms) as if they are identified with certainty. In many applications, this is not the case. On a visual line transect survey of discretely available cetaceans, for example, animals can only be detected for the short period when they surface. If both observers detect the same surfacing, there may be certainty about its duplicate status, but if they detect surfacings widely separated in time, there may be considerable uncertainty about whether or not they were from the same animal.

A simple if rather ad hoc way of dealing with the uncertainty associated with duplicate identification is to classify potential duplicates subjectively and then estimate abundance using increasingly liberal definitions of duplicates. For example, if they are classified as 'definite', 'probable', and 'possible', only definite duplicates might be treated as duplicates in the first analysis, then definite and probable, followed by definite, probable, and possible. This gives some indication of the effect of uncertainty in duplicate identification on estimated abundance. This method has been used for many years on annual surveys of Antarctic minke whales, for example, Butterworth and Borchers (1988).

A more rigorous approach is to specify an objective rule for duplicate classification. Detections by two observers might only be classified as duplicates if they occurred within some maximum time window, within some maximum horizontal separation, and so on. An advantage of this approach is that the properties of the method can be investigated by simulation. Cooke (1997) used this approach in a discrete-availability model analysis of Norwegian North Atlantic minke whale survey data. These data include resightings of animals by the same observer, and the resulting rule is fairly complex. Schweder *et al.* (1999) used the same kind of rule in their analyses of the 1995 Norwegian survey data, but integrated it into their estimation procedure. They developed a method they called the simulated likelihood method to do this. It allows likelihood-based inference in situations where the model is too complicated for analytic evaluation and maximization of the likelihood function.

The most conceptually satisfying method (to a statistician at least!) is to quantify the uncertainty due to duplicate identification using appropriate probability models. The simulated likelihood method of Schweder *et al.* (1999) is a way of doing this. Another was developed by Hiby and Lovell (1998). A notable feature of this method is that it does not use any rule for duplicate identification, and does not in fact require explicit duplicate identification at all. To understand the method, it is useful to consider why detections by two observers are not simultaneous and at the same location. Some of the most relevant explanations for this separation are:

1. The two detections are not of the same animal. The higher the animal density, the more likely different animals are to be seen in each others' vicinity.

2. The detections are of the same animal, but it moved between the time observer 1 detected it and the time observer 2 detected it. The greater or faster the movement, the more likely it is that duplicates are separated in space.

3. The detections are of the same animal, but one or both of the observers made some measurement error in locating it. The larger the measurement errors of each observer, the more likely it is that recorded locations of duplicates differ substantially.

The availability process is also relevant here. If animals are available at regular but infrequent intervals, for example, detections by different observers that are separated by a time much shorter than the mean inter-availability time are unlikely to be of the same animal. Depending on the application, there may be other relevant factors.

To incorporate the uncertainty associated with duplicate identification, probability models can be proposed for each of the relevant components (local density, animal movement, measurement error, availability process, etc.), each of which might have some unknown parameters associated with it. These can be incorporated into a likelihood function for the observed data. Note that in this case the observed data do not include capture histories, only times and locations of detections. The approach allows duplicate uncertainty to be incorporated into inference without subjective or objective duplicate identification. Hiby and Lovell (1998) developed the approach for analysis of aerial surveys of harbour porpoise. Their likelihood model includes components for animal density (animals are assumed to be distributed according to a Poisson process with unknown rate parameter), animal movement, and availability pattern.

6.9.2   *When should double-observer methods be used?*

Data gathered on distance sampling surveys always come from more than one detection function: detection probability is likely to vary by individual, attention level, fatigue, a host of sighting condition variables, survey platform, and many other things. If the vector z contains all variables that affect detection probability, there is a separate detection function for every combination of values of the variables in z that are encountered on the survey. If $p(0, z) = 1$, pooling robustness allows us to use CDS methods which ignore z, or to model the effect of some variables using MCDS methods and ignore other variables (except in extreme cases). When $p(0, z) < 1$, pooling robustness does not apply, and we have to take account of the effect of all variables that affect detection probability in order to estimate detection probability and abundance without bias. Doing this can be difficult and involve substantially more complex modelling than is required for CDS or MCDS methods.

If $p(0, z)$ can be made to be unity by adopting appropriate searching methods, increasing search effort, or by other means, this is the best solution. The best way to deal with the problem of $p(0, z) < 1$ is to avoid it! That said, (a) it is not always possible to avoid the problem, and (b) how do you know that $p(0, z) = 1$ after you have adopted appropriate searching methods and used what you believe is an appropriate level of survey effort? You may not know how bad the bias arising from use of CDS or MCDS methods is unless you use reliable methods to estimate $p(0, z)$.

The MRDS methods are therefore useful when $p(0, z)$ is known to be less than unity, or as a means of verifying that $p(0, z)$ is unity or at least nearly so. They are also useful when you are interested in the effects of the explanatory variables themselves on detection probability, when estimating trend, and when animals are not uniformly distributed in space. They can also be useful in the presence of substantial measurement error. We discuss each of these below.

6.9.2.1   *Estimation when $p(0, z) < 1$*

When $p(0, z)$ is less than unity, surveys should routinely contain a substantial double-observer component, or consist entirely of double-observer survey. If species identification or reliable cluster size estimation is problematic without leaving the transect line to 'close' on detections, conducting double-observer survey all the time may be inadvisable, because closing on the detection by one observer will alert the other to the presence of an animal.

6.9.2.2   *Verifying that $p(0, \underline{z}) = 1$*

When $p(0, \underline{z})$ is thought to be unity but this remains to be verified, one dedicated double-observer survey, or a double-observer component on an otherwise single-observer survey, may be adequate. When mixing double-observer and single-observer surveys, you should remember that the addition of a second observer will in itself increase the probability that an animal at $y = 0$ is detected. To estimate $p(0, \underline{z})$ for the single-observer component, you need to estimate $p_1(0, \underline{z})$ (where observer 1 is the observer operating in single-observer mode), not $p_.(0, \underline{z})$ (the combined observer detection probability).

6.9.2.3   *Effects of explanatory variables*

The MRDS methods are also useful for investigating the effects of explanatory variables on detection probability. You might be interested in investigating the effect of experience on detection probability in order to set criteria for acceptance of individuals as observers on future surveys, or to design training schedules, for example. You might be interested in discovering which are the important variables affecting detection probability, with the aim of cutting down the data recording load on future surveys to exclude unimportant variables. You might be interested in identifying individuals who are particularly good or poor observers. (This is easily done by including an individual identifier as a factor in $\underline{z}$.) Note that using MCDS methods to assess the efficiency of individual observers can be very misleading because MCDS methods can only tell you about the shape of individuals' detection functions, not their intercept. For example, from an MCDS analysis, an individual whose detection function has a wide shoulder will always appear more efficient than an individual whose detection function has a narrow shoulder; but if the wide shoulder was achieved by reducing search effort at $y = 0$ and hence reducing $p(0, \underline{z})$, the individual with the wide-shouldered detection function might have a substantially lower probability of detecting an animal than the individual with a narrow shoulder: $p(0, \underline{z})$ must be taken into account. For examples of the estimated effects of a variety of explanatory variables (including individuals) on detection probability, see Laake *et al.* (1997), Evans-Mack *et al.* (2001) and Southwell *et al.* (in preparation).

6.9.2.4   *Estimating trend*

One circumstance in which you might be tempted to ignore $p(0, \underline{z})$ is when interest focuses mainly on trend in density or abundance. In this case, it is in principle possible to obtain better precision in estimating trend by ignoring $p(0, \underline{z})$ than by estimating it, *provided $p(0, \underline{z})$ does not vary*. This is because fewer parameters need be estimated if $p(0, \underline{z})$ is ignored. However, the constancy of $p(0, \underline{z})$ is crucial to this argument, and it is something that is very unlikely to apply in practice because detection probability is likely to vary substantially between individuals, with conditions, etc. If $p(0, \underline{z})$ varies randomly between surveys, then inference about trend from CDS methods may be unbiased. Even in this instance, however, variance estimation must include the component of variance due to varying $p(0, \underline{z})$. This can be done without estimating $p(0, \underline{z})$ if the variance in trend is based on the inter-survey variance, and not the variance from the individual surveys alone (which are negatively biased because they neglect variance in $p(0, \underline{z})$). This requires an adequate number of survey occasions for reliable estimation; inferences about trend drawn from just a few surveys may be misleading.

If $p(0, \underline{z})$ varies systematically over time (between survey occasions), this effect will be completely confounded with trend in density or abundance when $p(0, \underline{z})$ is not estimated. In practice it is likely to be very difficult to ensure that $p(0, \underline{z})$ does not vary systematically over time (because people get more practised at surveying, methods evolve, survey platforms improve, etc.), so that estimation of $p(0, \underline{z})$ at regular intervals, if not on every survey, is advisable.

6.9.2.5 *Movement and non-uniform distribution*
The CDS and MCDS methods require animals to be uniformly distributed[8] with respect to distance from the line at the time of detection. In the absence of responsive movement, we can achieve this through appropriate design (Chapters 2, 7, and 10). As illustrated in Section 6.6.2, double-observer methods can be used to avoid bias when animal movement leads to violation of the uniformity assumption (by the primary observer in the case of this method). The method does, however, require one observer (the trial observer) to detect animals while they are uniformly distributed in space.

If animals are not uniformly distributed when detectable by any observer, double-observer methods can be used to estimate abundance and to estimate animal distribution within the covered region. This fact is not useful in most applications, where random location of the covered region results in uniform distribution within it. There may be circumstances in which it is useful though. An example is a species which inhabits a narrow strip (a river or coastline, for example) that can be contained entirely within the covered region. If it is not practical to lay transects across the strip, transects along the centre of the strip with double-observer survey methods could give unbiased estimates of abundance. Another example is migration counts (Section 11.8) in which the entire survey region is covered because all animals pass within view of the observer.

---

[8] Strictly, they need not be uniformly distributed, but their distribution must be known for unbiased estimation to be possible.

6.9.2.6 *Measurement error*

Double-observer methods to cope with the effects of measurement error should only be considered when measurement errors may be large and the errors are not quantified using estimated distance experiments. This use applies more to cue counting and point transect surveys, where the effects of measurement errors are much larger than is the case with line transect surveys. In this circumstance, double-observer data can be used to estimate measurement error (see Section 11.9.3) and to correct bias due to the error. With cue counting or point transect surveys, this bias can be large (Borchers *et al.* in preparation *b*).

6.10 Field methods

Providing guidance on field methods is a large undertaking, given the variety of techniques that can be used to collect data for MRDS. Buckland *et al.* (2001) provide nearly 60 pages describing various aspects of survey protocol and field methods for distance sampling, including several pages that focus on incomplete (uncertain) detection at zero distance. In this section, we will only consider aspects of field methods or MRDS not covered by Buckland *et al.* (2001). We will focus on the aspects of field methods that are important in satisfying two important assumptions for the mark-recapture aspect of an MRDS analysis: (1) required level of independence between observers, and (2) accurate recording of the capture history (e.g. determining duplicates between observers). While there are analytical techniques to handle uncertainty in duplicate identification (Hiby and Lovell 1998), we have focussed on methods that have assumed duplicates are determined without error. Even if the analysis can handle uncertain duplicate identification, it will be improved with accurate data collection to reduce the uncertainty.

6.10.1 *Marked animals*

Physical marking and tracking of animals with a radio, GPS or satellite location device to provide the initial sample is quite different from most of the other field techniques, so we describe it separately. This method has several advantages over MRDS methods in which both samples are based on visual location. Because the animal's location is potentially known at any time, availability bias can be removed without any concern about temporal separation of the samples as in visual samples. Also, duplicate identification may be confirmed by locating the marked animals.

Marked and tracked animals provide the initial sample, and the relevant observation configuration is the trial configuration. The marked sample of animals within the region covered during the conduct of the distance sampling provides a set of trials that are successes (animal observed) or failures. The assessment of whether a marked animal is observed must depend on the locations of the marked animals and the locations of any observations made during the distance sampling. This is likely to be done at the analysis stage, so it is important that the location data are as accurate as possible. If the marked animals can be tracked in real-time by a person not observing, it is preferable to make the determination in the field if possible. To achieve the one-way independence required by the trial configuration, it is essential that the marked animals are not made more or less visible by marking. This means that the marks cannot make the animal more visible; a bright orange collar say would not be a wise choice. The independence assumption also means that the marked animals must represent a random sample of animals at the time of the survey. For example, if a survey region contains two habitats that have differing visibility and the marked sample was in only one of the habitats, the detection of marked and unmarked animals would differ. Collection and incorporation of covariates (e.g. habitat) for the trials and observations can overcome the lack of randomness and induced dependence, as long as the marked sample is sufficiently large and contains some animals throughout the range of covariates in the population (e.g. some marked animals in both habitats).

6.10.2 *Observation configuration*

In visual surveys, the appropriate observation configuration and analysis will depend on the population of interest, the survey platform(s) and the availability of personnel. The investigator should consider carefully all of those aspects in making the choice. In some cases, the choice is obvious, while in others, two or more configurations may work.

The independent observation configuration has some important advantages over other configurations if duplicates can be determined accurately and responsive movement does not occur. In this case, it provides more information on detection probability (detection functions of all observers can be estimated), and analysis with an assumption of point independence (rather than the stronger assumption of full independence) is possible. However, it does require a platform large enough that observers do not cue each other while observing and recording. If the platform is too small, the trial configuration should be used if possible; the removal configuration is a last resort. The removal configuration generates less useful data and involves stronger assumptions about detection probabilities than the other two configurations. Every mark-recapture experiment contains within it a removal experiment on marked animals (see Borchers *et al.* 2002: 116–7), in addition to recapture data. The removal configuration lacks recapture data and is more limited as a result.

If animals are known or suspected to respond to the observers' presence while in detectable range, the trial-observer configuration should be used in preference to the independent observer configuration because it has the potential to correct for bias due to responsive movement. In this case, it is important that the observer who generates trials searches sufficiently far ahead that animals are unlikely to have responded when they are detected, and wide enough that no animals could enter the region covered by the other observer, from perpendicular distances greater than those searched by the trial observer. Use of high-powered binoculars or a different survey platform that is able to search far ahead should be considered. Buckland and Turnock (1992) used a helicopter to generate trials for a primary observer on a ship, for example. It is also important to gather data on as many variables as possible that might affect detection probability—because with this configuration, the data are inadequate to support a point independence assumption (and full independence is compromised if variables that affect detection probability are not taken into account).

The trial-observer configuration is also more able to deal with animals that are discretely or intermittently available. If it is possible to separate the regions searched by the two observers sufficiently so that there is no correlation in availability between them, this configuration can avoid bias due to the availability process. In principle, the independent observer configuration is equally able to avoid the bias, but when search regions are separated, duplicate identification becomes more uncertain and more difficult. This problem can be substantially reduced with the trial-observer configuration because the trial observer can track animals from detection through the other observer's detection region. The trial observer's detection function is not estimated, so he or she need not search continuously and can track detections without loss. For the same reason, the trial observer need not operate continuously—he or she could operate only for a fraction of the whole survey, as long as enough trials are generated to estimate the other platform's detection function. This means that the trial-observer configuration can be substantially less labour-intensive than the independent observer configuration. It does, however, beg the question of what constitutes enough trials? While we have not studied this thoroughly, and the answer will vary from survey to survey, from our experience we suggest that you aim at a design that will generate a minimum of around 30 duplicates.

All persons are not equally effective as observers. While you can try to assemble the best and most experienced survey crew, there will almost always be innate differences in their abilities. Thus, rotation of personnel through the observer roles is a consideration for each configuration. For independent observation, there is no need to rotate personnel between the roles because the roles are symmetric. If there are physically separate survey platforms for the observers, it is best to maintain the same personnel on each platform throughout the survey. If personnel are rotated and only platform effects are estimated, this can induce negative correlation between platforms: when platform 1 has a poor observer and platform 2 has a good observer, $p_1(y, \underline{z})$ is low and $p_2(y, \underline{z})$ is high, while when platform 1 has a good observer and platform 2 has a poor observer, $p_1(y, \underline{z})$ is high and $p_2(y, \underline{z})$ is low.

If personnel are not rotated, personnel and platform effects are confounded. If the goal was to estimate the separate effects of personnel and platform on detectability, the confounding would not be helpful. However, there is no need to estimate the effects separately for an MRDS analysis, and rotation of personnel between platforms allows for the possibility of a personnel–platform interaction and the need to estimate more parameters. If personnel and platform are confounded, there are fewer possible parameters that can and need to be estimated. Likewise, for the trial configuration, rotation may not be the best approach. Heterogeneity is the rule and not the exception in detection probability. If the observer generating the trials is a poor observer, he or she is likely to miss many of the animals that are hard to detect. The trials will then only consist of those that are easy to detect, and will not reflect the range of detectability in the population. Such an observer may also fail to generate an adequate number of trials. Thus, the trial observer should be the better observer. If the team consists of several equally experienced observers, they may be rotated through each role, but inexperienced or ineffective observers should not be used to generate trials. The removal configuration is the exception. Because it is necessary to assume equal detection probability for each observer role, there may be some advantage in having personnel rotate regularly. However, this runs the risk of inducing negative correlation (see above).

In some situations, lack of space or personnel or some other constraint may limit the amount of time that two observers can operate. For example, in an aircraft the range may be reduced by carrying four observers to provide a double-count on both sides of the aircraft. By carrying three observers, the range can be increased and a double-count can be conducted on only one side at a time. If the double-count cannot be conducted throughout the entire survey, if at all possible, conduct the double-count such that the observer(s) are unaware that they are being 'tested', as they may consciously or unconsciously change their searching behaviour. If they did change their searching behaviour, applying the double-count results to the entire survey may generate bias.

6.10.3 *Data collection and recording*

The focus of data collection and recording in MRDS should be to record sufficiently accurate data (e.g. distance, time, location) to enable duplicate determination, and secondly, to collect covariate data that will help model heterogeneity in detection probability. For MRDS, the quantity of data collected is much greater than for CDS, so a judicious choice of covariates and automated data recording (e.g. GPS linked to computer, audio/video recording with stereo channel input for each observer) are essential for quick and accurate data collection, and recording that does not interfere with observation. One such system proposed for MRDS was described by Southwell *et al.* (2002). If space is not a consideration, several persons can be used to record data from the various observers; otherwise, an automatic data recording system should be used. Observing personnel should not do their own recording unless the objects are immobile (e.g. nests) or they are very uncommon and widely spaced, such that data can be recorded without risk of missing a nearby observation.

It is quite easy to collect too many covariate data because 'that variable just might affect detection probability'. The number of covariates you record should depend on the expected number of observations and the amount of time available to record the data. Collect as many covariates as you can without compromising survey quality. Include those covariates that are obviously important, and only include lesser covariates if you expect a large sample size. Unless you are going to attempt a full-likelihood analysis by specifying distributions for the covariates in the population, you must only consider covariates that are independent of distance. Often, animal behavioural covariates are not independent of distance because the animal's behaviour is affected by the observer, and some behaviours are more likely to occur close to the observer (e.g. a flushing or diving bird). While you may want to record them, it will not be valid to include them in the kinds of analysis we have described, which assume independence between y and other variables.

7

Design of distance sampling surveys and Geographic Information Systems

S. Strindberg, S. T. Buckland, and L. Thomas

7.1 The potential role of GIS in survey design

Geographic Information Systems (GIS), together with advances such as the Global Positioning System (GPS) and remote sensing and satellite technology, mean that geo-referenced data of various sorts are readily available. This is starting to have a strong influence on how distance sampling surveys are designed, and on how data are collected and analysed. Line transect surveys especially, perhaps with randomly or systematically spaced lines within geographic strata, are not always straightforward to design by hand. Using GIS, survey region and stratum boundaries can readily be digitized, and survey lines positioned according to some automated algorithm within these boundaries. The design can then be printed off, or the coordinates of the lines downloaded directly into a GPS unit, without any need to plot lines by hand.

To take full advantage of GIS, we need to develop a range of automated design algorithms so that efficient designs for various circumstances can be generated. We also need the facility to generate a large number of realizations from a given design algorithm, so that for complex designs in which coverage probability varies spatially, we can estimate empirically coverage probability by location. If this probability varies appreciably, we will then need analysis methods that utilize the estimated coverage probabilities to allow asymptotically unbiased estimation of object density. Finally, we would like the capability to simulate populations of objects that mimic our population of interest. We can then generate a number of designs from each relevant automated algorithm, simulate survey data, and analyse them. From these analyses, we can assess which algorithm is likely to lead to optimal estimates of abundance; we would like these estimates to have high precision and low bias. Spatial covariates are useful for spatial modelling

(Chapter 4). They are also potentially useful for designing a survey, whether they are used simply to define better geographic strata, or to develop a more sophisticated sampling scheme, in which the probability of sampling a location is a function of covariate values at that location. The ability to try different options on simulated populations before designing the real survey allows more cost-effective options to be identified and implemented.

This chapter is based on the work of Strindberg (2001) and Strindberg and Buckland (in press). We consider a number of options for automated design algorithms, show how estimation can proceed if we use an algorithm that does not provide even coverage probability through the survey region (or at least within strata), and discuss the role of simulation in determining the optimal algorithm for any given survey.

7.2 Automated survey design

Distance sampling is a form of plot sampling in which the plots are generally either strips (line transect sampling) or circles (point transect sampling). We term these plots, prior to positioning them in the survey region, samplers, since they sample the region. For design purposes, to some extent, we can consider circular samplers to have no dimensions and strip samplers to have one dimension, with both sampling two-dimensional space. However, an area around each point or line of a design is sampled, so strictly, both are two-dimensional samplers. Typically, distances surveyed around the point or line are small relative to the dimensions of the survey region, allowing us to treat them conceptually as zero and one-dimensional samplers, respectively, for most purposes. In principle, this makes survey design straightforward, but practical considerations typically lead to modification of simple designs, in which case the effects of those modifications on coverage probability must be carefully considered. Line samplers lead to considerably more complications than point samplers, so we address point transect sampling first.

Area or polygon samplers are often of interest, for example, in quadrat sampling. However, in distance sampling, they are of relevance only to the extent that the samplers in point transect sampling are strictly small circles rather than points, and in line transect sampling, they are long, narrow strips rather than one-dimensional lines. Hence we do not consider polygon samplers separately.

When survey design is complex, it is often conducted subjectively, instead of relying on fully automated design algorithms. We stress here the advantages of automated design algorithms. First, subjective designs, and even random designs that are modified, for example, to prevent samplers falling very close to each other, do not provide even coverage probability. Indeed, bias can be surprisingly large when object density is variable through an irregularly shaped survey region. Second, if a simple random design is modified for practical reasons, provided the algorithm is automated, coverage probability can be readily evaluated by simulation, and estimation conducted as described in Section 7.3, if it is found to depart appreciably from uniformity. Third, survey design is quick and easy to carry out using one of the automated design algorithms and GIS functionality of Distance 4 (Thomas *et al.* 2003). Fourth, if we generate a population by simulation that has similar features to the population of interest, it is possible to generate designs using several algorithms, sample our simulated population with a number of realizations from each algorithm, and analyse the resulting data, to assess the performance of each algorithm in terms of bias and precision. The best-performing algorithm can then be used to generate the design for the real survey.

7.2.1 *Point transect design*

The simplest design is a set of K points randomly located through the survey region. To ensure that different realizations of this design do not result in widely varying estimates of object abundance, K should be at least 20, and preferably substantially more. This is generally easily achieved in point transect surveys.

A simple random sample of points has a range of possible problems, which we now examine.

7.2.1.1 *Edge effects*

In many designs, points are sampled only within the survey region, and objects beyond the survey region boundary are not considered. The distribution of objects with distance from the point does not then increase linearly out to the truncation distance w as assumed in standard point transect theory, because the proportion of area that falls outside the survey region typically increases with distance from the point for points within w of the boundary. There are several possible solutions to this problem. The first is simply to ignore it. The estimated detection function is then biased, because it is now a composite function, reflecting both probability of detection and availability of objects (Buckland *et al.* 2001: 214). However, provided this composite function is well approximated by our model for the detection function, abundance estimation remains asymptotically unbiased. If the effect is sufficiently extreme to render such modelling unreliable, other solutions are needed. This may occur for survey regions that are very narrow so that the edge-to-area ratio is high.

If a GIS is used, it is simple to evaluate area as a function of distance from the points, truncating at w from each point. The composite function can then be disaggregated into its constituent parts, allowing the probability of detection to be modelled (Buckland *et al.* 2001: 215–6).

If object density is similar either side of the survey region boundary, another option is to record objects detected beyond the boundary, and include them in the analysis. This ensures that object density increases linearly with distance from the point, as the standard theory requires. No upward bias in abundance estimation occurs provided abundance is estimated by multiplying estimated density by the size of the survey region; this excludes the area of the buffer zone, which extends a distance w beyond the region boundary even though detections in this zone are included in analyses.

If object density is likely to differ either side of the boundary, for example, when the boundary follows a habitat edge, then a buffer zone can be used in a different way. Random points are generated through both the survey region and the buffer zone that extends a distance w beyond the boundary. The area of survey region (excluding the buffer zone), as a function of distance from points in the buffer zone (and hence outside the survey region), increases faster than linearly, and compensates for the slower-than-linear increase for points just inside the boundary. The points in the buffer zone are surveyed, but only detections within the survey region are recorded (Fig. 7.1). The data are then analysed as normal, except that

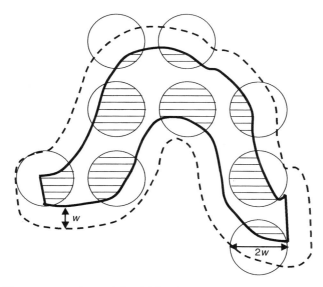

Fig. 7.1. For survey regions with a high edge-to-area ratio, such as nearly linear riparian habitats, a high proportion of surveyed circles may intersect the boundary of the survey region. In this circumstance, we recommend that a grid of points is laid down randomly over the survey region (delineated by the solid line) and a buffer zone (dashed line), extending a distance w beyond the region's boundary. Detected objects are only recorded if they fall within the survey region, indicated by the shaded sections of the surveyed circles of radius w, irrespective of whether they were recorded from a point that was inside or outside the boundary. For analysis of such data, see text.

care is required in recording effort correctly. The best option, which is implemented in Distance 4 using GIS functionality, is to record the effort for a point as the proportion of the circle of radius w and centred on the point that is in the survey region (assuming each point is surveyed just once). Thus points within a distance w of the boundary, or in the buffer zone, have effort less than one. To estimate abundance, estimated density is multiplied by the size of the survey region (excluding the buffer zone), as with a normal survey.

If the GIS functionality of Distance 4 is not used, another solution is to record no effort for points that lie in the buffer zone; detections within the survey region made from such points may be added to nearby points in the survey region for the purpose of analysis in Distance. Alternatively, effort can be recorded for these points the same as for other points, provided the estimated density of objects is multiplied by the combined area of the survey region and buffer zone to estimate abundance within the survey region; the reason for this is that the estimated density is now an average density for the survey region and buffer zone, under the assumption that there are no objects in the buffer zone (since such objects are not recorded). Both these solutions will tend to give a slightly higher variance than the above method, which weights points according to the proportion of the plot area within the survey region.

In most surveys, these edge effects are minor, but if perhaps around 25% or more of points are within w of the survey region, then one of the above solutions should be considered.

7.2.1.2 *Overlapping plots*

The surveyed plots are circles of radius w centred on the sampled points. For a simple random design, some of these circles may overlap. This compromises the assumed independence between points, so that standard variance estimation is biased. In fact, the standard methods in Distance are robust to this assumption failure, whether it is due to overlapping circles in a simple random design, or is due to use of a systematic design coupled with an analysis that assumes it was a simple random design (below). However, it is usually unsatisfactory, and an inefficient use of resources, to have a design in which the fieldworker is sometimes asked to survey points that are within $2w$ of each other.

The simplest solution to avoid overlapping plots is to use a systematic sample of points, which may be obtained by taking a grid of points, for example, the intersections of a square grid, and randomly placing the grid over the survey region. Spacing of the grid controls the number of points K in the design, and samples will not overlap, unless the grid spacing is less than $2w$.

Simple random sampling without overlap of plots can be achieved as follows. It is not possible to define a mutually exclusive and exhaustive set

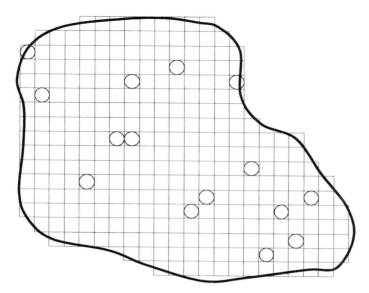

Fig. 7.2. If we wish to select a sample of circular plots at random, while avoiding sampler overlap, we can define a regular grid of polygons and randomly superimpose it on the survey region. Here, we show a grid of squares, each of side $2w$, so that a circle of radius w just fits inside a square. A random sample of $K = 15$ circles is shown. A grid of hexagons could also be used. Note that squares with centres falling outside the survey region are not included in the sampling frame. If a buffer zone were used as in Fig. 7.1, squares whose centres fall outside the survey region but within w of the boundary would also be included.

of circles over the survey region, but we can define a grid of squares of side $2w$, or a grid of hexagons such that a circle of radius w just fits within each hexagon. We can then randomly locate the grid over the survey region, and select a simple random sample of K squares or hexagons, taking as the sampled locations the centres of the sampled polygons. Only polygons whose centres lie within the survey region (plus buffer zone if relevant) are included in the sampling frame (Fig. 7.2).

7.2.1.3 *Uneven spatial distribution of samplers for any given realization*

Although random sampling (with an appropriate buffer zone) gives even coverage probability when averaged over many realizations, for any single realization the point samplers may exhibit a very uneven spatial distribution, especially if the number of points K is not large. In extreme cases, all points might fall at one end of the survey region. For the cases of line and quadrat samplers and when density varies through the survey region, Strindberg (2001) showed that systematic designs give smaller variation in density estimates from one realization to another, than do simple random designs. For this reason, and because samplers do not overlap for systematic designs, systematic samples are generally preferred to random samples in distance sampling.

Although systematic samples generally yield estimates with smaller variance, this benefit cannot be fully exploited, because variance is generally estimated assuming that the sample had in fact been random. In principle, the problem may be avoided by randomly placing more than one grid in the region, so that variance can be estimated from the among-grids variation. However, this solution is of little practical interest, because several grids would be required for reliable variance estimation, and the method has no obvious advantage over a simple random sample.

Another mechanism for ensuring a more even spatial distribution of sampler points relative to a simple random design is to define say a grid of polygons (usually squares or hexagons) that spans the survey region (Fig. 7.3). A point is then randomly located within each polygon. (For a polygon that spans the boundary of the survey region, the corresponding point may be included if it falls in that part of the polygon that lies within the region, and excluded otherwise.) The number of points K is controlled by the size of the grid. This ensures that the spatial distribution of samplers is more even through the region, but samplers in neighbouring polygons can still overlap, and as for systematic sampling, variances must be estimated assuming that the sample was an unrestricted simple random sample. Hence there is usually no reason to choose this method in preference to a systematic sample.

7.2.1.4 *Inefficient sampling when density is variable*

If density is known to vary appreciably through the survey region, simple random sampling is inefficient. If areas of high density are unpredictable in advance of the survey, adaptive distance sampling (Chapter 8) might be used. If areas of high density can be predicted, from past surveys, knowledge of the area, or identification of preferred habitats, then stratified sampling might be used to improve efficiency. Within strata, simple random, systematic, or cluster sampling of points might be adopted, depending on circumstances. More sophisticated designs might seek to allocate points with probability proportional to expected density (Underwood 2004). In this case, analyses would need to account for the uneven coverage probability (Section 7.3).

If simple random or systematic sampling is used when object density is variable, abundance estimation may be more efficient using spatial modelling (Chapter 4), which is based on a 'model-based' philosophy, whereas standard distance sampling is a hybrid philosophy, in which estimation of the detection function and its variance is model-based, but estimation of the encounter rate and its variance is design-based (Section 10.2).

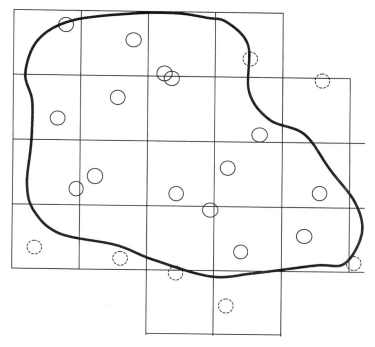

Fig. 7.3. To ensure that a sample of K points is spread fairly evenly throughout the survey region, we can define a grid of polygons (squares here) randomly superimposed on the survey region, and locate a single point randomly within each polygon. Note that in this realization, spread is more uniform than for the design in Fig. 7.2, but note also that two of the sampled circles of radius w overlap. In this example, $K = 15$ circles are sampled, although 23 polygons were required; the centres of the dashed circles fall outside the survey region, and so are not sampled (unless sampling is extended to a buffer zone as in Fig. 7.1). In practice, a systematic grid of points is usually preferred to this design. Such points could be defined for example by the intersections of the grid lines of this figure.

7.2.1.5 *Cost of travel between points*

If the survey region is large, a simple random or systematic sample of points may be inefficient, especially if it is costly to travel from one point to another. Cluster sampling is then a sensible option, so that, once a fieldworker reaches a sample location, he or she can sample several points around that location. The clusters may be located at random through the survey region, or on a systematic grid randomly superimposed over the region, and are the sampling units for variance estimation (e.g. the nonparametric bootstrap would be implemented by resampling clusters, not individual points). Within a cluster, the points might be positioned

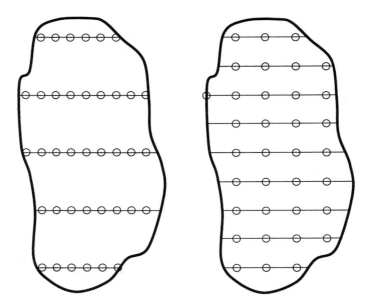

Fig. 7.4. Two examples of a point transect survey in which points are systematically spaced along lines, which are themselves systematically spaced. In the design on the left, separation between successive points is less than that between successive lines, so the points are not evenly distributed through the region. In this case, the lines are treated as independent sampling units (and are too few for reliable variance estimation here). In the design on the right, the points lie on a systematic grid of the whole region, so that individual points are treated as independent sampling units.

on a small grid centred on the sample location, with neighbouring points roughly $2w$ apart or slightly larger, to avoid overlapping plots. If small-scale spatial correlation in density is high, the separation distance between points within a cluster might be increased, so that the spatial autocorrelation between them is reduced. There is a balance to be met between improved precision and higher costs arising from larger travel times.

Point transect surveys are often designed as systematically spaced points along lines (Fig. 7.4). If the spacing between successive lines is the same as that between successive points on a single line, then the points fall on a regular grid, and are therefore a systematic sample of points. If, however, the spacing between lines is greater than that between points within lines, then the lines should be treated as clusters of points, because points within the same line cannot be considered independent. In other words, the lines must be treated as the sampling units, rather than the individual points within lines.

7.2.2 *Line transect design using lines of fixed length*

If the lines are very short relative to the size of the survey region, then design issues are largely the same as for points. At the other extreme, for shipboard and aerial surveys, lines typically pass from one boundary of the survey region to the opposite boundary. Intermediate between these two extremes are lines of fixed length, short enough that most do not intersect a boundary, but long enough that intersections with the boundary create problems that cannot be ignored. We consider this case first. We include in this class of designs the case that the line of fixed length is a circuit, such as the sides of a square or triangle. This has the advantage that the observer can start at any point on the circuit, so can choose the easiest access point, and finishes searching at this same point.

Denote the fixed line length by l. A simple random sample of K lines can be generated by selecting K points at random within the survey region, and drawing a line at a different random orientation from each point for a distance l. For lines that pass beyond the survey boundary, only the section of line within the survey region is sampled. Such a design is seldom if ever used. Practical problems are that some lines may cross, compromising independence and reducing survey efficiency; some lines that intersect a boundary may be very short, which is inefficient use of resources if it is costly to get to sample locations; and more resources are required in line transect surveys than in point transect surveys to ensure that K is sufficiently large to reduce variability arising from different stochastic realizations of the design to an acceptable level. We consider reasons why other designs may be preferred under the same headings as for point transects, but with an additional heading to cover the case of intersecting lines.

7.2.2.1 *Orientation of lines*

In practice, most designs use the same orientation for all lines. This ensures that lines cannot intersect. In a relatively homogeneous environment, or when lines are very short, this orientation can be selected at random. More usually, it is selected to take advantage of features of the survey region. Because variability in encounter rate is measured by among-line variability, greater precision is obtained if we orientate lines so that each line typically covers a range of habitats and hence of object densities. Thus if there is a known density gradient in the population, lines should normally be orientated roughly parallel to the gradient (i.e. perpendicular to the density contours). The longer the line length l, the greater the gain from doing this. Similarly, if habitat edges tend to be orientated in a given direction, lines should normally be orientated perpendicular to those edges. This can have an additional advantage if roads tend to follow such edges, as a higher proportion of lines will intersect a road, making access to the lines easier. Indeed, if roads are few and access problematic, orientation may be chosen to maximize the proportion of lines that intersect a road. If the survey region is split into strata, a different orientation can be used in each stratum, to improve efficiency further. (Note that such a strategy, in which line placement is random but orientation is not, does not bias sampling towards roads; rather, it ensures that the amount of ground covered near roads varies less among lines than it would if the line orientation was random.) If the transect lines are to be closed circuits, and there are reasons to prefer a particular orientation, a transect might be the sides of a long, narrow rectangle, with rectangle width chosen so that the two long sides are sufficiently separated to avoid overlap or disturbance effects.

7.2.2.2   Edge effects

If coverage of a line that intersects a boundary stops at the boundary, and if objects detected beyond the boundary are not recorded, then the distribution of objects with distance from the line is not in general uniform out to the truncation distance w. This is contrary to standard line transect theory, because the proportion of area that falls outside the survey region typically increases with distance from the line (Buckland *et al.* 2001: 215). As with point transect sampling, there are several possible solutions to this problem. Usually, we simply ignore it, and estimate the composite function, reflecting both probability of detection and availability of objects (Buckland *et al.* 2001: 214). In extreme cases with complex boundaries, such estimation may prove unreliable.

If a GIS is used, we can evaluate area as a function of distance from the line, truncating at w from each line. The composite function can then be disaggregated into its constituent parts, allowing the probability of detection to be modelled (Buckland *et al.* 2001: 215–16).

A simpler solution is to extend the line beyond the boundary through a buffer zone, until a line of width $2w$, perpendicular to the transect and centred on the observer, falls entirely outside the survey region. (If the transect would then exceed the pre-determined length l, it might be curtailed when that length is reached. To avoid bias, short lines that fall entirely outside the region, but for which the strip of width $2w$ falls partially within the region, should be covered. This solution is rather impractical, and omitting such lines, together with extending the sampled lines through the buffer zone, even if the occasional line is then longer than l, is approximately unbiased; the extra bits of line compensate for the short lines that should be sampled in the buffer zone but outside the survey region.) Detected objects are only recorded if they fall within the survey region (Fig. 7.5). Conventional analysis is then approximately unbiased for most shapes of survey region, provided the number of lines K is not very small, and the line length (effort) is recorded as the length of transect inside the survey region only; the additional line covered in the buffer zone is excluded.

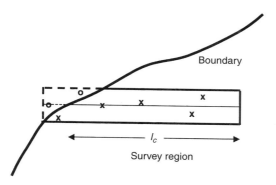

Fig. 7.5. A strategy for handling edge effects when using strip samplers. A strip that intersects the region boundary is extrapolated until it has completely emerged from the survey region (dashed lines). The whole of the centreline is walked (both the solid and the dashed sections), but only the length l_c of line inside the region (solid section) contributes to recorded effort. Animals detected within the survey region (**x**) are recorded, while those outside (**o**) are not. The area of the strip sampler within the survey region is the same on average as the area of a rectangular strip sampler of the same width $2w$ and length l_c, and the amount of ground surveyed is on average uniformly distributed with distance from the line up to distance w, as assumed in standard theory.

In most surveys, we can ignore these edge effects, because strip width $2w$ is typically small relative to the size of the survey region. Another edge effect is of greater concern for some surveys. Lines that intersect the boundary are truncated, and some of them may be too short to be worth visiting, if the costs of travel are high. One solution is to discard lines that, after truncation, are shorter than $l/2$, and to supplement those lines of length at least $l/2$ but less than l with additional effort, so that all retained lines are of length l. If this is done simply by moving lines that intersect the boundary back into the region until they are entirely within the boundary, the region's edge is undersampled, while the area a short distance in from the boundary (between $l/2$ and l from the boundary) is oversampled. A better option is to reflect the section of line beyond the boundary back into the survey region. That section of line may then need to be displaced a little along the boundary, to avoid overlap with the non-reflected part of the line (Fig. 7.6). This solution can also be used when lines are closed circuits (Buckland *et al.* 2001: 237).

7.2.2.3   *Overlapping plots*
The surveyed plots are strips of width $2w$ centred on the sampled lines. For a simple random design, some of these strips may overlap. As with point transect sampling, the methods of analysis are rather robust to this failure

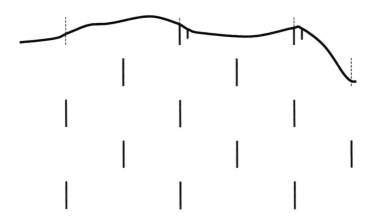

Fig. 7.6. A design based on a systematic sample of lines, showing how isolated short transects at the boundary can be avoided without undersampling near the boundary. In this example, four transects intersect the boundary. Two have their midpoints outside the region and are not sampled. The loss of effort near the boundary is compensated by increasing effort near the boundary in the locality of the two whose midpoints are inside the region. Conceptually, we can reflect these samplers at the boundary, and offset the reflected section along the boundary to avoid overlap of effort. In this example, the 'reflected' sections are chosen to be parallel to the original sections, and displaced a fixed distance to the right. If we regard the sample location to be the midpoint of the line, this algorithm ensures that a fixed length of transect is covered in the vicinity of every sample location that falls within the survey region.

of the independence assumption, but disturbance may become a significant issue, and a design that includes overlapping strips is not efficient. Thus systematic sampling is usually preferred.

Simple random sampling without the overlap of plots can be achieved as follows. Define a grid of rectangles, each of width $2w$ and length l, that covers the entire survey region. Select a simple random sample of K rectangles, and take the centreline of each sampled rectangle as the sampled lines. If any part of the centreline falls within the survey region, that line is sampled (Fig. 7.7). To avoid edge effects, a buffer zone could be added, so that all sections of sampled line that are either inside the survey region or within perpendicular distance w of the survey region are surveyed. This allows inclusion of objects that are within the survey region but detectable from sections of line outside the region.

7.2.2.4 *Uneven spatial distribution of samplers for any given realisation*
Uneven spatial distribution of samplers for a single realization of a design tends to be more problematic for line transect surveys than for point transect surveys because a single line requires more resources to survey than

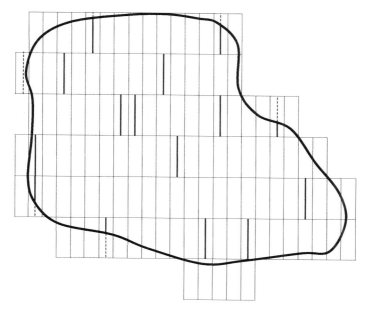

Fig. 7.7. To select a sample of strips of width $2w$ and length l at random without any sampler overlap, we can define a regular grid of strips with the required dimensions and randomly superimpose it on the survey region. The orientation of the grid might be random, or fixed to take account of practical considerations. A random sample of $K = 15$ strips is shown. Five of these intersect the boundary, so the corresponding transects (represented by the centrelines of sampled strips) have length $< l$. Note that strips that intersect the boundary but whose centrelines are entirely outside the boundary are not sampled under this scheme. (One such strip is shown with a dashed centreline.) If a buffer zone were used as in Fig. 7.5, all strips that intersect the boundary would be included in the sampling frame.

a single point, so that K is often rather small. Systematic designs have appreciably better spatial distribution properties than designs based on random, independent lines (Strindberg 2001). With appropriate spacing between line samplers, they also provide designs without overlap between neighbouring samplers. When transects are of a fixed length, a systematic design can be obtained by laying down a systematic grid of parallel lines over the survey region. A random point is chosen on one of the lines, and a transect of length l is drawn from that point along the line; the direction can be either predetermined or random. Having defined one transect, other transects on the same line are drawn such that the separation between successive transects on the same line is the same as the separation between successive parallel lines. Starting from the first line, transects along subsequent lines can be positioned systematically with respect to

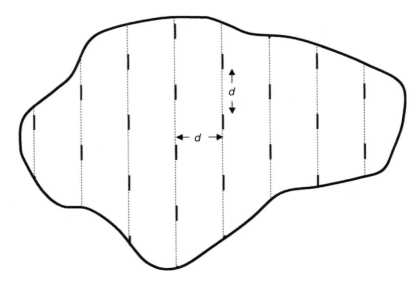

Fig. 7.8. A scheme for obtaining a systematic sample of transects by placing transects along parallel lines spanning the full survey region. Separation d between successive transects along a single line is the same as the separation between successive lines. In this example, transects on alternate lines are offset, which improves coverage probability properties.

those on the first line, possibly offset on alternate lines as illustrated in Fig. 7.8. (The 'offset' design of Fig. 7.8 can be obtained by generating lines on a rectangular grid, and then deleting alternate lines.) Sampling effort is then controlled by choice of l, and of line separation in the grid of parallel lines. The advantage of this design is that individual transect lines can be assumed to be independent when calculating variances (cf Section 7.2.1.3), whereas if separation of transects along a line is small relative to the separation of successive parallel lines, the transects do not provide even cover of the survey region, so that variance estimation must be based on variation among the long lines rather than the individual transects, and replication may then be insufficient for reliable estimation.

We could define a systematic grid of rectangles, and randomly position a line, parallel to two of the sides, in each rectangle (Fig. 7.9). The dimensions of the rectangles determine the number of lines K and total effort L. Typically, these rectangles would be substantially larger than those of length l and width $2w$ of the previous section. This is a compromise between a random and a systematic design, and shares the disadvantage of overlapping samplers with random designs, and the disadvantage of biased variance estimates with systematic designs.

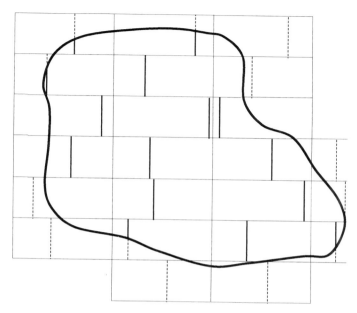

Fig. 7.9. To spread lines more evenly through a region, while retaining separate randomization for each line, we can superimpose a grid of rectangles on the survey region, and locate a single line in each rectangle, parallel to two of its sides and a random distance between those two sides. In this realization, spread is more uniform than for the design in Fig. 7.7, but two of the lines are sufficiently close that overlap of the associated strips of width $2w$ may occur. In this example, $K = 15$ lines are sampled, although 23 rectangles were required, and five of the lines are not of full length. Eight lines fell entirely outside the survey region and so are not sampled (unless sampling is extended to a buffer zone as in Fig. 7.5).

7.2.2.5 *Inefficient sampling when density is variable*

The same considerations apply as for point transect sampling (Section 7.2.1.4). Thus if density is known to vary appreciably, but areas of high density are not predictable in advance, then adaptive line transect sampling (Chapter 8) might be worthwhile. We often can predict such areas, from past surveys, knowledge of the area, or identification of preferred habitats. In that case, stratified sampling might improve efficiency. Within strata, lines might be defined according to the same range of sampling schemes as for the unstratified case. Usually, a systematic sample of lines or circuits, all with the same orientation, would be preferred. If a model-based approach is adopted, for example, the spatial models of Chapter 4, the choice of design scheme is less critical, although those that provide a fairly even spatial distribution of sampler lines over the survey area will tend to be more efficient and less biased than those that do not.

7.2.2.6 *Cost of travel between lines*

As with point transect sampling, if travel between sample locations is expensive, cluster sampling can be used, with several transects at each sample location. However, the gain is less clear for line transect sampling, because extra effort can be expended at each sample location simply by having longer lines. If there is a strong spatial correlation in object density over short distances, having several short lines at each sample location may be more efficient than having one long line, if the separation between lines is sufficient to reduce spatial correlation (and hence increase the precision of the abundance estimate) without adding substantially to travel costs. If variance in encounter rate is to be estimated by the sample variance (weighted by effort) of encounter rate across sampling units, or if variances are estimated by nonparametric bootstrapping of sampling units, the sampling units must be the clusters of transects, not individual transects. Theoretical considerations show that this strategy is better than two-stage bootstrap resampling of clusters, and then of transects within clusters (Davison and Hinkley 1997: 100–2).

7.2.3 *Line transect design using lines that span the full width of the survey region*

When each line spans the full width of the survey region, different design considerations apply relative to the case of lines of fixed length. Sometimes transects of fixed length are spaced along lines that span the survey region, in which case some of the issues addressed in this section are relevant to fixed-length transect designs. We examine issues under the same headings as in the previous section.

7.2.3.1 *Orientation of lines*

When lines span the full width of the survey region, it is almost a necessity to constrain the lines to be parallel (or to use zigzag samplers as in Section 7.2.4). If each line is independently randomly orientated, there are numerous intersections between lines, and unless the number of lines is large, individual realizations of such a scheme can have poor spatial distribution properties. If there is an identifiable density gradient through the survey region, precision of abundance estimates can be appreciably improved by ensuring that the grid of parallel lines is aligned along that gradient (i.e. perpendicular to density contours; Fig. 7.10). See Section 7.2.2.1 for a more detailed discussion. Another consideration is that many short lines give more reliable estimates of variance in encounter rate than do a few long lines. Hence, an orientation that achieves this is preferred, provided it does not conflict with the recommendation to avoid lines that are parallel to density contours.

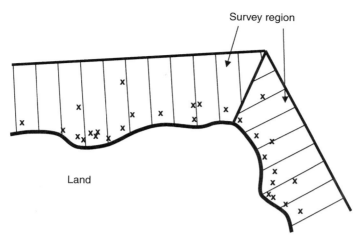

Fig. 7.10. A parallel-line marine survey of objects (indicated by crosses) whose density falls off with distance from the coast. The parallel lines are aligned to be approximately perpendicular to the coast (and hence to the density contours). Note how the survey region has been split into two strata so that this can be achieved.

7.2.3.2   *Edge effects*

The same considerations apply as for fixed-length transects (Section 7.2.2.2), except that we avoid the complication of requiring that lines are of fixed length, so that the solution of projecting lines beyond the boundary of the survey region is more straightforward (Fig. 7.11). There is also less of a problem with very short sections of transect at the region's boundary. For designs in which lines span the width of the region, only where that region is very narrow can lines that are impractically short occur.

7.2.3.3   *Overlapping plots*

The usual means of avoiding overlapping plots in this case is to use a grid of systematically spaced parallel lines, and randomly superimpose it over the survey region. For variance estimation, we assume that the lines in the design are independent.

Simple random sampling without overlap of plots can be achieved if we define a set of contiguous strips, each of width $2w$, that spans the entire survey region (Fig. 7.12). Select a simple random sample of K strips, and take the centreline of each sampled strip as the sampled lines. If any part of the centreline falls within the survey region, that line is sampled. As for fixed-length transects, we can avoid edge effects by adding a buffer zone, so that all sections of line that are either inside the survey region or within perpendicular distance w of the survey region are surveyed. This allows

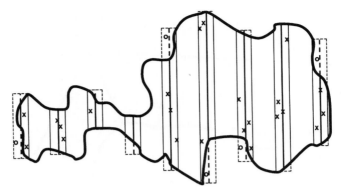

Fig. 7.11. If edge effects are severe, as here, the estimated detection function may be appreciably biased, since we would actually estimate the composite function reflecting fall-off with distance from the line both in detectability and in area within the survey region. By projecting each line beyond the boundary, until the associated strip emerges from the survey region, as shown here, such bias is reduced. Even though the dashed sections of transects are surveyed, only animals detected inside the region (**x**) are recorded; those outside the region (**o**) are ignored. Recorded effort must exclude the dashed segments of transect.

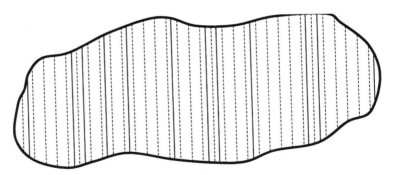

Fig. 7.12. To select a sample of strips of width $2w$ at random without any sampler overlap, we can define a regular grid of strips, each of which spans the full width of the region, and randomly superimpose the grid on the survey region. The strips with a solid centreline indicate a random sample of size $K = 12$.

inclusion of objects that are within the survey region but detectable from sections of line outside the region, reducing bias in the estimation of the detection function arising from edge effects.

7.2.3.4   *Uneven spatial distribution of samplers for any given realization*
Strindberg (2001) showed that a systematic grid of parallel lines, randomly superimposed on the survey region, has better spatial distribution

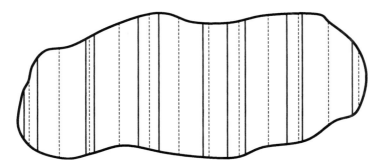

Fig. 7.13. To spread lines more evenly through a region, while retaining separate randomization for each line, we can superimpose a grid of strips on the survey region, and locate a single line (shown as a solid line) in each strip, parallel to its edges (shown as thin dashed lines) and a random distance between them. In this realization, spread is more uniform than for the design in Fig. 7.12, but two of the lines are sufficiently close that overlap of the associated strips of width $2w$ may occur. In this example, $K = 12$ lines are sampled, although 13 strips were required, as one of the lines fell entirely outside the survey region.

properties than single realizations of a random parallel line design for small numbers of lines. In particular, when there is a trend in density within the survey region, systematic samples show less variation between realizations of the design in the expectation of the abundance estimate, when that expectation is conditional on the realization obtained. See also Section 10.5 for an emphatic illustration of this effect.

A constrained random scheme could be implemented by defining K strips of equal width (equal to the line separation for the corresponding systematic design) to cover the whole survey region. Within each strip, a line can then be positioned parallel to the edges of the strip and at a random location between the two edges (Fig. 7.13). As with fixed-length transects, this is a compromise between a random and a systematic design, sharing the disadvantages of both, although it ensures a more even cover of the survey region than does a fully random design.

7.2.3.5 *Inefficient sampling when density is variable*
Typically when lines span the full width of the survey region, the number of lines tends to be small. Variable density can then lead to poor precision, which is improved somewhat by using systematic designs in preference to random designs. As for fixed-length transects, stratification can help if areas of high density can be predicted, or adaptive sampling (Chapter 8) if they cannot. Spatial models (Chapter 4) may lead to more precise abundance estimates, especially as they exploit the location of detected objects along the line, which standard analysis methods do not.

7.2.3.6 *Cost of travel between lines*

Shipboard surveys are often designed so that transects span the full width of the study area. Ship time is typically expensive, so that designs are required for which there is no dead time in travelling from one line to the next. Designs using continuous line samplers are therefore required. Such designs are also useful for aerial surveys if line separation in a parallel-line design would be large. Zigzag sampler designs (Strindberg and Buckland in press) offer a solution to this problem. These designs are similar to a systematic parallel-line design, but the lines are slanted so that the end of one line meets the start of the next (Fig. 7.14). Zigzag samplers raise a number of issues which we address below.

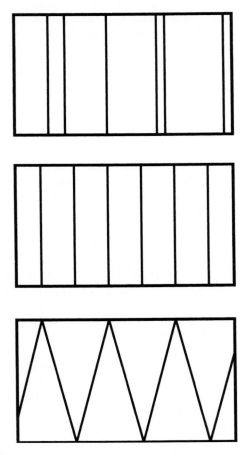

Fig. 7.14. Three commonly used design algorithms for line transect sampling: randomly spaced parallel lines (top); systematically spaced parallel lines (middle); and a continuous zigzag sampler.

7.2.4 *Zigzag samplers*

For survey designs that comprise a series of parallel lines, either randomly spaced or systematically spaced with a random start, the observer must travel from one line to the next 'off-effort', that is, without searching for the objects of interest. Sometimes, to make better use of resources, searching is conducted on these transit legs. However, the random design is then compromised, usually with too much effort along the boundaries of the survey region where the object density may be atypical, so that a design-based estimation of abundance is biased. A better solution is to use continuous zigzag samplers. If successive legs of the zigzag are assumed to be separate samplers, the sample of lines is effectively a systematic sample. Design-based analysis methods typically assume that the separate legs are independent, so that variance estimation proceeds as if the legs are random lines, independently placed within the survey region (Buckland *et al.* 2001: 238).

Generally, zigzag samplers do not provide uniform coverage probability, yet the standard analysis methods assume that they do. This can lead to substantial bias in abundance estimates if animal density varies appreciably through the study area. We consider the properties of commonly used zigzag samplers, and demonstrate their potential for bias in estimates of animal abundance. (In Section 7.3, we address how abundance estimation should be modified to take account of uneven coverage probability.) We also construct a zigzag sampler within a convex survey region that has even coverage probability with respect to distance along a design axis, and show how to modify the method for non-convex survey regions. The following is summarized from Strindberg and Buckland (in press).

7.2.4.1 *Properties of zigzag samplers*
In this section, we assume that the survey region is convex. Methods for non-convex regions are considered later.

A zigzag sampler is defined with respect to a design axis G that, without loss of generality, runs parallel to the x-axis of a Cartesian coordinate system. We thus use x to indicate distance along the design axis, and y for distance perpendicular to the design axis. Let the angle of the sampler to the design axis G be either $\theta(x)$ or $180° - \theta(x)$ at distance x along the axis, where $\theta(x)$ may vary in the range $[0, 90°)$. Let L denote the total length of the sampler, $2w$ its width, and A the area of the survey region R. Let P_c denote the mean coverage probability, where $P_c = 2wL/A$.

We would like a zigzag sampler to provide even coverage probability, so that coverage probability at point (x, y) is $P_c(x, y) = P_c$ for all (x, y) within the survey region, in which case conventional design-based estimation is unbiased. To date, survey designs based on zigzag samplers have been

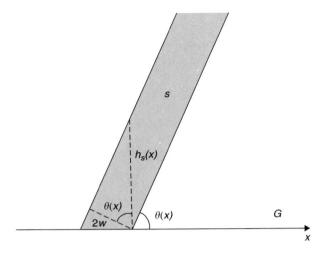

Fig. 7.15. The height $h_s(x)$ of the sampler s is determined by its width $2w$ and angle $\theta(x)$, where $h_s(x) = 2w/(\cos\theta(x))$.

ad hoc, with no proof that coverage probability is at least approximately uniform.

Let $h_s(x)$ denote the height of the zigzag sampler at location x along the design axis G. Fig. 7.15 shows that $h_s(x)$ is given by

$$h_s(x) = \frac{2w}{\cos\theta(x)}. \tag{7.1}$$

The coverage probability $P_c(x, y)$ achieved by a zigzag sampler typically varies spatially. We consider how it varies along the design axis, integrated over y: the 'horizontal component' $P_c(x) = \int P_c(x, y)dy$. Similarly, we use the term 'vertical component' to denote coverage probability at point y after integration over x: $P_c(y) = \int P_c(x, y)dx$. Let $H(x)$ denote the height of the survey region at x. For all types of zigzag design, the horizontal component is given by

$$P_c(x) = \frac{h_s(x)}{H(x)}. \tag{7.2}$$

This is illustrated in Fig. 7.16.

The vertical component $P_c(y)$ does not have an equivalent analytic form. Its value depends on the type of zigzag sampler, the survey region shape, and the total survey effort L. To ensure that coverage probability will be even on average, a possible solution is to select a random orientation for G, provided we select a zigzag sampler such that $P_c(x)$ as given by eqn (7.2) is

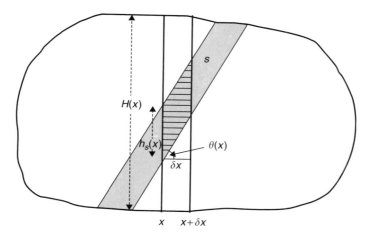

Fig. 7.16. Consider an infinitesimal strip perpendicular to the design axis G of width δx, height $H(x)$ and area (to first order) $H(x)\delta x$. The shaded parallelogram of area $h_s(x)\delta x$ indicates the covered proportion of the strip. The marginal coverage probability $P_c(x)$ is therefore given by $P_c(x) = h_s(x)\delta x / H(x)\delta x = h_s(x)/H(x)$.

constant. The algorithms for generating the zigzag samplers of this chapter are readily automated, so it is easy to estimate by simulation the combined vertical and horizontal coverage probability as a function of location. If this were considered to vary too much from a constant probability, a Horvitz–Thompson estimator of abundance can be constructed that does not require the assumption of an even coverage probability (Section 7.3).

Zigzag samplers are subject to constraints that must be taken into account by automated algorithms. Consider a survey region R. Any zigzag sampler s is orientated with respect to the design axis G. In general, the sampler angle $\theta(x)$ varies with distance x along G. Suppose the boundary of the survey region is defined by a set of straight edges $E = \{e_1, e_2, \ldots, e_M\}$, and let $\underline{\phi} = \{\phi_1, \phi_2, \ldots, \phi_M\}$ denote the set of angles formed by each of the e_k with G. Let E_U denote the set of 'upper' edges that can be intersected by s as it moves in the direction of increasing y. Let E_L denote the set of 'lower' edges that can be intersected by s as it moves in the direction of decreasing y.

Depending on whether an edge e_k $(k = 1, \ldots, M)$ belongs to the set E_U or E_L and the angle ϕ_k of e_k, there is a corresponding allowable range for the zigzag angle $\theta(x)$ that prevents the sampler s from 'bouncing outside' the survey region R. Figure 7.17 shows four categories of survey region edge along with the corresponding allowable sampler angle range. Thus, if we consider a survey region R with edge angles $\underline{\phi}$ for a selected design

Edge set	$e_k \in E_U$	$e_k \in E_L$	$e_k \in E_U$	$e_k \in E_L$
Edge angle	$0° \leq \phi_k < 90°$		$90° < \phi_k \leq 180°$	
Required angle	$\phi_k \leq \theta(x) < 90°$		$180° - \phi_k \leq \theta(x) < 90°$	
Access quadrant	IV	II	III	I
Diagram	1.	2.	3.	4.

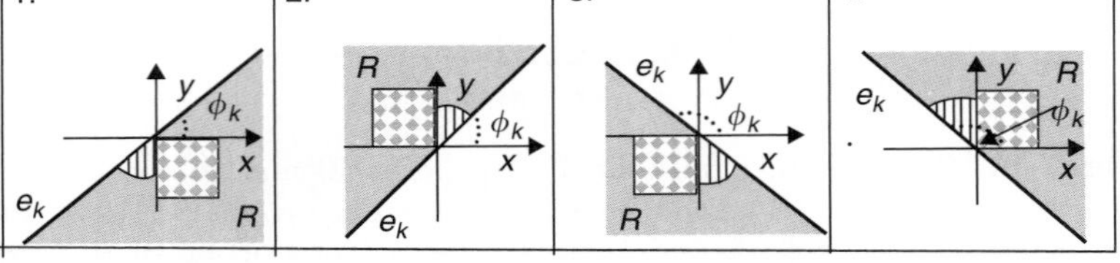

Fig. 7.17. A survey region edge e_k belongs to either the set of 'upper' or the set of 'lower' edges (E_U or E_L) that can be intersected by the zigzag sampler s as it moves in the direction of increasing or decreasing y, respectively. Consider a coordinate system whose origin O can be located at any point along e_k. The sampler angle (either $\theta(x)$ or $180° - \theta(x)$) may vary with distance x along the design axis G. The restrictions on $\theta(x)$, which ensure that s remains within the boundary of R as it intercepts e_k, are determined by its edge set and the angle ϕ_k (indicated by the dotted line in the diagrams) of e_k. As s approaches O from the left (as shown in diagrams 1 and 4) or from the right (diagrams 2 and 3), $\theta(x)$ must be such that s passes through the portion of the quadrant indicated by the vertical line fill, while the quadrant indicated by the diamond fill is completely accessible (i.e. s can take on any orientation in that quadrant and still remain within the boundary of R).

axis orientation, the smallest possible sampler angle (corresponding to the least effort L), which ensures that s remains within R, can be calculated by considering the edge that has the greatest absolute slope with respect to G (any edge e_k perpendicular to G, that is, $\phi_k = 90°$, is not considered, as the sampler is not required to 'bounce off' such edges). Let ϕ denote the angle that such an edge forms with G. Then the smallest possible allowable sampler angle is $\theta = \phi$ if $0 \leq \phi < 90°$ or $\theta = 180° - \phi$ if $90° < \phi \leq 180°$. (Note that the boundary is taken to be part of the survey region and the sampler is allowed to coincide with the boundary.) The minimum amount of effort, that is, sampler length L, required to avoid bouncing outside R can be estimated as

$$L = \frac{x_{\max} - x_{\min}}{\cos \theta}, \tag{7.3}$$

where $x_{\min}$ and $x_{\max}$ define the minimum and maximum values of x within the survey region R. When the ends of R are not lines perpendicular to G, eqn (7.3) is an approximation and will overestimate the amount of effort required.

For some survey regions, when the available survey effort is limited, it will be impossible to avoid a boundary and sampler orientation combination that forces the sampler out of R. This problem could be resolved by letting the sampler run along the boundary to minimize the discontinuity, while maintaining the same total line length, although this solution is likely to lead to over-sampling along the edge. The effect on coverage probability could be assessed by simulation. Depending on the shape of the survey region, it may be possible to avoid the problem by suitable choice of orientation of the design axis. Failing this, a pragmatic option is to approximate the shape of the survey region in a way that avoids the problem.

7.2.4.2   *The equal-angle zigzag sampler*

For the equal-angle zigzag sampler, $\theta(x) = \theta$ for all x within R. The design may be implemented by selecting a random (uniformly distributed) point along the design axis, and a sampler orientation with respect to the design axis of θ or $180° - \theta$, with equal probability. The sampler is extended from the selected point in both directions until it meets the upper and lower region boundary. At each point of intersection of the sampler with the boundary (termed a 'waypoint'), the sampler 'bounces off' the boundary. This is achieved by reflecting the sampler in a line perpendicular to G passing through the intersection of the sampler with the boundary. This procedure is continued until the sampler covers the entire x-value range of G within R. Designs based on an equal-angle zigzag sampler are denoted by S_{EAZ}.

An equal-angle zigzag sampler is generated using a specified value for either θ or L. We may estimate θ given L or vice versa using eqn (7.3). As θ is constant, eqn (7.1) yields

$$h_s(x) = \frac{2w}{\cos \theta} = h \text{ say}$$

from which eqn (7.2) gives

$$P_c(x) = \frac{h}{H(x)} \quad \forall x \in R.$$

Thus the coverage probability with respect to distance along the design axis is inversely proportional to the height of the survey region.

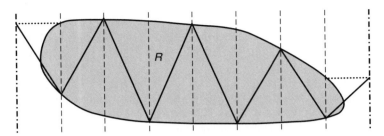

Fig. 7.18. The construction of an equal-spaced zigzag sampler (solid line) using the intersection points of a systematically spaced set of parallel lines (dashed lines) with the boundary of R. The sample end segments are constructed using the intersection of the dotted and dot-dashed line.

7.2.4.3   *The equal-spaced zigzag sampler*

To construct the equal-spaced zigzag sampler, a set of equally spaced parallel lines is constructed perpendicular to, and randomly positioned along, the design axis G and spanning its full length in R. Alternate intersections of these lines with the region boundary form the waypoints of the zigzag sampler, as illustrated in Fig. 7.18. Thus if the parallel lines pass north to south, the zigzag is formed by linking the north intersection of one line with the boundary to the south intersection of the next line, and so on. This fails to define the sections of sampler between $x_{\min}$ and the first parallel line, and between the last line and $x_{\max}$. We chose to define these sections as shown in Fig. 7.18. We denote designs based on an equal-spaced zigzag sampler by S_{ESZ}.

An equal-spaced zigzag sampler thus has waypoints equally spaced along the design axis. For a specified spacing c, it is possible to estimate the expected total sampler length for a realization of the design. The actual sampler length varies about this value, depending on the shape of the survey region and the particular realization of the design. Similarly, for a fixed total zigzag length L, an estimate of the spacing required to generate that total line length can be calculated. To estimate L given c or vice versa, the following formula is applied:

$$L = \frac{x_{\max} - x_{\min}}{c} \sqrt{\left(\frac{A}{x_{\max} - x_{\min}} \right)^2 + c^2}. \tag{7.4}$$

Again, the value for L or c is approximate unless the ends of R are lines perpendicular to G.

The equal-spaced zigzag sampler can be partitioned into K segments. For each sampler segment s_k, the sampler angle $\theta(x)$ remains constant.

That is, $\theta(x) = \theta_k \ \forall x \in s_k, k = 1, \ldots, K$. Thus for each sampler segment s_k, from eqn (7.1), the height h_k of s_k with respect to G is given by

$$h_s(x) = h_k = \frac{2w}{\cos \theta_k} \quad \forall x \in s_k, \ \ k = 1, \ldots, K.$$

Equation (7.2) yields

$$P_c(x) = P_{ck} \text{ say } = \frac{h_k}{H(x)} \quad \forall x \in s_k, \ \ k = 1, \ldots, K.$$

As for the equal-angle zigzag sampler, the coverage probability varies with distance along the design axis. It is inversely proportional to the height of the survey region within segment k. However, the constant of proportionality h_k differs between sampler segments. This allows a closer approximation to even coverage probability, with a smaller value for h_k in shorter segments.

7.2.4.4   The adjusted-angle zigzag sampler

The equal-angle zigzag and, to a lesser degree, the equal-spaced zigzag do not achieve an even coverage probability because the surface area of a sampler sub-section falling within any strip of R is not proportional to the surface area of that strip. We can construct a zigzag sampler that does have this property. From Fig. 7.16, we simply need to ensure that sampler height $h_s(x)$ is proportional to survey region height $H(x)$. Then $P_c(x) = P_c$ and from eqns (7.1) and (7.2), together with the approximation for mean coverage probability, this gives

$$\frac{2w}{\cos \theta(x) \times H(x)} = \frac{2w \times L}{A}.$$

Thus, the angle $\theta(x)$ must be continually adjusted in proportion to the survey region height $H(x)$. At any position along G, the angle of the sampler is given by

$$\theta(x) = \cos^{-1}\left(\frac{A}{LH(x)}\right), \quad \forall x \in R_s \subset R, \quad \frac{A}{LH(x)} \leq 1 \ \forall x, \qquad (7.5)$$

where R_s denotes the sampled sub-region of R.

Note that even coverage probability for the above continuous sampler is achievable only if $H(x) \geq A/L$ for all $x \in R$. As for other zigzag samplers, the procedure relies on being able to 'bounce' off a survey region boundary at an angle $\theta(x)$ to G. This is possible only when the absolute value of the slope of the boundary of R is smaller than the absolute value of the slope of the sampler. This condition always holds when $\theta(x) \to 90°$, which occurs as $L \to \infty$. For smaller values of $\theta(x)$, sub-regions may occur in which a continuous sampler with constant coverage probability along the design axis is not achievable. When $H(x) = A/L$, $\theta(x) = 0°$ and at smaller values of $H(x)$, even a sampler that is parallel to the design axis over-samples an infinitesimal strip at x. For most surveys, bias from this source is likely to be minimal, but if it is considered to be a significant problem, a solution is to reduce search effort along this section of transect, for example, by travelling 'off-effort' for short segments.

In the special case of a rectangular survey region of height H and width W, the sampler angle is given by $\theta(x) = \cos^{-1}(H \times W/L \times H) = \cos^{-1}(W/L)$ $(L \geq W)$, independent of x. In this case, all three zigzag samplers considered here are equivalent.

To implement the adjusted-angle zigzag sampler, first select a starting point at random within the survey region, and determine the corresponding angle $\theta(x)$ from eqn (7.5). Select from $\theta(x)$ and $180° - \theta(x)$ with equal probability to determine the angle of the sampler to the design axis G. Extend the sampler from the selected point in both directions along G, while continually adjusting the angle according to eqn (7.5). When it intersects the upper or lower boundary of R, it is reflected in a similar manner to the equal-angle zigzag, except that the sampler angle is determined by the height of R at that x-coordinate. This procedure is continued until the sampler covers the entire x-value range of G. We denote designs based on an adjusted-angle zigzag sampler by S_{AAZ}.

7.2.4.5 *Comparison of zigzag samplers*

As noted previously, general practice is to estimate animal abundance assuming that the zigzag sampler design attains even coverage probability P_c. Under that assumption, and given the knowledge of animal density $D(x, y)$ within the area covered by the sampler, we could obtain a Horvitz–Thompson estimator (Horvitz and Thompson 1952) of abundance as

$$\widehat{N} = \frac{1}{P_c} \int_{x_{\min}}^{x_{\max}} \int_{y_s(x)}^{y_s(x) + h_s(x)} D(x, y) \, dy dx,$$

where $x_{\min}$ and $x_{\max}$ define the full range of x within the survey region, and $y_s(x)$ is the lower y-extreme of the sampler at distance x along the design axis. This equation can equivalently be written as

$$\widehat{N} = \frac{1}{P_c} \int_R D(\underline{u}) z(\underline{u}) \, d\underline{u}, \tag{7.6}$$

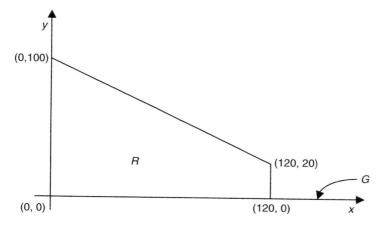

Fig. 7.19. A trapezoidal survey region R, whose design axis G coincides with the x-axis.

where $z(\underline{u})$ is an indicator function, equal to one if $\underline{u} = (x, y) \in R_s$, and zero otherwise.

These results are not useful for analysis of real survey data, because $D(x, y)$ (equivalently $D(\underline{u})$) cannot be observed. However, they allow us to evaluate bias arising from assuming that the equal-angle and equal-spaced zigzag samplers provide even coverage probabilities, given an assumed form for $D(\underline{u})$, without the need to simulate. Strindberg (2001) evaluated the bias in $\widehat{N}$ for three cases: $D(\underline{u}) = \alpha x + \beta \ \forall \underline{u} \in R$, where α and β are positive constants; $D(\underline{u}) = \alpha x + \beta y + \gamma xy + \delta \ \forall x, y \in R$, where α, β, γ, and δ are positive constants; and $D(\underline{u}) = \alpha(x)y^2 + \beta(x)y + \gamma(x) \ \forall x, y \in R$, where $\alpha(x) = 4[D_{\max}(x) - D_{\min}(x)]/[H(x)]^2$, $\beta(x) = -\alpha(x)H(x)$, $\gamma(x) = D_{\max}(x)$ and $D_{\min}(x)$ and $D_{\max}(x)$ are the minimum and maximum density with respect to y for a given x, to be specified as functions of x. The functional form of $D_{\max}(x)$ and $D_{\min}(x)$ is such that $D_{\max}(x) > D_{\min}(x) > 0 \ \forall x \in R$. The survey region was taken to be a trapezium with vertices at $(0, 0)$, $(0, 100)$, $(120, 20)$, and $(120, 0)$, respectively (Fig. 7.19).

Each of the three zigzag samplers was constructed with the same total length L. To compare them visually, we start all three from the origin in Fig. 7.20, although in practice, their location within the survey region would be random. The curve of the adjusted-angle zigzag becomes especially apparent in Fig. 7.20 as the height of the survey region becomes small. Also apparent is the much better approximation of the equal-spaced zigzag sampler to the adjusted-angle sampler than that of the equal-angle sampler. This is reflected in Fig. 7.21, which shows that the equal-spaced sampler provides less variable coverage probability than does the equal-angle sampler.

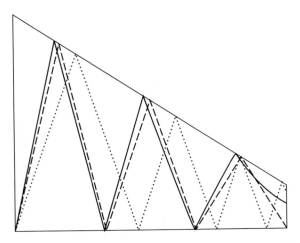

Fig. 7.20. A trapezoidal survey region illustrating three types of zigzag sampler: equal-angle (dotted line), equal-spaced (dashed line), and adjusted-angle (solid line). The design axis runs parallel to the base of the trapezium.

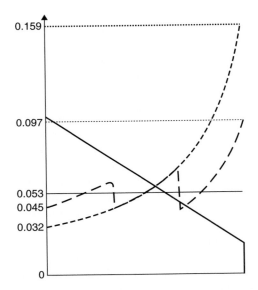

Fig. 7.21. Coverage probability $P_c(x)$ plotted against x for the three zigzag samplers in Fig. 7.20. Also shown is the height of the trapezium (not to scale) as a function of distance along the design axis, which indicates that the equal-angle zigzag (short dashes) has too low coverage probability where the survey region is wide, and too high where it is narrow. For the equal-spaced zigzag (long dashes), the coverage probability changes for each change in sampler angle. The coverage probability for the adjusted-angle zigzag (solid line) remains constant, because the angle varies as a smooth function of the trapezium height.

Table 7.1. Percentage bias in $\widehat{N}$ arising from use of the equal-angle S_{EAZ} and equal-spaced S_{ESZ} zigzag samplers under the model $D(\underline{u}) = \alpha x + \beta$

α	β	N	% bias, S_{EAZ}	% bias, S_{ESZ}
2	5	708,000	27.1	5.8
0.2	5	103,200	18.6	4.0
0.02	5	42,720	4.5	1.0
2	0.5	675,600	28.4	6.1
0.2	0.5	70,800	27.1	5.8
0.02	0.5	10,320	18.6	4.0
2	0.05	672,360	28.6	6.1
0.2	0.05	67,560	28.4	6.1
0.02	0.05	7,080	27.1	5.8

Table 7.2. Percentage bias in $\widehat{N}$ arising from use of the equal-angle S_{EAZ} and equal-spaced S_{ESZ} zigzag samplers under the model $D(\underline{u}) = \alpha x + \beta y + \gamma xy + \delta$ with $\delta = 0.005$

α	β	γ	N	% bias, S_{EAZ}	% bias, S_{ESZ}
0.2	0.03	0.01	165,876	16.8	3.5
0.2	0.03	0.001	83,796	22.9	4.9
0.2	0.003	0.01	159,180	18.0	3.8
0.2	0.003	0.001	77,100	26.0	5.6
0.02	0.03	0.01	105,396	10.0	2.1
0.02	0.03	0.001	23,316	8.2	1.7
0.02	0.003	0.01	98,700	11.6	2.4
0.02	0.003	0.001	16,620	16.8	3.5
0.002	0.03	0.01	99,348	8.9	1.8
0.002	0.03	0.001	17,268	1.1	0.2
0.002	0.003	0.1	913,452	10.5	2.2
0.002	0.003	0.01	92,652	10.5	2.1
0.002	0.003	0.001	10,572	10.0	2.1

Using eqn (7.6), the zigzag samplers generate the biases of Tables 7.1, 7.2, and 7.3 under the three models listed above. These evaluations conform with expectation, with the equal-spaced sampler (S_{ESZ}) yielding much lower bias under the assumption of even coverage probability than the equal-angle sampler (S_{EAZ}).

Table 7.3. Percentage bias in $\widehat{N}$ arising from use of the equal-angle S_{EAZ} and equal-spaced S_{ESZ} zigzag samplers under the model $D(\underline{u}) = \alpha(x)y^2 + \beta(x)y + \gamma(x)$ with $D_{\min}(x) = \alpha_d x + \beta_d$, $D_{\max}(x) = \gamma_d D_{\min}(x)$, $\forall x \in R$

α_d	β_d	γ_d	N	% bias, S_{EAZ}	% bias, S_{ESZ}
0.02	3	10	113,280	6.8	1.5
0.02	3	0.1	19,824	6.8	1.5
0.02	3	0.01	18,974	6.8	1.5
0.02	0.3	10	35,520	21.6	4.6
0.02	0.3	0.1	6216	21.6	4.6
0.02	0.3	0.01	5950	21.6	4.6
0.02	0.03	10	27,744	27.7	5.9
0.02	0.03	0.1	4855	27.7	5.9
0.02	0.03	0.01	4647	27.7	5.9
0.002	3	10	89,088	0.9	0.2
0.002	3	0.1	15,590	0.9	0.2
0.002	3	0.01	14,922	0.9	0.2
0.002	0.3	10	11,328	6.8	1.5
0.002	0.3	0.1	1982	6.8	1.5
0.002	0.3	0.01	1897	6.8	1.5
0.002	0.03	10	3552	21.6	4.6
0.002	0.03	0.1	622	21.5	4.7
0.002	0.03	0.01	595	21.7	4.7

7.2.4.6 *Non-convex survey regions*

We illustrate strategies for non-convex regions using a complex ice-edge stratum from a survey of minke whales in the Antarctic. The ice-edge is very irregular, and can move by up to 100 km in a single day, so it is not possible to lay down the design in advance. Instead, an algorithm must be defined that allows the transects to be determined as the survey progresses.

For a non-convex region, some areas may be inaccessible to a zigzag sampler, and the properties explored above do not hold. When departure from the convexity requirement is mild, the problem may be overcome by designing the survey within a convex hull around the survey region. Sections of sampler that fall outside the survey region are then not sampled. However, Fig. 7.22 shows that this solution is far from satisfactory for our example, which represents rather extreme departure from convexity. Another solution to the problem is to divide the stratum into sub-regions, each of which is convex. This strategy is often successful, but again, for the ice-edge stratum of Fig. 7.22, far too many sub-regions would be required. However, by combining these two strategies and defining convex hulls around almost convex sub-regions of the original survey region, a reasonable design can be found (Fig. 7.23). In this case, seven sub-regions were defined, and a convex hull was established around each one. The solution is not ideal, but fortunately survey regions are seldom this complex.

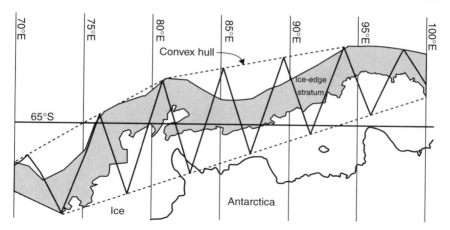

Fig. 7.22. A complex ice-edge stratum in the Antarctic. A single realization of an adjusted-angle zigzag sampler is shown in the convex hull of the stratum. The sampler is discontinuous and thus inefficient.

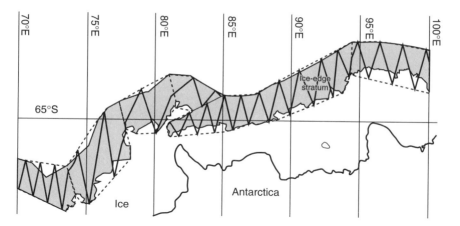

Fig. 7.23. The ice-edge stratum of Fig. 7.22 has been divided into seven almost convex sub-regions, and a zigzag sampler placed within the convex hull of each.

7.2.4.7 *Discussion*

For populations known to exhibit a density gradient, the zigzag sampler orientation should usually be chosen so that each leg as far as possible cuts across all density levels. This minimizes variation in encounter rate among lines (Buckland *et al.* 2001: 238–9), so that better precision in the estimated abundance is achieved. For zigzag designs, this is accomplished by placing the design axis approximately parallel to the density contours. In our example, minke whales have highest density along the ice-edge, so that the design axis is chosen to be roughly parallel to the ice-edge.

If nothing is known about density gradients within the population, and zigzag legs are used as sampling units, then a sampler orientation that maximizes the number of legs is advantageous. This increases the sample size for estimating variance in encounter rate (assuming that the sampling unit is taken to be a single leg of the zigzag sampler) and yields a more precise estimate of variance.

Zigzag samplers have an associated width, so overlap at the boundaries of the survey region occurs between successive legs of the sampler. This overlap is not a problem provided detected animals falling within the area of overlap are recorded for just one of the legs and that the effort in the area of overlap contributes to total effort only once. Another problem is that part of the sampler may fall outside the survey region at the boundary. This can lead to a coverage probability that is smaller than expected. Usually, the sampler width is very small compared to the sampler length, so that overlap and other edge effects are negligible.

7.3   Estimation for uneven coverage probability designs

Standard line transect analyses assume that samplers have even coverage probability throughout the survey region. (An unrealistic alternative is to assume that the objects of interest are uniformly distributed through the survey region.) Provided coverage probability is known or can be estimated, analysis proves to be relatively straightforward for samplers that do not have even coverage probability (Strindberg and Buckland in press). However, estimation becomes less precise as variability in coverage probability increases, and biased if any parts of the survey region have zero coverage probability. We allow for uneven coverage probability using a Horvitz–Thompson estimator of abundance. Cooke (1985) was the first to suggest this approach for line transect sampling. When coverage probability is uneven, it is also usually unknown. However, given automated algorithms for generating the samplers, we can readily estimate coverage probability at any chosen point by simulation. To do this, we repeatedly generate realizations of a given survey design, and observe the proportion of realizations for which the resulting samplers cover the chosen point. For estimating object abundance, we must obtain this proportion for all points at which objects were detected. These coverage probabilities are estimated, whereas the Horvitz–Thompson estimator assumes that they are known. However, this presents no practical difficulty as we can estimate them to any desired precision, simply by increasing the number of realizations of the survey design that we generate. In practice, except for strip transect surveys in which all objects within w of the line are counted, the effective half-width of search μ must also be estimated, and its precision is determined by the survey data, so that variance estimation should take this into account (below).

7.3.1   *Objects that occur singly*

In conventional line transect sampling, object abundance N is estimated by

$$\widehat{N} = \frac{An}{2\hat{\mu}L}, \tag{7.7}$$

where A is the area of the survey region, n is the number of animals detected, $\hat{\mu}$ is the estimated effective strip half-width, and L is the total length of line (Buckland *et al.* 2001). This may be obtained as a Horvitz–Thompson-like estimator as follows.

   If an object i is located outside the sampled strips, it cannot be sampled. If it is inside the strip, and we do not condition on its distance from the line, the probability that it is detected is estimated by $\hat{\mu}/w$ where w is the half-width of the strip (Buckland *et al.* 2001: 37–8). Let P_c denote the mean coverage probability, where $P_c \approx 2wL/A$ for small w. (Due to edge effects, the exact value includes an additional term in w^2 that depends on the angle of the sampler to the design axis and on the shape of the survey region. Modified designs with mean coverage probability exactly equal to $2wL/A$ are possible, but they introduce bias if mean animal density at the boundary differs from that inside the region.) Assuming even coverage probability, the probability that a sampled strip includes a given animal is P_c. Hence the estimated inclusion probability $\widehat{P}_i$ of animal i is the product of these two quantities, and

$$\widehat{N} = \sum_{i=1}^{n} \frac{1}{\widehat{P}_i} = \sum_{i=1}^{n} \frac{1}{(\hat{\mu}/w) \cdot P_c} \approx \sum_{i=1}^{n} \frac{1}{(\hat{\mu}/w) \cdot (2wL/A)} = \frac{An}{2\hat{\mu}L}. \tag{7.8}$$

We now allow the probability that a sampled strip covers an object to vary by location. Denote the coverage probability at the location of detected object i by P_{ci}. Then

$$\widehat{N} = \sum_{i=1}^{n} \frac{1}{\widehat{P}_i} = \sum_{i=1}^{n} \frac{1}{(\hat{\mu}/w) \cdot P_{ci}} = \frac{w}{\hat{\mu}} \sum_{i=1}^{n} \frac{1}{P_{ci}}. \tag{7.9}$$

If $P_{ci} = P_c$ for all i (even coverage probability), this gives the same result as before. This equation may be further generalized by allowing animals within the strip to have their own individual effective strip half-widths μ_i:

$$\widehat{N} = \sum_{i=1}^{n} \frac{1}{\widehat{P}_i} = \sum_{i=1}^{n} \frac{1}{(\hat{\mu}_i/w) \cdot P_{ci}} = w \sum_{i=1}^{n} \frac{1}{\hat{\mu}_i P_{ci}}. \tag{7.10}$$

The μ_i may be modelled as a function of their distance from the line and of other covariates (Chapter 3).

7.3.2 *Objects that occur in clusters*

If objects occur in clusters, and the ith detected cluster is of size s_i objects, then the formulae of the previous section yield estimates of cluster abundance. Object abundance is estimated simply by replacing $\sum 1/\widehat{P}_i$ by $\sum s_i/\widehat{P}_i$ in these formulae. Hence eqn (7.9) becomes

$$\widehat{N} = \sum_{i=1}^{n} \frac{s_i}{\widehat{P}_i} = \frac{w}{\hat{\mu}} \sum_{i=1}^{n} \frac{s_i}{P_{ci}}. \tag{7.11}$$

Note that this assumes that probability of detection is independent of cluster size s_i. If it is not, we need to use eqn (7.10), modified similarly:

$$\widehat{N} = \sum_{i=1}^{n} \frac{s_i}{\widehat{P}_i} = w \sum_{i=1}^{n} \frac{s_i}{\hat{\mu}_i P_{ci}}. \tag{7.12}$$

7.3.3 *Variance estimation*

Variances for the above estimators may be estimated by bootstrapping transects (Buckland *et al.* 2001: 83–4). For zigzag survey designs, these are usually taken to be the separate legs of the zigzag sampler. This tends to be more robust than the analytic variance for the Horvitz–Thompson estimator, because it does not require the assumption that different detections on the same transect are independent, and uncertainty due to estimating μ is readily incorporated into the bootstrap variance. The simple percentile method is generally adequate for estimating confidence limits (Buckland *et al.* 2001: 82).

If analytic variances are preferred, then care is needed to include components to allow for estimation of inclusion probabilities (Huggins 1989; Borchers 1996), and for extrapolation from the surveyed strips to the entire survey region (Innes *et al.* 2003), as noted in Chapter 3.

7.4 Choosing between survey designs by simulation

Given GIS capability, together with automated design algorithms and analysis methods for when coverage probability is uneven, we now need methods to help choose between what could be a bewildering array of options. We would like to know which is the optimal design for our own population of interest, but in practice seek to identify an option that is a reasonable approximation to this. To proceed, we need to be able to generate simulated populations of objects that mimic our population of interest. We could generate a few populations with differing properties, in an attempt to span the properties of our real population, and identify an algorithm that performs well across the range of populations. We need also to identify criteria to measure how well algorithms perform.

If we have a past survey of the population, we can use the spatial modelling methods of Chapter 4 to estimate a density surface, and then simulate objects using the fitted spatial model. In that case, a subset of bootstrap replicates could be used to allow generation of a range of populations. The model could also be modified subjectively, to take account of known or suspected changes in population abundance or distribution since the last survey.

In the absence of a past survey, a spatial model could be selected, and its parameter values chosen to match what is known of the population of interest. The less that is known, the wider the range of parameter sets that will be needed if we are to be confident that at least one of our sets provides an adequate approximation to the true density surface. Objects can be simulated from each of the selected parameterizations as for the case in which the spatial model is fitted to past survey data.

The above approach allows smooth changes in density through the study area to be generated. With sufficiently flexible models such as generalized additive models, 'patches' of higher density can also be created. However, for many purposes, we may wish to simulate populations of objects that are patchily distributed, for example, when the presence of one object at a given location enhances the chance that other objects will occur nearby. For this purpose, we might use the Neyman–Scott process. Given past survey data, we might fit the process, for example, using the methods of Hagen and Schweder (1995) and Schweder et al. (1997) or of Brown and Cowling (1998) and Cowling (1998) (but note the correction to this latter paper in Aldrin et al. 2003). Otherwise, we might assign suitable values to the parameters of the Neyman–Scott process to provide a range of levels of patchiness of objects that is expected to span the level of patchiness of the real population.

In the Neyman–Scott process, 'parents' (patches here) are assumed to be distributed according to a Poisson distribution. Objects are then generated by simulating 'offspring' for each parent, whose number is determined by a second Poisson distribution with rate λ_o, and whose locations are determined by simulating values from a bivariate normal density. We assume that we have an estimate $\hat{\lambda}_o$ of λ_o, and that the 'offspring' represent the objects of interest (so that 'parents' are unused). We can therefore combine a model for spatial trend with one for patchiness by assuming that the spatial model for density has Poisson errors. If the estimated number of objects in the survey region under that model is $\hat{N}$, we start by generating a number from a Poisson distribution with parameter $\hat{N}/\hat{\lambda}_o$. We then generate this number of 'parents' through the survey region, by simulating observations from the assumed spatial model. For each 'parent', the number of 'offspring' is obtained by simulating values from a Poisson distribution with parameter $\hat{\lambda}_o$. The objects and their locations are then obtained by distributing these 'offspring' according to the bivariate normal distribution, whose variance–covariance matrix must be estimated or specified. When fitting to real data, the distribution of 'offspring' about the 'parents' might be assumed to be radially symmetric (*e.g.* Cowling 1998), so that a single variance need be estimated. More generally, the 'offspring' could be distributed elliptically about the 'parents' by suitable choice of the variances and covariance of the bivariate normal. If required, greater generality could be achieved by use of distributions other than the Poisson for the 'parents' and/or the 'offspring'.

Having simulated a range of populations, we can generate a number of realizations from each competing design algorithm, implement each realization on each population, and sample from that population using simulation. The data from each of these simulations can then be analysed using the methods that are to be used on the real data (but probably with a greater degree of automation!), and summaries of the population abundance estimates extracted.

We would like to select an algorithm with high precision and low bias, so perhaps the most useful summary statistic would be the root mean squared error (RMSE)

$$\text{RMSE}_a = \sqrt{\sum_{j=1}^{J} (\widehat{N}_{ja} - N_j)^2 / J}, \tag{7.13}$$

where subscript a denotes which design algorithm is being used, J is the number of simulated data sets analysed, N_j is the size of the population for data set j, and $\widehat{N}_{ja}$ is estimated abundance of the population corresponding to data set j, using data generated under design algorithm a. We can tabulate RMSE_a for each population separately, and across all populations. The algorithm with the lowest overall RMSE might then be selected, although an algorithm with a slightly worse performance might be preferred if its performance varied less between populations. It may also be useful to tabulate variance ($s^2 = [\sum_{j=1}^{J}(\widehat{N}_{ja} - \overline{\widehat{N}}_a)^2]/(J-1)$, where $\overline{\widehat{N}}_a$ is the sample mean of the J estimates $\widehat{N}_{ja}$) and bias ($\widehat{B} = \overline{\widehat{N}}_a - \overline{N}$ where $\overline{N}$ is the mean over the J data sets of the N_j) separately; note the relationship $\text{RMSE} = \sqrt{(J-1)s^2/J + \widehat{B}^2}$.

8

Adaptive distance sampling surveys

J. H. Pollard and S. T. Buckland

8.1 Introduction

Distance sampling is especially effective for surveying objects that are
sparsely distributed through a large region. However, if those objects tend
to occur in a relatively small number of patches, standard distance sampling
may yield both a low number of detections (because much effort is expended
away from these patches) and high variance (because the main contributor
to variance of abundance estimates from distance sampling surveys is generally variance in encounter rate; patchiness increases this component of
variance appreciably).[1]

Adaptive sampling (Thompson 1990, 2002; Thompson and Seber 1996)
was designed for surveying objects whose spatial distribution is patchy.
It provides a mechanism for conducting additional survey effort in local-
ities where objects have been found. If objects tend to occur in patches,
this ensures that a substantially higher proportion of total effort is conducted in areas of high object density, thus significantly enhancing sample
size. Careful design and analysis ensures that subsequent estimators are
design-unbiased. By developing adaptive distance sampling methods, we
should therefore be able to extend the applicability of distance sampling to
populations that are distributed patchily as well as sparsely.

In adaptive sampling, an initial sample of units is selected at random.
If the number of observations in a sampled unit satisfies some *trigger condition*, then units in a pre-defined *neighbourhood* of the triggering unit are
also sampled. If any of the adaptive units in the neighbourhood meet the
condition, then the neighbourhood of each of these units is also sampled.
The process repeats until no newly sampled units meet the condition.

The combination of an initial unit and its associated adaptive units is
termed a *cluster*. Within the cluster, units which do not meet the trigger

---

[1] In spatial modelling, patchy distributions are often referred to as clustered. In this
chapter, we use the term 'patchy' in preference to 'clustered' because the term 'cluster'
in distance sampling is generally taken to be a well-defined group of objects.

condition are termed *edge units*, while units which meet the condition form a *network*. (Below, we do not use the term 'cluster' for a set of units, to avoid confusion with the use of 'cluster' to mean a group of animals.) Any initial unit that does not meet the condition is also a network, consisting of a single unit. The neighbourhood must have the property that, if any unit within a network is sampled, then all other units within the same network must be sampled.

The final sample will therefore consist of a network for each of the initially sampled units. However, networks from two or more separate initial units may merge into one larger network.

This chapter is based on the work of Pollard (2002) and Pollard *et al.* (2002). We first apply Thompson's (1990) methods to obtain design-unbiased estimators for adaptive point transect sampling. We then show how this is extended to line transect sampling. A practical limitation of adaptive surveys is that the amount of survey effort required is not known in advance. We therefore include a section on fixed-effort adaptive line transect sampling, in which the design-unbiased property of estimators is relaxed.

8.2   Design-unbiased adaptive point transect surveys

8.2.1   *Survey design*

For simplicity, we will assume that an initial random sample of points is selected. In point transect surveys, a systematic grid of points is often randomly located over the study area, but subsequent analyses assume that points were selected randomly. Strindberg (2001) shows that a similar strategy in line transect sampling generates abundance estimates with appreciably smaller root mean square errors (RMSE) than if lines are actually positioned at random. Because point transect surveys typically involve large numbers of points, the benefit of such a strategy is likely to be less, but still worthwhile.

Commonly, the design of a point transect survey comprises a systematic or random sample of lines, with points systematically spaced along each line. In this case, the lines are considered to be the initial random sample of units, and the adaptive methods described here are readily extended. Pollard (2002) discusses various design options in detail, and how analysis proceeds under each.

There are many ways in which units can be defined for point transect surveys. Three options, based on triangular, square, and hexagonal units, are illustrated in Figs 8.1–8.3.

Suppose that a grid of squares is located over the survey region, and that a systematic or random sample of squares is selected. The points are taken to be at the centres of the selected squares. We use the term *plot* for the circle of radius w centred on a selected point; we denote plot area

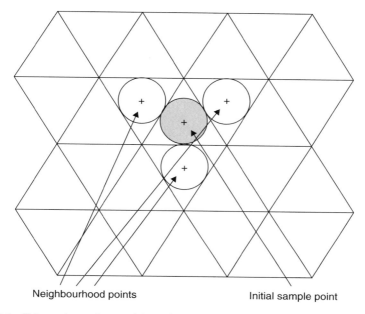

Neighbourhood points   Initial sample point

Fig. 8.1. Triangular units combine a low number of edge units with good spread to detect networks, at the expense of comparatively large gaps between plots.

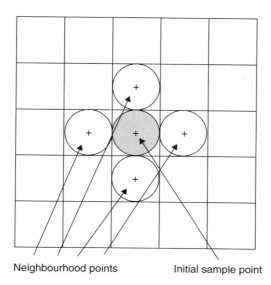

Neighbourhood points   Initial sample point

Fig. 8.2. Typically, square units are used, and the neighbourhood is defined to be the north, south, east, and west units, although many other combinations are possible.

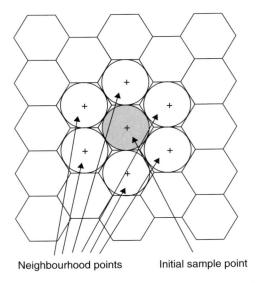

Neighbourhood points   Initial sample point

Fig. 8.3. Hexagonal units provide a good spread to detect patches, with minimal gaps between circles, although there is a comparatively high number of edge units.

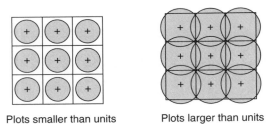

Plots smaller than units   Plots larger than units

Fig. 8.4. Point transect plots may be larger or smaller than the units they are located within. Thus the covered area, given by the total number of plots multiplied by the area of each plot, may be larger or smaller than the survey region. Shown here is part of a survey region in which plots are located on a grid of square sampling units.

by $A_p = \pi w^2$. Squares might be chosen to have side $2w$, so that adaptive plots touch the initial plot with which they are associated (as in Fig. 8.2), although there is no necessity for this. The grid squares are the sampling units. We assume that there are K such units in total. We define the covered area A_s as the area of each plot multiplied by K: $A_s = KA_p$. This will generally be smaller than A, the size of the survey region, but if plots overlap, it could be larger (Fig. 8.4).

8.2.2   *Estimation*

8.2.2.1   *Homogeneous detection probabilities*

In the following, we assume that animals occur in well-defined clusters, so that a detection refers to the detection of a cluster of animals. If animals occur singly, then estimates of cluster density and abundance derived below become estimates of animal density and abundance. We use adaptive sampling methodology (Thompson 1990) to estimate two key parameters: m_c, the expected number of clusters detected per plot, and m, the expected number of animals detected per plot. (In each case, this applies to the full set of plots; the sampled plots will be expected to yield more detections per plot, since the adaptive plots will tend to occur in areas of high density.) However, we pool detection distances across all sampled plots to estimate the detection function. This assumes that, given that a cluster is present in a unit, its probability of detection is independent of whether the unit was selected in the initial sample or was an adaptive unit.

We define

N_c = number of clusters in the population

$D_c = \dfrac{N_c}{A}$ = density of clusters

s_h = size of cluster $h, h = 1, \ldots, N_c$

$N = \displaystyle\sum_{h=1}^{N_c} s_h$ = number of animals in the population

$D = \dfrac{N}{A}$ = density of animals

M_c = mean number of clusters (whether detected or not) per plot

M = mean number of animals (whether detected or not) per plot

Hence, if we can obtain estimates $\widehat{M_c}$ and $\widehat{M}$, then $\widehat{D_c} = \widehat{M_c}/A_p$ and $\widehat{D} = \widehat{M}/A_p$, where $A_p = \pi w^2$. Further, $\mathrm{cv}(\widehat{D_c}) = \mathrm{cv}(\widehat{M_c})$ and $\mathrm{cv}(\widehat{D}) = \mathrm{cv}(\widehat{M})$.

Standard adaptive sampling methods assume that all animals are observed within a sampled unit. Thompson and Seber (1996: 211–33) extend the methods so that an animal within a sampled unit is assigned a probability of detection, as in distance sampling. We start by assuming that probability of detection depends only on distance from the point. Then, unconditional on the specific design, all clusters have the same probability of detection. Let P_a denote the probability of detection of a cluster, given that it lies within one of the sampled plots, but unconditional on its location in that plot. Then we can write

$$m_c = P_a M_c,$$

where m_c is the expected number of clusters detected for a randomly selected plot. Hence

$$\widehat{M}_c = \frac{\hat{m}_c}{\widehat{P}_a}. \tag{8.1}$$

Assuming independence between $\hat{m}_c$ and $\widehat{P}_a$, the delta method (Seber 1982: 7–9) yields the approximate result

$$\left[\mathrm{cv}(\widehat{M}_c)\right]^2 = [\mathrm{cv}(\hat{m}_c)]^2 + \left[\mathrm{cv}(\widehat{P}_a)\right]^2. \tag{8.2}$$

Similarly,

$$\widehat{M} = \frac{\hat{m}}{\widehat{P}_a} \tag{8.3}$$

with approximately

$$\left[\mathrm{cv}(\widehat{M})\right]^2 = [\mathrm{cv}(\hat{m})]^2 + \left[\mathrm{cv}(\widehat{P}_a)\right]^2, \tag{8.4}$$

where m is the expected number of animals detected on a randomly selected plot, estimated by $\hat{m}$.

If $g(r)$ is the probability that a cluster at distance r from a point is detected, then conventional point transect theory yields $P_a = \frac{\nu}{A_p}$, where $\nu = 2\pi \int_0^w r g(r) dr$ (Buckland *et al.* 2001: 55), and the corresponding estimators $\widehat{P}_a$ and $\mathrm{cv}(\widehat{P}_a)$. It remains therefore to find estimators $\hat{m}_c$ and $\hat{m}$. Note that, in the absence of size-biased detection, mean cluster size in the population $E(s)$ is estimated by $\widehat{E}(s) = \hat{m}/\hat{m}_c$. (If large clusters are more detectable than small clusters, this should be reflected in a model for P_a, which will then be a function of cluster size; $E(s)$ is then estimated by $\widehat{E}(s) = \widehat{M}/\widehat{M}_c$.)

Thompson (1990) considers two forms of estimator. The first uses the draw-by-draw probabilities that the initial sample will intersect a unit's network and is formed from the Hansen–Hurwitz estimator (Hansen and Hurwitz 1943). The second uses probabilities that the initial sample will intersect each detected network, and is based on the Horvitz–Thompson estimator (Horvitz and Thompson 1952). Pollard (2002) develops both estimators, with either a systematic or a random initial sample, and sampling either with replacement or without. In the following, we use the Horvitz–Thompson estimator, assuming that the initial sample is selected at random and without replacement.

The expected number of clusters detected per plot is estimated by

$$\hat{m}_c = \frac{1}{K} \sum_{i=1}^{\eta} \frac{\sum_{j=1}^{m_i} n_{ij}}{\alpha_i}, \tag{8.5}$$

where

$$\alpha_i = 1 - \frac{\binom{K - m_i}{k}}{\binom{K}{k}}$$

α_i = probability that network i is included in the sample

K = number of plots in the survey region

k = number of initial plots

η = number of distinct networks in the sample

m_i = number of plots in network i

n_{ij} = number of clusters detected on plot j within network i

The variance is estimated by

$$\widehat{\text{var}}(\hat{m}_c) = \frac{1}{K^2} \sum_{i=1}^{\eta} \sum_{i'=1}^{\eta} \frac{(\alpha_{ii'} - \alpha_i \alpha_{i'}) \sum_{j=1}^{m_i} n_{ij} \sum_{j'=1}^{m_{i'}} n_{i'j'}}{\alpha_i \alpha_{i'} \alpha_{ii'}} \tag{8.6}$$

with $\alpha_{ii} = \alpha_i$ and

$$\alpha_{ii'} = 1 - \frac{\binom{K - m_i}{k} + \binom{K - m_{i'}}{k} - \binom{K - m_i - m_{i'}}{k}}{\binom{K}{k}}$$

and $\text{cv}(\hat{m}_c) = (\sqrt{\widehat{\text{var}}(\hat{m}_c)})/(\hat{m}_c)$.

An estimate of the expected number of animals observed per plot is given by

$$\hat{m} = \frac{1}{K} \sum_{i=1}^{\eta} \frac{\sum_{j=1}^{m_i} \sum_{h=1}^{n_{ij}} s_{ijh}}{\alpha_i}, \tag{8.7}$$

where s_{ijh} is the number of animals in the hth detected cluster on the jth plot of the ith network.

The variance is estimated by

$$\widehat{\mathrm{var}}(\hat{m}) = \frac{1}{K^2} \sum_{i=1}^{\eta} \sum_{i'=1}^{\eta} \frac{(\alpha_{ii'} - \alpha_i \alpha_{i'}) \sum_{j=1}^{m_i} \sum_{h=1}^{n_{ij}} s_{ijh} \sum_{j'=1}^{m_{i'}} \sum_{h'=1}^{n_{i'j'}} s_{i'j'h'}}{\alpha_i \alpha_{i'} \alpha_{ii'}}$$

(8.8)

with $\mathrm{cv}(\hat{m}) = (\sqrt{\widehat{\mathrm{var}}(\hat{m})})/(\hat{m})$.

8.2.2.2 *Heterogeneous detection probabilities*

Thompson and Seber (1994, 1996: 228) present results for the case that each object has its own probability of detection. Adapting their results to distance sampling, we can extend the results of the previous section as follows.

$$\widehat{M_c} = \frac{1}{K} \sum_{i=1}^{\eta} \frac{\hat{u}_i}{\alpha_i},$$

(8.9)

where $\hat{u}_i = \sum_{j=1}^{m_i} \sum_{h=1}^{n_{ij}} 1/\hat{p}_{ijh}$ and $\hat{p}_{ijh}$ is the estimated probability of detection for the hth detected cluster on plot j of network i. The methods of Chapter 3 (or of Chapter 6 if detection on the line is uncertain) provide this estimate. The corresponding variance estimate is

$$\widehat{\mathrm{var}}(\widehat{M_c}) = \frac{1}{K^2} \left[\sum_{i=1}^{\eta} \sum_{i'=1}^{\eta} \frac{(\alpha_{ii'} - \alpha_i \alpha_{i'}) \hat{u}_i \hat{u}_{i'}}{\alpha_i \alpha_{i'} \alpha_{ii'}} + \sum_{i=1}^{\eta} \frac{1}{\alpha_i} \sum_{j=1}^{m_i} \sum_{h=1}^{n_{ij}} \frac{1 - \hat{p}_{ijh}}{\hat{p}_{ijh}^2} \right.$$
$$\left. + \sum_{i=1}^{\eta} \sum_{i'=1}^{\eta} \frac{1}{\alpha_{ii'}} \sum_{j=1}^{m_i} \sum_{h=1}^{n_{ij}} \sum_{j'=1}^{m_{i'}} \sum_{h'=1}^{n_{i'j'}} \frac{\widehat{\mathrm{cov}}(\hat{p}_{ijh}, \hat{p}_{i'j'h'})}{\hat{p}_{ijh}^2 \hat{p}_{i'j'h'}^2} \right]$$

with notation as before.

The expected number of animals per plot (whether detected or not) is estimated by

$$\widehat{M} = \frac{1}{K} \sum_{i=1}^{\eta} \frac{\hat{w}_i}{\alpha_i},$$

(8.10)

where $\hat{w}_i = \sum_{j=1}^{m_i} \sum_{h=1}^{n_{ij}} s_{ijh}/\hat{p}_{ijh}$. Estimated variance is

$$
\widehat{\text{var}}(\widehat{M}) = \frac{1}{K^2} \left[\sum_{i=1}^{\eta} \sum_{i'=1}^{\eta} \frac{(\alpha_{ii'} - \alpha_i \alpha_i')\hat{w}_i \hat{w}_{i'}}{\alpha_i \alpha_{i'} \alpha_{ii'}} + \sum_{i=1}^{\eta} \frac{1}{\alpha_i} \sum_{j=1}^{m_i} \sum_{h=1}^{n_{ij}} \frac{(1 - \hat{p}_{ijh}) s_{ijh}^2}{\hat{p}_{ijh}^2} \right.
$$

$$
\left. + \sum_{i=1}^{\eta} \sum_{i'=1}^{\eta} \frac{1}{\alpha_{ii'}} \sum_{j=1}^{m_i} \sum_{h=1}^{n_{ij}} \sum_{j'=1}^{m_{i'}} \sum_{h'=1}^{n_{i'j'}} \frac{s_{ijh} s_{i'j'h'} \widehat{\text{cov}}(\hat{p}_{ijh}, \hat{p}_{i'j'h'})}{\hat{p}_{ijh}^2 \hat{p}_{i'j'h'}^2} \right].
$$

8.2.3 *Simulated example*

We illustrate the methods using simulation. A population was created in a square area of 130 units by 130 units, using a Poisson cluster process (Diggle 1983). The number of parents (termed patches here) in the whole population followed a Poisson (15) distribution, and the number of objects in each patch a Poisson (40) distribution. The vertical and horizontal coordinates of the centre of each patch were selected from a uniform (0, 130) distribution. Finally the position of each object within a patch, relative to the centre, was generated using a radial distance following a normal (0, 4) distribution and an angle following a uniform (0, 2π) distribution. For this example, cluster size was set to one; that is each object represented a single animal only.

A single population was created containing 722 objects. An adaptive point transect survey was then simulated, using 100 initial points. The points were equally spaced in a 10 by 10 grid at 12-unit intervals. A 5-unit 'buffer-zone' was used to reduce edge effects, so that coordinates of the initial points were restricted to the range 5–125 both horizontally and vertically. The grid was then randomly located to fit within this inner boundary, using deviates from uniform (0, 10) to determine the horizontal and vertical offsets.

Detections were simulated using a half-normal detection function $g(r) = \exp\left(-r^2/2\sigma^2\right)$, where r is distance from the point, with $0 \le r \le 2.0$, and $\sigma = 1.0$.

The unit pattern was a grid of squares and the neighbourhood was defined using the four adjacent units above, below, left, and right of the triggering unit, as in Fig. 8.2. The squares were each four units by four, so that the sampling circle of each point fitted exactly inside the square. Any adaptive plots that overlapped the population edge were included, but no adaptive plots were added outside the edge. The resulting survey is shown in Fig. 8.5.

The adaptive survey added 82 adaptive plots, making a total of 182 plots surveyed, from which 79 animals were detected. Although there are some large networks, careful inspection of Fig. 8.5 reveals that no two initial plots joined into a single network. For example, the large network towards the

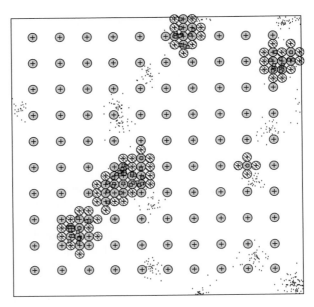

Fig. 8.5. Simulation of an adaptive point transect survey on a very patchy population. There were 100 initial sample plots in a systematic grid and 82 adaptive plots. The crosses identify the point transects and the circles the corresponding plots. The solid circles identify points which belong to a network and the diagonally hatched circles are edge units. Objects are black dots and detected objects are surrounded by a square.

middle of the population emanates from the fifth initial plot of the sixth row down. Although this network touches both the fourth initial plot in the sixth row and the fourth and fifth initial plots in the seventh row, none of these initial plots have any detections, and so each is a distinct network of size 1.

For comparison, a conventional survey was run using a 14 by 14 grid of points on the same population, giving a total of 196 survey points. In this case, there were 33 observations in total, rather small for reliable estimation of the detection function. In this simulation exercise, the true half-normal model was fitted to the data, so that results will be less variable than might be expected when the true model is unknown.

The surveys used a randomly positioned, systematic grid of points, but analyses assumed that it was a random initial sample. The results are summarized in Table 8.1. Both confidence intervals comfortably include the true density. Relative to the conventional survey, the adaptive survey yields a substantial increase in number of detections, and consequently more precise estimation of the detection probability P_a, but less precise

Table 8.1. Simulation results for conventional and adaptive sampling of the population of 722 objects. True density is $D = 0.038$. $\widehat{P}_a$ is the estimated probability of detection of an object, given that it is in one of the sampled plots, $\hat{m}$ is the estimated expected number of detections per plot, $\widehat{D}$ is the estimated animal density, with corresponding lower and upper 95% confidence limits $\widehat{D}_l$ and $\widehat{D}_u$. Coefficients of variation are expressed as percentages

	Conventional	Adaptive
Number of plots	196	182
Number of detections	33	79
$\widehat{P}_a$	0.440	0.460
$\text{cv}[\widehat{P}_a]$	23	14
$\hat{m}$	0.168	0.143
$\text{cv}(\hat{m})$	24	28
$\widehat{D}$	0.0304	0.0247
$\text{cv}(\widehat{D})$	34	31
$\widehat{D}_l$	0.0160	0.0136
$\widehat{D}_u$	0.0577	0.0450

estimation of the expected number of animals detected per plot m. Overall, the adaptive method yields higher precision for estimating animal density in this example, with around 7% fewer plots.

8.2.4 *Discussion*

Point transect surveys are typically multi-species, so that the trigger condition needs careful consideration. One option is to trigger additional effort only for relatively rare species, for which the number of detections might otherwise be insufficient. This may reduce efficiency for common species, unless densities tend to be higher at locations where the rarer species occur. Another possibility would be to use a combination trigger. This could be as simple as the total count of observations at the point across all species, or a total count across a subset of rarer species.

Bootstrapping is frequently used for robust estimation of variances and confidence intervals (e.g. Buckland *et al.* 2001: 161). Here, the bootstrap sample is created by sampling with replacement from the k points in the initial sample, and including any associated networks in the bootstrap resample. For points systematically distributed along lines, the resampling units should be the lines.

In line transect sampling, the main component of the density estimate variance is usually variance in sample size, while in point transect sampling, the variance of $\widehat{P}_a$ is often the larger component. The adaptive method outlined here provides a means of substantially increasing the number of sightings for animals that are patchily distributed, giving the potential for improved precision. This is borne out by the simulated example, where for a very patchy population, the adaptive approach gave more than double the number of observations compared with a conventional survey in which a larger number of plots was surveyed. The gain comes at the cost of greater survey complexity.

8.3   Design-unbiased adaptive line transect surveys

8.3.1   *Survey design*

Suppose the survey region is covered by a grid of units, each of which will typically be a long, narrow rectangle. Suppose further that the centreline of each unit comprises the line transect to be surveyed, if that unit is sampled. In this section, we assume that an initial random sample of units is selected. Strindberg (2001) shows that a systematic sample of units, combined with a line transect analysis that assumes the sample of units was random, generates abundance estimates with appreciably smaller RMSE than if lines are actually positioned at random. We recommend the same strategy here. Figure 8.6 shows two options for the survey design. The adaptive units would typically be the strips either side of a strip in which the trigger condition was met. If lines are long, efficiency of this approach may be poor, as the density may vary appreciably along the line. In that case, the line may be split into segments, each segment of which corresponds to a unit. Thus if the trigger condition is met within a segment, the adaptive

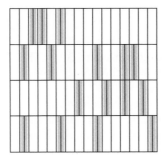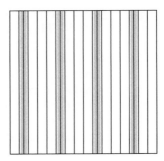

Fig. 8.6. Two designs for an adaptive line transect survey. In the left-hand design, a random sample of rectangles forms the initial sample of units. In the right-hand design, the rectangles extend right across the survey region, and a systematic sample of units is selected, with a random start. The transect line is shown down the centre of units in the initial sample.

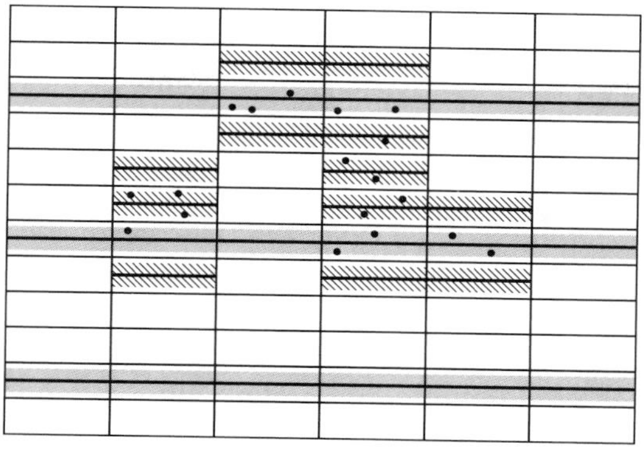

Fig. 8.7. An adaptive line transect survey in which the initial sample is a systematic sample of *primary* units. Each primary unit is divided into shorter *secondary* units. If the trigger condition is met within a secondary unit (indicated by a detection •), the secondary units on either side are sampled. The adaptive units are shown here by cross-hatching.

units sampled correspond to the two segments adjacent to it (Fig. 8.7). Under this design, the term 'line' in the following would refer to a line segment, rather than the full line.

Thus line i passes down the centre of the corresponding rectangular unit. We define a strip of half-width w and length l_i with the line i as the centreline. The strip might be chosen to coincide with the corresponding rectangular unit, but there is no need for the strip and the unit to have the same width. The strip is the plot, and has area $A_{pi} = 2wl_i$. Typically, strip lengths l_i will all be the same, except perhaps for strips truncated at the boundary. There are K units in total, so that the covered area is $A_s = \sum_{i=1}^{K} A_{pi}$. If all units have the same length l, $A_s = KA_p$ where $A_p = 2wl$. If the strips and units have the same width, then $A_s = A$, the size of the survey region. (Implicit in this formulation is the assumption that the survey region can be approximated by a set of K rectangles. This may require some redefinition of boundaries of the survey region, but on a scale that is unlikely to have any practical significance.)

8.3.2 Estimation

8.3.2.1 Homogeneous detection probabilities

We again assume that animals occur in well-defined clusters, so that a detection refers to the detection of a cluster of animals. If animals occur singly, then all 'clusters' are of size one, and cluster density is animal density. For simplicity, we assume that all units have the same length of transect l. As for point transect sampling, we define the two key parameters as follows: m_c is the expected number of clusters detected for a randomly selected plot, and m is the corresponding expected number of animals detected. As for point transect sampling, we will pool detection distances across all sampled plots to estimate the detection function. Thus probability of detection of a cluster is assumed independent of whether the cluster is in a plot from the initial sample, or from the additional adaptive units.

As for point transect sampling, define

N_c = number of clusters in the population

$D_c = \dfrac{N_c}{A}$ = density of clusters

s_h = size of cluster $h, h = 1, \ldots, N_c$

$N = \displaystyle\sum_{h=1}^{N_c} s_h$ = number of animals in the population

$D = \dfrac{N}{A}$ = density of animals

M_c = mean number of clusters (whether detected or not) per plot

M = mean number of animals (whether detected or not) per plot

Thus given estimates $\widehat{M_c}$ and $\widehat{M}$, $\widehat{D}_c = \widehat{M_c}/A_p$ and $\widehat{D} = \widehat{M}/A_p$, where $A_p = 2wl$. Further, $\mathrm{cv}(\widehat{D}_c) = \mathrm{cv}(\widehat{M_c})$ and $\mathrm{cv}(\widehat{D}) = \mathrm{cv}(\widehat{M})$.

The methods of Thompson and Seber (1996: 211–33) allow uncertain detection of clusters in a sampled plot. We first assume that probability of detection depends only on distance from the line. Then, unconditional on the specific design, all clusters have the same probability of detection. Let P_a denote the probability of detection of a cluster, given that it lies within one of the sampled plots, but unconditional on its location in that plot. Then

$$m_c = P_a M_c$$

so that

$$\widehat{M_c} = \frac{\hat{m}_c}{\widehat{P}_a}. \tag{8.11}$$

Assuming independence between $\hat{m}_c$ and $\widehat{P}_a$, the delta method (Seber 1982: 7–9) yields the approximate result

$$\left[\mathrm{cv}(\widehat{M_c})\right]^2 = [\mathrm{cv}(\hat{m}_c)]^2 + \left[\mathrm{cv}(\widehat{P}_a)\right]^2. \tag{8.12}$$

Similarly,

$$\widehat{M} = \frac{\hat{m}}{\widehat{P}_a} \qquad (8.13)$$

with approximately

$$\left[\mathrm{cv}(\widehat{M})\right]^2 = [\mathrm{cv}(\hat{m})]^2 + \left[\mathrm{cv}(\widehat{P}_a)\right]^2, \qquad (8.14)$$

where m is the expected number of animals detected on a randomly selected plot, estimated by $\hat{m}$.

If $g(x)$ is the probability of detection of a cluster at distance x from the line, conventional line transect theory yields $P_a = \frac{\mu}{w}$, where $\mu = \int_0^w g(x)dx$ (Buckland *et al.* 2001: 53), and the corresponding estimators $\widehat{P}_a$ and $\mathrm{cv}(\widehat{P}_a)$. We therefore need estimators $\hat{m}_c$ and $\hat{m}$.

In the absence of size-biased detection, mean cluster size in the population $E(s)$ is estimated by $\widehat{E}(s) = \hat{m}/\hat{m}_c$. If large clusters are more detectable than small clusters, P_a should be modelled as a function of cluster size; $E(s)$ is then estimated by $\widehat{E}(s) = \widehat{M}/\widehat{M}_c$.

As for point transect sampling, Pollard (2002) considers both a Hansen–Hurwitz estimator and a Horvitz–Thompson estimator, with either a systematic or a random initial sample, and sampling either with replacement or without. We consider just the Horvitz–Thompson estimator, first assuming that the initial sample is selected at random and without replacement, as in the first design of Fig. 8.6, then assuming an initial systematic sample, in which lines are split into legs, as shown in Fig. 8.7.

Random initial sample of transects, units correspond to transects   Formulae follow exactly as for point transect sampling. Thus m_c, the expected number of clusters detected on a random plot (where a plot is now a strip of length l and width $2w$), is estimated using eqn (8.5). The variance of this estimate is given by eqn (8.6). The corresponding formulae for m, the expected number of animals detected on a random plot, are eqns (8.7) and (8.8).

Systematic initial sample of transects, units correspond to transect legs   If a systematic sample of transects is selected, and each transect is divided into legs of length l, as in Fig. 8.7, then different formulae apply. Given the findings of Strindberg (2001), and in common with Pollard (2002), we assume that a grid of systematic lines, randomly positioned, may be treated as a random sample of lines for variance estimation purposes.

The expected number of clusters detected on a random plot (corresponding now to a transect leg) is estimated by

$$\hat{m}_c = \frac{1}{K} \sum_{i=1}^{\eta} \frac{\sum_{j=1}^{m_i} n_{ij}}{\alpha_i}, \tag{8.15}$$

where

$$\alpha_i = 1 - \frac{\binom{R - t_i}{r}}{\binom{R}{r}}$$

α_i = probability that network i is included in the sample

R = number of potential transects in the survey region

K = total number of plots (legs) across all R transects

η = number of distinct networks in the sample

m_i = number of transect legs in network i

n_{ij} = number of clusters detected on plot (leg) j within network i

t_i = number of initial transects that intersect network i

r = number of initial transects in the sample

The variance is estimated by

$$\widehat{\mathrm{var}}(\hat{m}_c) = \frac{1}{K^2} \sum_{i=1}^{\eta} \sum_{i'=1}^{\eta} \frac{(\alpha_{ii'} - \alpha_i \alpha_{i'}) \sum_{j=1}^{m_i} n_{ij} \sum_{j=1}^{m_{i'}} n_{i'j}}{\alpha_i \alpha_{i'} \alpha_{ii'}}, \tag{8.16}$$

with $\alpha_{ii} = \alpha_i$ and

$$\alpha_{ii'} = 1 - \frac{\binom{R - t_i}{r} + \binom{R - t_{i'}}{r} - \binom{R - t_i - t_{i'} + t_{ii'}}{r}}{\binom{R}{r}},$$

where

$t_{ii'}$ = number of initial transects that intersect networks i and i'

As before, $\mathrm{cv}(\hat{m}_c) = (\sqrt{\widehat{\mathrm{var}}(\hat{m}_c)})/(\hat{m}_c)$.

An estimate of the expected number of animals observed per plot is given by

$$\hat{m} = \frac{1}{K} \sum_{i=1}^{\eta} \frac{\sum_{j=1}^{m_i} \sum_{h=1}^{n_{ij}} s_{ijh}}{\alpha_i},$$

(8.17)

where s_{ijh} is the number of animals in the hth detected cluster on the jth plot of the ith network.

The variance is estimated by

$$\widehat{\mathrm{var}}(\hat{m}) = \frac{1}{K^2} \sum_{i=1}^{\eta} \sum_{i'=1}^{\eta} \frac{(\alpha_{ii'} - \alpha_i \alpha_{i'}) \sum_{j=1}^{m_i} \sum_{h=1}^{n_{ij}} s_{ijh} \sum_{j'=1}^{m_{i'}} \sum_{h'=1}^{n_{i'j'}} s_{i'j'h'}}{\alpha_i \alpha_{i'} \alpha_{ii'}}$$

(8.18)

with $\mathrm{cv}(\hat{m}) = (\sqrt{\widehat{\mathrm{var}}(\hat{m})})/(\hat{m})$.

8.3.2.2   *Heterogeneous detection probabilities*

As for point transect sampling, we can extend the above results for the case that each cluster has its own probability of detection. These probabilities may be estimated using the methods of Chapter 3. Formulae for the random initial sample case are as for point transect sampling, and we do not repeat them here. Below, we give formulae for the systematic initial sample case, in which transects are split into legs of length l.

$$\widehat{M}_c = \frac{1}{K} \sum_{i=1}^{\eta} \frac{\hat{u}_i}{\alpha_i},$$

(8.19)

where $\hat{u}_i = \sum_{j=1}^{m_i} \sum_{h=1}^{n_{ij}} 1/\hat{p}_{ijh}$ and $\hat{p}_{ijh}$ is the estimated probability of detection for the hth detected cluster on plot j of network i. The corresponding variance estimate is

$$\widehat{\mathrm{var}}(\widehat{M}_c) = \frac{1}{K^2} \left[\sum_{i=1}^{\eta} \sum_{i'=1}^{\eta} \frac{(\alpha_{ii'} - \alpha_i \alpha_{i'}) \hat{u}_i \hat{u}_{i'}}{\alpha_i \alpha_{i'} \alpha_{ii'}} + \sum_{i=1}^{\eta} \frac{1}{\alpha_i} \sum_{j=1}^{m_i} \sum_{h=1}^{n_{ij}} \frac{1 - \hat{p}_{ijh}}{\hat{p}_{ijh}^2} \right.$$
$$\left. + \sum_{i=1}^{\eta} \sum_{i'=1}^{\eta} \frac{1}{\alpha_{ii'}} \sum_{j=1}^{m_i} \sum_{h=1}^{n_{ij}} \sum_{j'=1}^{m_{i'}} \sum_{h'=1}^{n_{i'j'}} \frac{\widehat{\mathrm{cov}}(\hat{p}_{ijh}, \hat{p}_{i'j'h'})}{\hat{p}_{ijh}^2 \hat{p}_{i'j'h'}^2} \right]$$

with notation as before.

The expected number of animals per plot (leg) is estimated by

$$\widehat{M} = \frac{1}{K} \sum_{i=1}^{\eta} \frac{\hat{w}_i}{\alpha_i}, \tag{8.20}$$

where $\hat{w}_i = \sum_{j=1}^{m_i} \sum_{h=1}^{n_{ij}} s_{ijh}/\hat{p}_{ijh}$. Estimated variance is

$$\widehat{\text{var}}(\widehat{M}) = \frac{1}{K^2} \left[\sum_{i=1}^{\eta} \sum_{i'=1}^{\eta} \frac{(\alpha_{ii'} - \alpha_i \alpha_{i'}) \hat{w}_i \hat{w}_{i'}}{\alpha_i \alpha_{i'} \alpha_{ii'}} + \sum_{i=1}^{\eta} \frac{1}{\alpha_i} \sum_{j=1}^{m_i} \sum_{h=1}^{n_{ij}} \frac{(1 - \hat{p}_{ijh}) s_{ijh}^2}{\hat{p}_{ijh}^2} \right.$$

$$\left. + \sum_{i=1}^{\eta} \sum_{i'=1}^{\eta} \frac{1}{\alpha_{ii'}} \sum_{j=1}^{m_i} \sum_{h=1}^{n_{ij}} \sum_{j'=1}^{m_{i'}} \sum_{h'=1}^{n_{i'j'}} \frac{s_{ijh} s_{i'j'h'} \widehat{\text{cov}}(\hat{p}_{ijh}, \hat{p}_{i'j'h'})}{\hat{p}_{ijh}^2 \hat{p}_{i'j'h'}^2} \right]$$

8.3.3 *Discussion*

These methods may have less benefit for line transects than for point transects, as for line transects, the contribution to the overall variance of estimating detection probability is typically smaller. The methods should be readily implementable in the field, although the additional time off-effort, travelling to and from the adaptive transects, may reduce the efficiency of adaptive line transect sampling. Where movement between transects or transect legs is fast or low cost, and surveying is comparatively slow, then the overhead of the adaptive process is reduced. For example, in the deer dung pellet surveys of Marques *et al.* (2001), a systematic grid of short lines was used. For these surveys, many transects yield low encounter rates while some have high rates; observers can easily move to a neighbouring unit; and responsive movement does not occur. Hence adaptive surveys would be expected to work well.

The Thompson-based adaptive line transect approach shares problems with adaptive point transects: the complex notation and unwieldy estimation formulae; the potential to disturb animals which may bias estimates; identification of a suitable trigger; and the complications arising from multi-species surveys.

As with the point transect estimators, it is possible to use bootstrap methods (Efron and Tibshirani 1993) to estimate variances and confidence intervals. For a simple random design, the sampling units will be the initial transects plus any adaptive units they include. If the adaptive units form a network, then the complete network and any edge units should be added.

Although not used in the estimates of the mean number of animals and animal clusters per transect or transect leg, observations made on edge units may be included for estimating the detection function.

8.4 Fixed-effort adaptive line transect surveys

8.4.1 *Survey design*

Often a survey must be completed within a fixed time period, such as when a survey ship, observers and crew are available for a fixed period only. The above adaptive designs are then impractical. We now consider an adaptive line transect sampling approach with fixed total effort (Pollard and Buckland 1997; Pollard 2002; Pollard *et al.* 2002). Design-unbiased adaptive line transect sampling leads to time off-effort, while the observer moves between units in a neighbourhood. The methods described here can be implemented so that there is no loss of survey effort, which enhances survey efficiency.

As for design-unbiased methods, survey effort is increased in areas of high abundance using a trigger condition (e.g. when encounter rate rises above some value). However, the amount of adaptive effort is a function of the degree to which the survey is ahead of or behind schedule. This increase in sampling intensity is measured by the effort factor. Thus if effort in the next sampling unit is doubled when the trigger is activated, the effort factor is two for that section of the survey. The effort factor is then used to weight sightings to reduce bias. The approach conditions on these effort factors, which are data-dependent, and thus the method is not design-unbiased. However simulations have shown that bias is negligible.

The ability to vary adaptive effort as a function of the available effort remaining is useful for at least two reasons: survey time is often predetermined; and survey effort is often lost due to poor weather conditions. For conventional surveys, the second problem may lead to gaps in the nominal effort, or a failure to cover the entire study area. In fixed-effort adaptive surveys, adaptive effort would be reduced, so that lost time is made up. Only if so much effort is lost that the survey cannot be completed on schedule even without any adaptive effort, does the fixed-effort adaptive survey have gaps in the effort.

Although many adaptive patterns are feasible, we consider a zigzag track (Fig. 8.8), as this avoids time off-effort, allows the effort factor to take any desired value ≥ 1 (by varying the angle of the zig-zag legs), and is easily implemented.

8.4.2 *Estimation*

Using conventional line transect estimators, systematically increasing the effort in areas of higher animal density to increase detections would lead to abundance overestimation. The fixed-effort adaptive line transect approach downweights the data from the adaptive sections to compensate for the increased effort. The weighting used is inversely proportional to the effort factor, so that each section of transect is weighted in

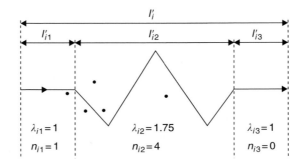

Fig. 8.8. When the trigger condition is met within a line segment (as in l'_{i1} here), a zig-zag search path commences in the next segment. The effort factor λ for the segment is actual track length of the zig-zag divided by the nominal track length of the straight transect that would have been surveyed had adaptive effort not been triggered.

proportion to the length of straight-line (nominal) effort through that section.

8.4.2.1 *Notation*

Each transect is divided into a number of sub-transects or legs, where the start and finish of each leg occurs at a change in effort.

L = total line length (total effort)

l = length of a single transect or transect leg

λ = the effort factor

n = number of clusters detected

e = encounter rate (number of detections per unit length of transect)

s = cluster size (number of animals in cluster)

D = animal density

$f(0)$ = pdf of detection distances from the line, evaluated at zero distance.

Subscript i is used to refer to the transect, $i = 1, \ldots, k$, and subscript j refers to the leg within the transect, $j = 1, \ldots, m_i$. One leg is deemed to end and the next to start whenever adaptive effort is triggered, or whenever normal effort is resumed. Thus l_{ij} is the actual distance travelled, or effort, for the jth leg of the ith transect. Subscript h refers to the observation within a leg, $h = 1, \ldots, n_{ij}$. Thus s_{ijh} refers to the hth observation of the jth leg of the ith transect (see Fig. 8.8).

Nominal values refer to the values expected if a conventional straight-line transect is followed. Nominal effort is signified by a prime, such as L', the total nominal effort, whereas the corresponding actual effort is L. The expected sample size, if only the nominal effort had been used, is represented by $E(n \mid L')$. The same approach is also used for both the expected encounter rate and expected cluster size if only the nominal effort had been used, giving for example $E(e_i \mid l_i')$ and $E(s_i \mid l_i')$.

8.4.2.2   *Assumptions*

In deriving the estimating equations, the following standard line transect assumptions are made:

1. Probability of detection on the line is certain: $g(0) = 1$.
2. There is no size bias (the probability of detection is independent of cluster size).
3. There is no responsive movement of animals in advance of detection, and any non-responsive movement is slow relative to the speed of the observers.

These assumptions can be weakened or removed using similar strategies as for conventional line transect sampling. In addition the following assumptions are made specifically for adaptive line transect sampling:

1. The expected encounter rate for an adaptive track is the same as the expected encounter rate for the corresponding nominal track.
2. The expected cluster size for an observation on an adaptive track is the same as the expected cluster size for an observation when following the corresponding nominal track.
3. Conditional on the location of the actual (as distinct from the nominal) track line, each observation is an independent event. That is, the probability of an observation is only a function of its perpendicular distance from the actual line (although the position of the line itself may depend on past observations).

Approaches to dealing with heterogeneity in the detection function estimate, and thus weakening assumption 6, are considered later.

8.4.2.3   *Effort factor calculation*

The effort factor λ is the ratio of the actual effort to the nominal effort. Thus the effort factor for the jth leg of the ith transect is

$$\lambda_{ij} = \frac{l_{ij}}{l_{ij}'}. \tag{8.21}$$

Suppose additional effort is triggered, so that we need to calculate the effort factor, as a function of the remaining effort available. Let $L_E(t)$ (measured in units of time or distance) be the total excess effort remaining at time t. This quantity is calculated as total effort available at the start of the survey, less the actual effort used up to time t, less the nominal effort required to complete the survey (without any further adaptive effort). Let ξ be the expected number of times the effort will increase above the nominal level for the remainder of the survey. Then the increase in effort, following an observation, is given by the excess effort available divided by the expected number of times the effort will increase plus one (for the current increase). So the increase in effort for a leg is given by

$$l_{ij} - l'_{ij} = \frac{L_E(t)}{1 + \xi}.$$

Thus from eqn (8.21) we get

$$\lambda_{ij} = 1 + \frac{L_E(t)}{l'_{ij}(1 + \xi)}.$$

If each effort increase is applied for the same fixed distance along the nominal trackline, then ξ can be calculated from an estimate of the encounter rate. Let l'_Z be the nominal effort over which the effort increase occurs; $L'_R(t)$ be the nominal effort remaining at time t; and γ be an estimate of the encounter rate, where γ might be obtained from previous survey data or be a best guess provided by the user. Then

$$\xi = \gamma \left\{ L'_R(t) - \xi l'_Z \right\}$$

so that

$$\xi = \frac{\gamma L'_R(t)}{1 + \gamma l'_Z}.$$

8.4.2.4   *Estimating equations*

Conceptually, we estimate animal density separately for each transect line, using formulae from conventional line transect sampling. To avoid bias arising from concentrating more effort in areas of high density, weighted means of encounter rate and cluster size are found, weighting by the reciprocal of the effort factor. To simplify the methodology, we assume that $f(0)$ is independent of animal density, and use a single pooled estimate of $f(0)$.

The density for conventional line transect sampling, assuming detection on the line is certain, is given by (Buckland *et al.* 2001: 54)

$$D = \frac{E(n)f(0)E(s)}{2L'}.$$

Assuming $f(0)$ is constant across transects, in the absence of adaptive effort, the density corresponding to the ith transect is

$$D_i = \frac{E(n_i)f(0)E(s_i)}{2l'_i}.$$

Replacing the parameters by their estimators, we have

$$\widehat{D}_i = \frac{\widehat{E}(n_i|l'_i)\hat{f}(0)\widehat{E}(s_i|l'_i)}{2l'_i}, \qquad (8.22)$$

where $\widehat{E}(n_i|l'_i)$ and $\widehat{E}(s_i|l'_i)$ are, respectively, estimates of the sample size and mean cluster size for transect i if the nominal track line had been followed. For conventional surveys, where the effort factor is unity, these estimators are simply n_i and, assuming no size bias, $\bar{s}_i$ respectively, where $\bar{s}_i$ is the mean size of clusters detected from transect i.

Thus the overall density estimate is (Buckland *et al.* 2001: 80)

$$\widehat{D} = \frac{\sum_{i=1}^{k} l'_i \widehat{D}_i}{L'}. \qquad (8.23)$$

The estimate of $f(0)$ is based on pooled data across all transects, thus an estimate of the variance of the density estimate, $\widehat{\text{var}}(\widehat{D})$, can be separated into two components, assumed to be independent: $\widehat{\text{var}}(\widehat{H})$, with $\widehat{H} = \widehat{D}/\hat{f}(0)$, and $\widehat{\text{var}}\{\hat{f}(0)\}$. Using the delta method (Seber 1982: 7–9), an estimate of the variance of the density estimate is

$$\widehat{\text{var}}(\widehat{D}) = \widehat{D}^2 \left[\frac{\widehat{\text{var}}(\widehat{H})}{\widehat{H}^2} + \frac{\widehat{\text{var}}\{\hat{f}(0)\}}{\{\hat{f}(0)\}^2} \right], \qquad (8.24)$$

where

$$\widehat{\text{var}}(\widehat{H}) = \frac{\sum_{i=1}^{k} \left\{ l'_i (\widehat{H}_i - \widehat{H})^2 \right\}}{L'(k-1)}$$

with $\widehat{H}_i = \widehat{D}_i/\hat{f}(0)$.

By definition, the nominal effort for the jth leg of the ith transect is $l'_{ij} = l_{ij}/\lambda_{ij}$, with the nominal transect effort and nominal total survey effort given by $l'_i = \sum_{j=1}^{m_i} l'_{ij}$ and $L' = \sum_{i=1}^{k} l'_i$.

An estimate of the sample size if only the nominal effort had been used for the jth leg of the ith transect is given by $\widehat{E}(n_{ij}|l'_{ij}) = n_{ij}/\lambda_{ij}$, with the corresponding transect and survey estimates of

$$\widehat{E}(n_i|l'_i) = \sum_{j=1}^{m_i} \widehat{E}(n_{ij}|l'_{ij}) \tag{8.25}$$

and

$$\widehat{E}(n|L') = \sum_{i=1}^{k} \widehat{E}(n_i|l'_i). \tag{8.26}$$

An estimate of the variance of estimated expected sample size if only the nominal effort had been used is given by

$$\widehat{\mathrm{var}}\{\widehat{E}(n|L')\} = \frac{L'}{k-1} \sum_{i=1}^{k} \left[l'_i \{\widehat{E}(n_i|l'_i) - \widehat{E}(n|L')\}^2 \right]. \tag{8.27}$$

The encounter rate for the jth leg of the ith transect is given by $e_{ij} = n_{ij}/l_{ij}$. So from assumption 4, an estimate of the expected encounter rate, if only the nominal effort had been used for the jth leg of the ith transect, is given by $\widehat{E}(e_{ij}|l'_{ij}) = n_{ij}/l_{ij} = \widehat{E}(n_{ij}|l'_{ij})/l'_{ij}$. Using weighted averages, an estimate of the expected encounter rate, if only the nominal effort was used for the ith transect, is

$$\widehat{E}(e_i|l'_i) = \frac{\sum_{j=1}^{m_i} l'_{ij} \widehat{E}(e_{ij}|l'_{ij})}{\sum_{j=1}^{m_i} l'_{ij}} = \frac{\sum_{j=1}^{m_i} \widehat{E}(n_{ij}|l'_{ij})}{\sum_{j=1}^{m_i} l'_{ij}} = \frac{\widehat{E}(n_i|l'_i)}{l'_i} \tag{8.28}$$

and an estimate of the survey encounter rate, if only the nominal effort was used, is

$$\widehat{E}(e|L') = \frac{\sum_{i=1}^{k} l'_i \widehat{E}(e_i|l'_i)}{\sum_{i=1}^{k} l'_i} = \frac{\sum_{i=1}^{k} \widehat{E}(n_i|l'_i)}{\sum_{i=1}^{k} l'_i} = \frac{\widehat{E}(n|L')}{L'}. \tag{8.29}$$

Thus an estimate of the variance of the expected survey encounter rate if only the nominal effort had been used is

$$\widehat{\mathrm{var}}\left\{\widehat{E}(e|L')\right\} = \frac{\widehat{\mathrm{var}}\left\{\widehat{E}(n|L')\right\}}{(L')^2}. \tag{8.30}$$

In the absence of size bias, an estimate of the expected cluster size for the jth leg of the ith transect is given by the observed mean cluster size for the leg. So from assumption 5, an estimate of the expected cluster size if only the nominal effort had been used is $\widehat{E}(s_{ij}|l'_{ij}) = \widehat{E}(s_{ij}) = \sum_{h=1}^{n_{ij}} s_{ijh}/n_{ij}$. Using weighted averages, an estimate of the expected cluster size for the ith transect, if only the nominal effort had been used, is given by

$$\widehat{E}(s_i|l'_i) = \frac{\sum_{j=1}^{m_i} \widehat{E}(n_{ij}|l'_{ij})\widehat{E}(s_{ij}|l'_{ij})}{\sum_{j=1}^{m_i} \widehat{E}(n_{ij}|l'_{ij})} = \frac{\sum_{j=1}^{m_i} \widehat{E}(n_{ij}|l'_{ij})\widehat{E}(s_{ij}|l'_{ij})}{\widehat{E}(n_i|l'_i)}. \quad (8.31)$$

Similarly, an estimate of the overall expected cluster size, if only the nominal effort had been used, is

$$\widehat{E}(s|L') = \frac{\sum_{i=1}^{k} \widehat{E}(n_i|l'_i)\widehat{E}(s_i|l'_i)}{\sum_{i=1}^{k} \widehat{E}(n_i|l'_i)} = \frac{\sum_{i=1}^{k} \widehat{E}(n_i|l'_i)\widehat{E}(s_i|l'_i)}{\widehat{E}(n|L')}. \quad (8.32)$$

An estimate of the variance of the expected cluster size, if only the nominal effort had been used, is

$$\widehat{\text{var}}\{\widehat{E}(s|L')\} = \frac{\sum_{i=1}^{k} \left[\widehat{E}(n_i|l'_i) \left\{ \widehat{E}(s_i|l'_i) - \widehat{E}(s|L') \right\}^2 \right]}{\widehat{E}(n|L')(k-1)}. \quad (8.33)$$

It is assumed there is no correlation between density and $f(0)$ and so observation data are pooled across all transects to produce a single estimate of $f(0)$ using conventional techniques.

8.4.2.5   Modelling heterogeneity in $f(0)$

The above methods do not allow for heterogeneity between clusters in the probability of detection due to cluster size, weather conditions, etc. In practice, adaptive effort is more likely to be triggered in good sighting conditions, and so the probability of detection on the adaptive leg may be enhanced. Since such observations will be over-represented in the sample, $\hat{f}(0)$ (which is the reciprocal of the estimated effective strip half-width) will be negatively biased.

We can seek to model the heterogeneity; however, if we have not measured the relevant covariates, or if probability of detection of further animals changes following an observation (because observers become more alert, or because they continue to watch detected animals), this approach may not be wholly effective. Another approach is to downweight the influence of observations made during adaptive legs on the likelihood. Adopting the principle that a single observation at distance y from the line when the effort factor is one should have the same contribution to the likelihood as λ observations at distance y when the effort factor is λ, we obtain the modified likelihood $\mathcal{L}(\underline{\theta}) = \prod_{i=1}^{n} \{f(y_i)\}^{1/\lambda_i}$ where y_i is the perpendicular distance from the line of the ith observation, $i = 1, \ldots, n$; λ_i is the effort factor corresponding to the ith observation; and $\underline{\theta}$ represents the parameters of $f(y_i)$.

We now maximize this modified likelihood function with respect to the parameters of $f(\cdot)$. For example, consider the half-normal model with ungrouped data and no truncation, so that $\underline{\theta}$ is the scalar σ^2 and $f(y) = \exp\left(-y^2/2\sigma^2\right)$, $y \geq 0$. Then a weighted estimate of $f(0)$ is given by

$$\hat{f}_w(0) = \sqrt{\frac{2}{\pi \hat{\sigma}_w^2}}, \tag{8.34}$$

where

$$\hat{\sigma}_w^2 = \frac{\sum_{i=1}^{n} y_i^2/\lambda_i}{\sum_{i=1}^{n} 1/\lambda_i} = \frac{\sum_{i=1}^{n} y_i^2/\lambda_i}{\widehat{E}(n|L')} \tag{8.35}$$

and the variance estimate is given by

$$\widehat{\mathrm{var}}\{\hat{f}_w(0)\} = \frac{\{\hat{f}_w(0)\}^2}{2\widehat{E}(n|L')}. \tag{8.36}$$

8.4.3 *Simulation*

To investigate the efficiency of the adaptive approach, populations exhibiting patchy distributions were simulated. Both conventional and fixed-effort adaptive line transect surveys of these populations were simulated. We assumed that animals occurred singly; that is, cluster size was always one. The following 'key + adjustment' models were fitted to each simulated data set: half-normal + cosine; half-normal + hermite polynomial; hazard-rate + cosine; hazard-rate + simple polynomial; and uniform + cosine. Akaike's information criterion (AIC) was used to select between them.

8.4.3.1 *Population models*

Three types of population were simulated: a population exhibiting complete spatial randomness (CSR); and two patchy populations, one exhibiting a medium degree of patchiness (referred to here as 'patchy'), and the other a high degree (referred to as 'highly patchy'). Each population had an expected size of 600 and was created within a square area (population frame) of 100 by 100 units. The two patchy populations were simulated using a Poisson cluster process (Diggle 1983). The number of patches was simulated using a Poisson (40) distribution for the patchy population and a Poisson (15) distribution for the highly patchy population. The number of animals within each patch was then simulated using a Poisson (15) distribution for the patchy and a Poisson (40) distribution for the highly patchy population. The location of the centre of each patch was simulated using a uniform (0, 100) distribution for the vertical coordinate and another uniform (0, 100) for the horizontal coordinate. Finally the position of each animal within each patch was calculated relative to the patch centre. The radial distance to each animal was simulated from a normal (0, 4) distribution and the radial angle using a uniform (0, 2π). If following this, the animal lay outside the population frame, the distance to the animal was wrapped around to the opposite edge, horizontally or vertically as necessary, until the animal was within the population frame.

8.4.3.2 Survey simulation

Each simulated population was sampled first using a conventional line transect survey and then using an adaptive line transect survey. The transects were systematically spaced with a random start for the first transect, and a buffer zone was defined in which transects could not be located, to avoid edge effects. For each survey, the total effort was set at 1500 units, and for the adaptive surveys, the nominal effort was set at 1300 units, leaving 200 units for additional, adaptive effort. The detection function was simulated using a half-normal model with parameter $\sigma = 0.3$. Perpendicular distances were truncated at two units (effectively no truncation).

The trigger to start adapting was a single observation in the previous 0.66 length of transect, after which zigzagging occurred for 12 steps (three complete zigzag cycles), spanning approximately six units of nominal line, with the angle of the zigzags determined by the effort factor. If there was an observation during the last step of an adaptive section, the adapting continued for another 12 steps. Each transect started in straight-line mode, irrespective of whether the survey was still adapting at the end of the previous transect.

Additional simulations were performed using the highly patchy populations to investigate the effects of heterogeneity in the detection function. First, to simulate an increase in observer awareness following an observation, the detection function for the adaptive surveys was changed. For the conventional surveys and the straight-line sections of the adaptive surveys, a half-normal detection function with $\sigma = 0.3$ was used as before. However to simulate an increase in observer awareness, σ was increased to 0.4 on the adaptive (zigzag) sections of the adaptive survey. Second, to simulate heterogeneity introduced by changes in weather, simulation of the highly patchy population was again re-run, this time using a half-normal detection function with $\sigma = 0.15$ for the first 400 units of survey effort, reverting to $\sigma = 0.3$ for the remainder of the survey. This heterogeneity was applied to both the adaptive and the conventional surveys. For these simulations, the detection function was estimated, assuming the half-normal model, using eqns (8.34) and (8.36).

8.4.3.3 *Results*

For each population type, 1000 populations were generated, with both a conventional and an adaptive survey run on each population. Two additional runs of 1000 highly patchy populations were generated to test heterogeneity in the detection function, due to increased observer awareness and changes in weather conditions. The comparative efficiencies of estimators were calculated by dividing the mean estimator variance from 1000 conventional survey simulations by the mean estimator variance of the corresponding 1000 adaptive survey simulations. Hence efficiencies greater than one indicate that the adaptive surveys give better precision than the conventional surveys. The efficiencies for the expected nominal encounter rate, $f(0)$ and density estimators for the three population types are given in Table 8.2.

For the patchily-distributed populations, adaptive sampling indicated improved density estimate precision with an efficiency of 1.050 for the patchy and 1.059 for the highly patchy populations. As expected, adaptive sampling was less efficient, at 0.99, than conventional sampling for the CSR population type. This is because all animals are randomly located, so increasing the search effort following an observation does not increase the probability of detecting another animal. Thus, with a CSR population, the expected total number of sightings for an adaptive survey is the same as the expected total for a conventional survey. However the sightings in the adaptive survey are then weighted, to account for any adaptive bias, and so there is a slight decrease in efficiency.

Table 8.3 gives 95% confidence intervals for the mean percent relative bias of the encounter rate, $f(0)$ and density estimators. Overall there appears to be no or minimal bias. There is a small negative bias in the $f(0)$

Table 8.2. Efficiencies of adaptive surveys, where efficiency is measured as mean variance of conventional estimator from 1000 simulations divided by mean variance of corresponding adaptive estimator

Population	Estimator adaptive efficiency		
	$\widehat{\text{var}}\{\widehat{E}(e\|L')\}$	$\widehat{\text{var}}\{\hat{f}(0)\}$	$\widehat{\text{var}}(\widehat{D})$
CSR	0.959	1.032	0.990
Patchy	0.993	1.265	1.050
Highly patchy	1.035	1.349	1.059

Table 8.3. Estimated 95% confidence intervals for mean percent relative bias over all 1000 simulations. For each estimator, the top confidence interval is for the adaptive survey simulations, and the lower one is for the conventional survey simulations

	Estimator		
Population	$\widehat{E}(e|L')$	$\hat{f}(0)$	$\widehat{D}$
CSR	$(-0.91\%, 0.52\%)$	$(-1.45\%, 0.83\%)$	$(-1.96\%, 0.07\%)$
	$(-0.56\%, 0.86\%)$	$(-1.41\%, 0.16\%)$	$(-1.54\%, 0.56\%)$
Patchy	$(-1.46\%, 0.32\%)$	$(-1.84\%, -0.35\%)$	$(-2.86\%, -0.58\%)$
	$(-1.32\%, 0.38\%)$	$(-1.34\%, 0.25\%)$	$(-2.15\%, 0.15\%)$
Highly patchy	$(-2.00\%, 0.19\%)$	$(-0.67\%, 0.74\%)$	$(-2.16\%, 0.44\%)$
	$(-0.50\%, 1.55\%)$	$(-2.01\%, -0.32\%)$	$(-1.90\%, 0.78\%)$

Table 8.4. RMSE estimated from 1000 simulations. For each estimator, the top value corresponds to the adaptive survey simulations and the bottom value to the conventional survey simulations

	RMSE		
Population	$\widehat{E}(e|L')$	$\hat{f}(0)$	$\widehat{D}$
CSR	0.00520	0.330	0.0098
	0.00519	0.336	0.0101
Patchy	0.00648	0.321	0.0111
	0.00618	0.341	0.0112
Highly patchy	0.00804	0.304	0.0127
	0.00752	0.362	0.0131

estimate for the adaptive surveys of the patchy populations and the conventional surveys of the highly patchy populations. This is probably because AIC selected the Fourier series model 70% of the time, whereas the true model was the half-normal. There was also a small negative bias for the density estimate of the adaptive surveys of patchy populations, presumably largely due to the negative bias in the $f(0)$ estimate.

Root mean square errors for the encounter rate, $f(0)$ and density estimators are given in Table 8.4. The encounter rate estimators under adaptive sampling do not perform well. However, the improvement in the $f(0)$ estimate outweighs this, leading to an overall improvement in the precision of density estimates.

Table 8.5. Percentage of occasions the true value is below, above a 95% confidence interval for the estimator for 1000 simulations. For each estimator, the top values are for the adaptive survey simulations, and the lower ones for the conventional survey simulations

Population	Estimator (%)		
	$\widehat{E}(e\mid L')$	$\hat{f}(0)$	$\widehat{D}$
CSR	3.0, 3.2	2.9, 13.2	2.8, 5.7
	3.0, 2.7	4.1, 13.2	3.4, 4.9
Patchy	1.2, 1.7	5.3, 16.2	1.1, 3.5
	0.4, 1.2	3.8, 12.9	0.8, 2.3
Highly patchy	0.4, 1.1	6.3, 13.1	0.6, 2.1
	0.3, 0.6	5.0, 14.0	0.4, 2.5

Table 8.5 shows the coverage of a log-normal 95% confidence interval for the encounter rate, $f(0)$ and density estimators. Values are presented as the percentage of occasions the true value is below or above the estimated confidence interval. Coverage for the $f(0)$ estimates was poor, with the true value being larger than the upper confidence limit for 13–16% of the time. Much of this can be explained by the tendency for AIC to select the Fourier series model rather than the (true) half-normal model, and the Fourier series model tended to underestimate $f(0)$. In general the confidence interval coverage was very similar for the two approaches.

The simulations of heterogeneity in $f(0)$ gave improved adaptive density variance estimator efficiencies relative to the conventional estimates, of 1.068 for the simulation of increased observer awareness and 1.036 for the simulation of bad weather. However in the case of the bad weather simulation, there was an improvement in the encounter rate efficiency and a decrease in the $f(0)$ efficiency. This was borne out in RMSE for the density variance estimators. In the increased observer awareness simulation, the adaptive RMSE (0.0115) improved on the conventional estimator RMSE (0.0119), while for the bad weather simulation, the adaptive RMSE (0.0129) was larger than the conventional RMSE (0.0120). There was a small negative bias in the adaptive density estimate for the increased observer awareness simulation, while the conventional survey indicated a small positive bias: 95% confidence intervals for percent relative bias were $(-3.35, -1.01)$ and $(0.36, 2.79)$ respectively. The corresponding bad weather simulation confidence intervals were $(-8.62, -6.16)$ and $(-6.28, -3.90)$ for the adaptive and conventional surveys; thus, both density estimators showed negative bias.

8.4.4 *Discussion*

The methods of this section were tested out on harbour porpoise in the Gulf of Maine and Bay of Fundy region. Results are reported by Palka and Pollard (1999) and Pollard *et al.* (2002). Adaptive and conventional methods gave very similar estimates of porpoise density, with slightly improved precision for the adaptive method. Perhaps the most significant gain from the adaptive method however was a substantial increase in the number of detections using the adaptive method (551, compared with 313 for the conventional survey). In surveys where sample size is problematic, because a species is scarce or resources limited, adaptive methods may prove a cost-effective way of increasing sample size.

Overall the simulation results indicate that conditioning on the effort factors only introduces small bias and that adaptive sampling offers potential for improving density estimator precision for patchy populations. They also indicate a correlation between the degree of patchiness and the adaptive efficiency.

Efficiency of the method is dependent on many factors, including appropriate selection of the trigger and stopping function; effort factor calculation; adaptive pattern; and amount of excess effort available. If the adaptive track is too large, so that it frequently steps outside a patch, some bias is introduced into the encounter rate estimate. This is due to violation of the assumption that the expected encounter rate for the adaptive track is the same as the expected encounter rate for the corresponding nominal track (assumption 4). In reality, extra effort is more likely to be triggered when passing near the centre of a patch, so that adaptive legs may tend to have a slightly lower expected encounter rate than the corresponding nominal legs.

One possible use of the methods developed here is for effort factors that are less than one. There may be poor survey coverage in certain areas of a large survey due to insufficient time or poor weather. It may be possible to treat these incomplete areas as sections with an effort factor less than one, and so compensate for the lack of uniform cover of a survey region.

Fixed-effort point transect surveys are also possible. In this case, the number of neighbouring points to be surveyed when extra effort is triggered can be made a function of whether the survey is ahead of or behind schedule. Pollard (2002) provides estimating equations for such surveys.

9

Passive approaches to detection in distance sampling

P. M. Lukacs, A. B. Franklin, and D. R. Anderson

9.1 Introduction

Most distance sampling methods require active detection of animals from a line or point. Sometimes, active detection of animals is impractical or even impossible. Consider estimating abundance of a species of salamander. The species is small, cryptic, and spends much of the time under litter on the forest floor, so that active detection is not feasible, yet the salamander can be trapped in a pitfall trap. This example provides the impetus for considering passive approaches to distance sampling. These approaches are passive in that the animal detects itself by entering a trap whose distance from the sample point or line is known.

In general, a passive distance sampling method is any method in which the animal marks its own distance from the point or line (Lukacs 2001). The best known passive method, the trapping web, uses traps arranged around a random point. The relevant distance is the distance from the centre of the web to the trap in which the animal was encountered. Passive methods can use any form of trap (e.g. pitfall traps, snap traps, live traps, sticky boards), hair or feather snags for DNA extraction, or remote sensing (e.g. remote camera systems). Passive approaches are not restricted to terrestrial vertebrate species. Insects can be sampled with passive distance methods. Aquatic species such as crabs, lobsters, and crayfish fit naturally into a trap sampling procedure.

Two approaches to passive distance sampling are developed in this chapter: trapping webs and trapping line transects. The trapping web is a special case of point transect theory, and trapping line transects are a special case of line transect theory. Passive methods usually involve multiple sampling occasions, whereas active methods usually involve a single sampling occasion. Therefore, the use of passive sampling tends to result in some individuals being encountered more than once (i.e. recaptured).

Theory now exists to allow the use of recaptures in the estimation of detection probability, leading to increased precision (Lukacs *et al.* in preparation *b*). We note that an alternative approach that also uses recapture information has been developed by Efford (in press) and Efford *et al.* (in press); this approach has been field-tested by Efford *et al.* (in preparation).

The assumptions for passive approaches are similar to those for active distance sampling. They are restated here in the context of passive sampling:

1. All animals at the centre of the trapping web or along the line of a trapping line transect must be encountered with certainty or with a probability that can be estimated from empirical data (i.e. $\hat{g}(0)$ and $\widehat{\text{var}}[\hat{g}(0)]$).

2. No directional movement of animals occurs. Therefore, animals cannot be more likely to move toward or away from the centre of the web or transect. The increasing trap density towards the centre of a trapping web has the effect of causing this assumption to be violated, but in a well-designed web the impact on estimated density is small.

3. Distances from the centre of the web or from the transect line to each trap are accurately measured or placed within the appropriate distance ring. This assumption is easy to uphold because distance between two inanimate objects is easy to measure accurately.

Other issues common to all sampling methods are important for passive distance sampling also. Webs or trapping transects must be objectively placed within a predefined study area according to some probabilistic scheme (Anderson 2001). Trapping line transects could be systematically set with a random starting point as long as the pattern in the transects does not follow any pattern on the landscape. There is no need to assume a specific spatial distribution of animals across the study area. The variance of n is estimated empirically, and is therefore robust to animal spatial distribution. Passive distance sampling methods, in common with other distance sampling methods, are robust to heterogeneity in capture probabilities, provided detection is certain at the line or point. Detection probability unconditional on distance from the line or point, P_a, is estimated as an average quantity over a large, unknown set of covariates, and is therefore estimated well even when detection probability varies. All distance sampling methods assume instantaneous sampling—a 'snapshot' of animal locations. Passive methods violate this assumption to a greater degree than well-designed active methods. Therefore, attention must be paid to the behaviour of the animal of interest to ensure that the assumption can be met reasonably well.

Software exists to design and analyse trapping web and trapping line transect surveys. WebSim (Lukacs 2002) provides simulation and response surface methods to guide researchers in design issues. Distance 4 (Thomas *et al.* 2003) computes density and variance estimates from passive designs just as it does for line and point transect surveys.

9.2   Trapping webs

The trapping web was originally developed to allow density estimation from trapping data without having to estimate separately the size of the area being sampled and the number of animals in that area (Anderson *et al.* 1983). The trapping web analysis also avoids complications of individual heterogeneity in capture probabilities, which cause substantial bias in closed population mark-recapture models. In addition, data from multiple trapping webs can be pooled to estimate P_a. Basic trap set up and analysis issues are addressed in Buckland *et al.* (2001: 216–23); field trials are presented in Parmenter *et al.* (2003). Here we focus on new advances in trapping web analysis and design.

The trapping web is a special case of point transect theory. With a trapping web, trapped animals provide their own detection distance from the web centre. Trapping webs are often analysed as grouped distance data because traps are usually arranged in defined concentric rings. The data can be analysed as exact distance data when traps are arranged irregularly within a distance ring, perhaps because vegetation prevents perfect placement of traps.

9.2.1   *Density estimation*

The density estimator for the trapping web follows that of point transect sampling:

$$\widehat{D} = \frac{n}{k\pi w^2 \widehat{P_a}},\tag{9.1}$$

where n is the number of different animals trapped, k is the number of webs, w is the radius of a web, and $\widehat{P_a}$ is the estimated probability that an animal, located within a web, is trapped. Standard point transect sampling analysis of grouped distance data is conducted to estimate P_a. For example, if there are rings of traps at 5 m, 15 m, 25 m, ... from the point, and an animal is caught in the ring at 15 m, then it is assigned to the distance interval 10–20 m.

Cluster size s is usually difficult to determine accurately with passive methods, therefore it is rarely used; if three animals are caught in the same trap, these would be considered as three independent observations at that distance, and we then have $E(s) \equiv 1$ (and $\text{var}(s) \equiv 0$). With camera traps, for example, it may be possible to estimate the mean cluster size, $E(s)$, in which case the above equation must be multiplied by this estimate.

The sampling variance of $\widehat{D}$ also follows from point transect theory:

$$\widehat{\mathrm{var}}[\widehat{D}] = \widehat{D}^2 \left[\{\mathrm{cv}(n)\}^2 + \{\mathrm{cv}(\widehat{P}_a)\}^2 \right]. \tag{9.2}$$

The coefficient of variation for n usually contributes most to the variance of $\widehat{D}$ and therefore stratification and other approaches should be considered to reduce this variance component, just as in active methods.

9.2.2 Including data from recaptures

In active distance sampling, animals are usually encountered only once. Even if a line or point is visited more than once, it is not known if the same individuals are detected more than once. In passive methods using traps or remote sensing devices, animals are often encountered more than once. Information about detection probability exists in the subsequent encounters of animals. Detections in distance sampling are assumed to come from a distribution of animals, which is uniform with respect to the placement of the trapping web. After an animal is captured once, subsequent captures are sampled from a distribution that follows the shape of the detection function of the previous detection. This is similar to size bias in regression estimates of mean cluster size (Buckland et al. 2001: 74).

Lukacs et al. (in preparation b) developed a method to incorporate recaptures in a distance sampling analysis. This method assumes that recaptures are sampled from the same distribution as the first captures. The method requires a few additional assumptions:

1. Capture probability is constant across the duration of the study.
2. Animals do not exhibit a behavioural response to being encountered.
3. Capture probability is constant across individual animals.

Anyone familiar with mark-recapture analysis would be skeptical of an analysis making these three assumptions, but it turns out that distance sampling is robust to violations of these assumptions.

Some additional notation is required to develop the recapture estimator of detection probability:

$u_i =$ distance interval cutpoint $i, i = 0, \ldots, U (u_0 = 0)$
$n_{kij} =$ number of kth captures in ring i and web j
$n_{1..} =$ total number of first captures
$n_{...} =$ total number of captures, including recaptures
$n_{.i.} =$ total number of captures in ring i
$\widehat{P}_a^* =$ probability of detection, estimated from both first captures and recaptures.

Note that 'first captures' are defined as captures of animals caught for the first time, so that $n_{1..}$ is the number of different animals caught. Density can be estimated using recaptures by first estimating the detection probability. For the trapping web, detection probability is estimated as

$$\widehat{P}_a^* = \frac{2}{w^2} \int_0^w r \, \hat{g}(r, \hat{\underline{\theta}}) \, dr, \tag{9.3}$$

where r represents distance from the web centre. The counts $n_{.i.}$ follow a multinomial distribution when the analysis is based on grouped distance data:

$$\Pr[n_{.1.}, \ldots, n_{.U.}, \underline{\theta}] = \frac{n_{...}!}{\prod_{i=1}^U n_{.i.}!} \prod_{i=1}^U p_i(\underline{\theta})^{n_{.i.}}, \tag{9.4}$$

where

$$p_i(\underline{\theta}) = \frac{\int_{u_{i-1}}^{u_i} r \, g(r, \underline{\theta}) \, dr}{\int_0^w r \, g(r, \underline{\theta}) \, dr}, \tag{9.5}$$

which gives us the likelihood:

$$\mathcal{L}(\underline{\theta} \mid n_{.1.}, \ldots, n_{.U.}) \propto \prod_{i=1}^U [p_i(\underline{\theta})]^{n_{.i.}}. \tag{9.6}$$

Maximum likelihood estimates (MLEs) $\hat{\underline{\theta}}$ then map back into $\widehat{P}_a^*$ using eqn (9.3). The variance of $\widehat{P}_a^*$ can be computed using the delta method on the maximum likelihood variances and covariances of $\hat{\underline{\theta}}$.

Density is then estimated as

$$\widehat{D} = \frac{n_{1..}}{k\pi w^2 \widehat{P}_a^*}. \tag{9.7}$$

Note that the sample size in this estimator is the number of first captures, $n_{1..}$. Its variance is estimated as in Buckland *et al.* (2001: 154). The variance of $\widehat{D}$ is estimated using the delta method:

$$\widehat{\mathrm{var}}[\widehat{D}] = \widehat{D}^2 [\{\mathrm{cv}(n_{1..})\}^2 + \{\mathrm{cv}(\widehat{P}_a^*)\}^2]. \tag{9.8}$$

The additional assumptions required to use recaptures in a trapping web analysis turn out to be not too troubling. Assumption 1 is largely satisfied because the duration of a trapping web study is short, usually only a few days. Therefore, seasonal changes in animal behaviour are unlikely to influence capture probability, although careful design may be required if the target species is short-lived, as for some insect species. Assumption 3 becomes irrelevant if all animals at the centre of the web are captured. Distance sampling is robust to heterogeneity in capture probability and this also pertains when using recaptures. Assumption 2 is more problematic, especially if animals become 'trap happy'. If animals are attracted to traps after being encountered once, then too many animals will be captured near the centre of the web. That will bias the density estimate high. Animals which become 'trap-shy' cause less trouble because the estimation problem reduces to the analysis of first captures only. Some species show very little behavioural response. These species are more suited to use with the recapture estimator than species which have a substantial behavioural response to capture.

The assumption that recaptures are sampled from the same distribution as first captures will result in the point estimate of detection probability being biased low. Consequently, density estimates will be biased high. The extent of this bias is often small relative to the standard error of the estimate. Therefore, the estimator can be useful. Simulations suggest that confidence interval coverage is near the nominal level when a well-designed web is used.

Using recaptures in trapping web data is similar to repeated sampling of a point transect. When points are visited multiple times, animals seen during previous visits are likely to be seen again, but because the animals are unmarked, it is unknown how many are encountered more than once. Density is estimated for multiple visits to the same point by adding an additional term c to the denominator of the estimator, which is the sampling effort (Buckland *et al.* 2001: 294). With recaptures, the total number of encounters divided by the number of first captures is analogous to estimating c, except that recaptures are sampled proportionately less as distance from the web increases. Drawing a parallel with repeat visits to a point transect, an alternative way to use recaptures would be to pool all captures (whether recaptures or not) over all trapping sessions, and to record the effort for the web as the number of trapping sessions. The reason this method is not adopted is that it is seldom reasonable to assume that all animals at or near the web centre are trapped with certainty ($g(0) = 1$) in a single session, and this method is only valid when the assumption holds.

It may be possible to drop the assumption that $g(0) = 1$ when recaptures are used with trapping webs. Several methods may be used to estimate the population size of animals at the centre of the web. For example, one could use one of the many closed mark-recapture models to estimate population size (Otis *et al.* 1978). Unfortunately, these methods bring more assumptions with them to which the methods are particularly sensitive. Nonetheless, the combination of distance sampling and mark-recapture methodologies is a field ripe for exploration. For more on this topic, see Chapter 6 and Section 11.5.

9.2.3 *Design of trapping webs*

Design of a trapping web study is important for reliable density estimation. Lukacs *et al.* (in preparation *a*) explore design issues through simulation. The trapping web is an effective way to sample populations of animals which have home ranges. Webs should have at least 90 traps. The number of traps per web may seem large, but when compared to a 12×12 trapping grid, the number of traps is not unreasonable. Often 5–7 trapping occasions are needed to ensure all animals at the centre of the web are encountered.

The most important design issue for a trapping web study is the use of multiple webs. Just as active distance sampling approaches require multiple lines or points, trapping web studies require multiple webs. At least 15 webs are needed for reliable estimation and 25 webs is better.

In general, at least 60–80 encounters in total across all webs are required to obtain a reliable estimate of density. Several ways exist to increase the number of encounters. First, increasing the number of webs is the best way to increase encounters because it also leads to better estimates of $\mathrm{var}(n)$. Second, the number of traps per web can be increased. This method will only increase the number of encounters away from the web centre, so it has limited impact on reducing the sampling variance of $\widehat{D}$. Increasing the number of traps to at least 90 has a substantial benefit in reducing bias in $\widehat{D}$. Third, traps can be baited to attract more animals into the traps. If bait is used, careful consideration must be taken to ensure that animals are not lured from long distances to the traps, as this generates upward bias in estimated densities. Finally, recaptures can be used to increase the number of encounters. The use of recaptures is most effective when the number of first captures is small and recaptures are frequent.

WebSim provides a tool for researchers to use to plan their own trapping web study (Lukacs 2002). It allows an animal population to be simulated and a trapping web designed to sample the animals. The user can set the animal characteristics to mimic the species for which the web is intended. The web can also be set up with varying trap spacing, numbers of traps, trapping occasions, and numbers of webs. WebSim allows the user to explore whether only first captures should be used, or whether recaptures can usefully be included; in some designs and levels of animal movement, inclusion of recaptures in the estimator can result in marked upward bias in the estimate of density. The output of WebSim allows the researcher to determine the amount of effort needed for a study and whether the precision of the result will be enough to answer a desired question.

9.2.4 *An example*

The flat-tailed horned lizard (*Phrynosoma mcallii*) is a relatively rare species inhabiting the desert in the southwestern United States and Mexico

Table 9.1. Trapping web data from a simulated flat-tailed horned lizard population, using 15 trapping webs

Trap ring	1st captures	2nd captures	3rd captures	4th captures
1	1	1	1	0
2	4	2	0	1
3	4	4	2	2
4	8	3	1	0
5	7	1	1	1
6	3	1	0	0
7	5	2	2	0
8	3	0	0	0
9	6	1	0	0
10	6	1	1	0
11	7	0	0	0
12	5	2	0	0
Total	59	18	8	4

(Rorabaugh *et al.* 1987; Turner and Medica 1982). The species is small, cryptic, uncommon, and moves very quickly. Therefore, an observer only rarely sees their original location if the lizard is seen at all. For these reasons an active distance sampling survey will fail, yet the species can be trapped in pitfall traps. Therefore, we simulated a trapping web study to estimate the density of the lizards. Standard Monte Carlo methods were used and program WebSim simulated data under known parameters. For each individual lizard, movement distances are generated as a uniform (0, 25 m) random variable and in a random direction. Each lizard was confined to a 0.785-ha home range.

Fifteen trapping webs are set with 96 traps per web. Each web has eight lines of 12 pitfall traps per line and are trapped for 5 days. The true density was 3 lizards/ha. The data in Table 9.1 were obtained from the simulated study.

Two issues arise from these data which are typical for trapping web data. First, more animals are encountered in the outer 2–3 rings than naively expected. This is likely to occur because animals come from an unknown distance outside the web and are encountered in one of the outer rings. This does not cause a problem because the outer rings can be truncated. Second, a higher proportion of recaptures occurs at the centre of the web than in the outer rings. This is expected because of the higher trap density at the web centre. By the last day of trapping, we would expect most captures near the centre of the web to be recaptures if the assumption that $g(0) = 1$ has been achieved.

Table 9.2. Models, model selection statistics, detection probability $(\widehat{P}_a)$, and density estimates for trapping web data for the flat-tailed horned lizard. K is the number of parameters and w_i is the Akaike weight of the ith model. The data are first captures only

Model	K	AIC_c	$\Delta\mathrm{AIC}_c$	w_i	$\widehat{P}_a$	$\mathrm{cv}[\widehat{P}_a]$	$\widehat{D}$	$\mathrm{cv}[\widehat{D}]$
Half-normal	1	141.621	0.020	0.461	0.4101	0.2201	2.833	0.291
Uniform + cosine	1	141.601	0.000	0.471	0.4026	0.1577	2.882	0.247
Hazard-rate	2	143.540	1.939	0.068	0.4711	0.3033	2.463	0.358

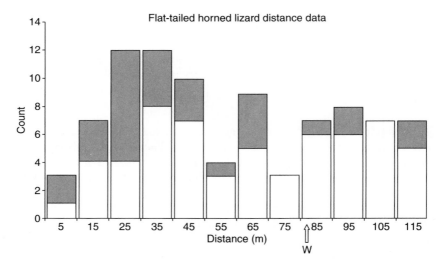

Fig. 9.1. Histograms of the untruncated trapping web data for the flat-tailed horned lizard example. Unshaded bars represent first captures, and shaded bars recaptures; w is the truncation point.

Examining a histogram of the data for first captures suggests truncation is needed at 80 m (Fig. 9.1). After truncation, a detection function was fitted to the first-capture data. The half-normal, uniform + cosine adjustment, and hazard-rate models have similar model fits and density estimates (Table 9.2). Density estimates from each model were model averaged (Burnham and Anderson 2002) to obtain a density estimate of 2.831 animals/ha and 95% confidence interval of [1.665, 4.814].

Examining the data including both first captures and recaptures shows a histogram similar to that without recaptures (Fig. 9.1). Using the method

Table 9.3. Models, model selection statistics, detection probability ($\widehat{P}_a^*$), and density estimates for the flat-tailed horned lizard trapping web first capture and recapture data. K is the number of parameters and w_i is the Akaike weight of the ith model

Model	K	AIC_c	$\Delta\mathrm{AIC}_c$	w_i	$\widehat{P}_a^*$	$\mathrm{cv}[\widehat{P}_a^*]$	$\widehat{D}$	$\mathrm{cv}[\widehat{D}]$
Half-normal	1	241.999	0.000	0.552	0.3297	0.1572	3.519	0.214
Uniform + cosine	1	242.489	0.490	0.338	0.3559	0.0945	3.260	0.147
Hazard-rate	2	243.610	1.611	0.110	0.3672	0.2285	3.160	0.279

of Lukacs *et al.* (in preparation *b*), the half-normal, uniform + cosine, and the hazard-rate models fit the data well and model selection results are similar to those without recaptures (Table 9.3). As with the first capture analysis, the density estimates were model averaged. The model-averaged density estimate was 3.392 with a 95% confidence interval of [2.286, 5.034]. Using the alternative recapture estimator, the estimate was 3.221 lizards/ha.

Whether just first captures or all captures are used, confidence intervals are obtained that cover the true value of 3 lizards/ha. The confidence interval using all captures was shorter than that for only first captures. Moreover, the recaptures increased the data set to 60 detections which begins to be large enough for reliable inference. In this example, 55% of the variance in $\widehat{D}$ came from variance in encounter rate and 45% came from estimating P_a. In a field situation, the proportion of variance due to encounter rate would typically be larger because animals tend to be patchily distributed, so that density varies by web, whereas in the simulation, a homogeneous Poisson distribution was assumed. In wild populations, $\mathrm{var}(n)$ is often large, perhaps of the order of $3E(n)$, while in the simulation, $\mathrm{var}(n)$ was equal to $E(n)$.

9.2.5 *A critique of the trapping web*

The trapping web is a useful method for sampling populations of species which are difficult to sample actively. It is especially useful for species which are rare, fast-moving, or cryptic. Data can be pooled across webs to allow a detection function to be estimated even for species at very low abundance.

Trapping webs face two potential downfalls. First, the analysis theory for data from trapping webs uses point transect theory. This theory focuses attention on the data near the centre of the web. Unfortunately, most of the data are at the middle and outer edge of the web. Focusing attention towards the centre avoids some of the problem of movement of animals onto the web, but it limits the use of the whole data set. Second, unconstrained movement of animals will cause positive bias in trapping web estimates. Lukacs *et al.* (in preparation *b*) demonstrate that movement within home ranges up to roughly 10 times the trap spacing has little impact on density estimation, whereas unconstrained movement results in substantial bias.

Recaptures are most beneficial when the number of first captures is low and recaptures are frequent. In the horned lizard example, the number of first captures is small, roughly half of that recommended for reliable inference. The recaptures provide additional information for fitting the detection function and reduce the variance in density estimates. In our example, the confidence interval length decreased by about 13%. In addition, the cost of getting the recapture data is minimal.

9.3   Trapping line transects

The trapping line transect is a new approach to passive distance sampling. Here traps are placed on each side of a transect line with declining density following a hazard-rate or half-normal distribution (Fig. 9.2). The distance from the transect line to the trap containing an animal of interest is the detection distance.

Analysis of trapping line transect data is similar to that of line transects in that data can be grouped or ungrouped. Detection at traps at the beginning and end of the line can be truncated if a strong end effect is suspected. Captures at far distances may be truncated to reduce the need to fit additional adjustment terms in the detection function. Recaptures may be used as for trapping webs.

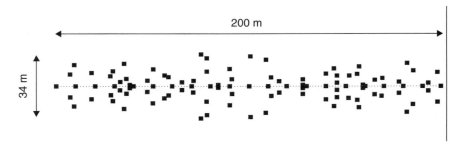

Fig. 9.2. Layout of traps in a trapping line transect. In this example, trap density follows a hazard-rate distribution with $\sigma = 9$ and $b = 3$ for distance away from the centreline and a uniform distribution along the line.

9.3.1 *Density estimation*

The density estimator for trapping line transects follows that of active line transects,

$$\widehat{D} = \frac{n}{2wL\widehat{P_a}} \tag{9.9}$$

with notation as before, but with L equal to total transect length. In the absence of cluster size data, each trapped animal is analysed as a separate detection, so that $E(s) \equiv 1$. If clusters and their sizes can be recorded, for example using remote cameras, the above equation is multiplied by $\widehat{E}(s)$. The variance of estimated density also follows from line transect theory:

$$\widehat{\mathrm{var}}[\widehat{D}] = \widehat{D}^2[\{\mathrm{cv}(n)\}^2 + \{\mathrm{cv}[\widehat{P_a}]\}^2]. \tag{9.10}$$

Trapping line transects have the same advantage over trapping webs as line transects do over point transects: a relatively high proportion of the detected animals in a trapping line transect are trapped at or near the line. This allows better estimation of P_a. Also, animal movement affects line transect estimation less severely than it affects point transect estimation.

9.3.2 *Including data from recaptures*

Recaptures can be used in a trapping line transect analysis in much the same way as they are used with trapping webs. All captures are used in the likelihood function assuming recaptures are sampled from the same distribution as the first captures:

$$\mathcal{L}(\theta|y_1, \ldots, y_n) = \prod_{i=1}^{n_{\ldots}} f(y_i), \tag{9.11}$$

where y_i is the distance of the trap that caught the ith animal from the transect line and $f(y) = g(y, \underline{\theta}) / \int_0^w g(y, \underline{\theta}) dy$ for ungrouped data; and

$$\mathcal{L}(\underline{\theta} \mid n_{.1.}, \ldots, n_{.U.}) \propto \prod_{i=1}^{U} \left(\frac{\int_{u_{i-1}}^{u_i} g(y, \underline{\theta}) \, dy}{\int_0^w g(y, \underline{\theta}) \, dy} \right)^{n_{.i.}} \tag{9.12}$$

for grouped data.

Having maximized the likelihood, P_a is estimated as

$$\widehat{P_a^*} = \frac{1}{w\hat{f}(0)}. \tag{9.13}$$

Density is then estimated with the number of first captures in the numerator and $\widehat{P}_a^*$ in the denominator

$$\widehat{D} = \frac{n_{1..}}{2wL\widehat{P}_a^*}. \tag{9.14}$$

Thus this is just eqn (9.9), but with P_a estimated by $\widehat{P}_a^*$ instead of $\widehat{P}_a$. The variance of $\widehat{P}_a^*$ can be computed using the covariance matrix corresponding to $\hat{\theta}$ and the delta method. The variance of $n_{1..}$ is estimated as it would be when recaptures are not used. The variance of estimated density is estimated by

$$\widehat{\text{var}}[\widehat{D}] = \widehat{D}^2[\{\text{cv}(\widehat{P}_a^*)\}^2 + \{\text{cv}(n_{1..})\}^2]. \tag{9.15}$$

It again should be noted that $\widehat{D}$ of eqn (9.14) is biased. The magnitude of the bias is typically small relative to the standard error of estimated density. In addition, a trap-happy response will also bias the density estimate high if recaptures are used. Therefore, recaptures should be used cautiously, or not used if the species is expected to be trap-happy.

9.3.3   *Design of trapping line transects*

WebSim allows simulation of trapping line transects. This enables a researcher to gain insights into design before attempting a field study. Four major design features specific to the trapping line transect can be manipulated by the user. First, the length of the transect can be set. The transect length determines in part the density of the traps. Second, the maximum distance from the transect line at which traps are placed is defined by the user. Third, the user can define a half-normal or hazard-rate distribution for the traps and set the parameter(s) of the distribution. All of the other options in WebSim that exist for trapping webs also exist for trapping line transects.

9.3.4   *An example*

A field study was undertaken in 2002 to determine the density of dusky-footed woodrats (*Neotoma fuscipes*) and deer mice (*Peromyscus* spp.) in two forest seral stages in California and to explore the usefulness of the trapping line transect. Particular interest was on feasibility and field considerations. The sampled area contained oldgrowth coniferous forest (referred to as older forest) and clearcut areas that had few trees but substantial shrub cover. Between the clearcut and the oldgrowth was a transition area or ecotone. The forest seral stages (old clearcut, ecotone, and oldgrowth) act as strata because density was expected to be different in each type. Sampling was conducted in each forest type four times during the summer.

The first sampling period was considered a pilot study to improve the survey design. The analysis here was based on the last three surveys. Two types of traps were used: Tomahawk live traps for the woodrats and Sherman live traps for the mice. Animals were individually marked with monel ear tags.

The design of the trapping line transect was based on a hazard-rate function (Fig. 9.2). For the pilot study, a hazard-rate function with parameters $\sigma = 9$ and $b = 3$ as defined in Buckland et al. (2001: 47) was used to assign trap distances from the centreline. In addition to the traps placed off of the line following a hazard-rate function, traps were uniformly distributed 5 m apart along the 100-m length of the centreline. In total, 105 trap stations were established per trapping line transect (21 stations on the centreline and 42 stations on each side), and 158 traps were used; 105 Sherman traps were placed one at each trap station, and 53 Tomahawk traps were placed at randomly selected stations that still maintained the hazard-rate detection function distribution off the centreline. We set up the line transect in the field by using a surveyor's compass and tape measures to maintain straight centrelines and to establish trap stations at exact and perpendicular distances from the centreline. Three people were most efficient at setting up transects, with one person on the centreline using the compass to establish perpendicular bearings to trap stations and one person on each side of the centreline with measuring tapes to mark off the exact distances from the centreline to each trap station.

After the pilot study was complete, results of trapping were compared to simulation results from WebSim and new simulations were run. The simulations suggested that the trap density, especially on the centreline, was higher than needed. This consideration led to a redesign of the trapping line transect for the remainder of the field study. The transect was extended to 200 m and trap spacing on the centreline was increased to 10 m. This change necessitated just 12 additional Tomahawk traps. The new design thus used 170 traps/trapping line transect. Lines were trapped for 6 days, but only the first 3 days of each trapping session will be used in the distance analysis. (The remaining days were part of another study.)

Over all three trapping sessions, 63 woodrats were captured, with 50 recaptures of those individuals. In addition, 38 deer mice were captured with six recaptures. Distance 4 was used to fit detection functions from all captures. Detection functions were fitted globally across strata and trapping sessions for each species. Half-normal, uniform + cosine, and hazard-rate functions were fitted (Fig. 9.3). Data were grouped into four equal size bins. No truncation of data from the outer traps was necessary.

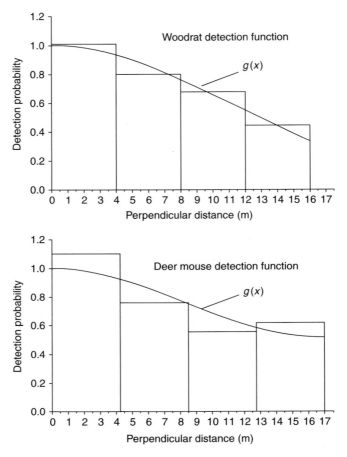

Fig. 9.3. Histograms and fitted detection functions for (a) dusky-footed woodrats ($n = 113$) and (b) deer mice ($n = 44$) from trapping line transects in three forest seral stages in California. Both first captures and recaptures are used to estimate the detection functions.

Detection probability and its variance were extracted from Distance 4 and used in eqn (9.14) to estimate density for each detection model. Density estimates were model averaged using Akaike weights (Table 9.4). The half-normal is the best model as judged by AIC_c for the woodrats, and the uniform + cosine for the deer mice despite use of a hazard-rate model for trap placement. This is not surprising because all three models are similar in shape at the parameter values used. In addition, the half-normal and uniform + cosine models have only one parameter to fit while the hazard-rate model has two parameters. Given models of nearly identical shape,

Table 9.4. Detection function model selection statistics and estimated detection probability, $\widehat{P}_a^*$, for the dusky-footed woodrat and deer mouse from trapping line transect data including recaptures. K is the number of parameters and w_i is the Akaike weight of the ith model

Species	Model	K	AIC_c	ΔAIC_c	w_i	$\widehat{P}_a^*$	$cv[\widehat{P}_a^*]$
Woodrat	Half-normal	1	306.23	0.00	0.443	0.733	0.0929
	Uniform + cosine	1	306.34	0.11	0.420	0.710	0.0908
	Hazard-rate	2	308.57	2.34	0.137	0.696	0.2092
Deer mouse	Half-normal	1	122.10	0.44	0.370	0.790	0.1549
	Uniform + cosine	1	121.66	0.00	0.461	0.756	0.1590
	Hazard-rate	2	123.66	2.01	0.169	0.699	0.3408

Table 9.5. GOF of the hazard-rate model used to generate trap placements for the woodrat data

Cell (m)	$E(n)$	n	χ^2	p
0–4	41.81	39	0.189	0.664
4–8	39.02	31	2.075	0.149
8–12	22.14	26	0.672	0.412
12–16	10.03	17	4.848	0.027
Total			7.784	0.100

AIC_c will favour those with fewer parameters. Moreover, $\widehat{P}_a^*$ is similar for each of the three models.

To examine model fit and potential assumption violations, we performed a goodness-of-fit (GOF) test on the model that was used to generate trap placements. Traps were arranged following a hazard-rate model with $\sigma = 9$ and $b = 3$. The total χ^2 value for the woodrat data suggests the model fits the data, $\chi_4^2 = 7.784$, $p = 0.10$ (Table 9.5). The most distant bin does suggest some lack of fit with more detections observed than expected. This is consistent with the possibility that animals are moving from outside the trapping line transect and being trapped in the outer traps. GOF of the model used to generate trap placements for the deer mice also suggests lack of fit in the outermost distance category, which contributes most to the apparent lack of fit of the model, $\chi_4^2 = 10.07$, $p = 0.04$ (Table 9.6). Despite some lack of fit to the trap placement generating model, we do not believe there is a large cause for concern about the estimated densities. Model fit was good along the centreline where data contribute most to the estimate of density. Poor fit only occurred away from the centreline where

Table 9.6. GOF of the hazard-rate model used to generate trap placements for the deer mouse data

Cell (m)	$E(n)$	n	χ^2	p
0–4.25	17.06	16	0.066	0.797
4.25–8.5	15.44	11	1.301	0.254
8.5–12.75	8.01	8	0.000	0.997
12.75–17	3.49	9	8.780	0.003
Total			10.067	0.040

Table 9.7. Numbers of first captures, model-averaged density estimates, coefficients of variation, and 95% confidence limits for each trapping session ($i = 1, 2, 3$), forest type, and species. Clearcut (CC), ecotone (EC), and oldgrowth (OG) data were fitted assuming a global detection function

Species	Forest seral stage and trapping session	$n_{1..}$	$\widehat{D}$	$\mathrm{cv}[\widehat{D}]$	Confidence interval Lower	Upper
Woodrat	CC 1	19	41.35	0.257	25.20	67.85
	CC 2	18	39.17	0.262	23.62	64.97
	CC 3	16	34.82	0.275	20.49	59.16
	EC 1	2	4.35	0.717	1.23	15.38
	EC 2	2	4.35	0.717	1.23	15.38
	EC 3	2	4.35	0.717	1.23	15.38
	OG 1	2	4.35	0.717	1.23	15.38
	OG 2	1	2.18	1.010	0.42	11.22
	OG 3	1	2.18	1.010	0.42	11.22
Deer mouse	CC 1	9	17.47	0.390	8.35	36.55
	CC 2	7	13.61	0.431	6.06	30.57
	CC 3	11	21.39	0.365	10.69	42.78
	EC 1	1	1.94	1.022	0.37	10.20
	EC 2	2	3.89	0.738	1.07	14.15
	EC 3	1	1.94	1.022	0.37	10.20
	OG 1	4	7.59	0.539	2.82	20.44
	OG 2	1	1.94	1.022	0.37	10.20
	OG 3	2	3.89	0.738	1.07	14.15

data contribute little to the density estimate. If a lot of movement into the surveyed strip is suspected, outer detections could be truncated.

Density estimates were computed for each stratum, trapping session, and species based on the model-averaged global detection functions (Table 9.7). Low woodrat and deer mouse densities in the oldgrowth and

Table 9.8. Comparisons in effort required to set up passive transects and trapping grids

Characteristic	Trapping design	
	Grid	Transect
Size (m)	135 × 60	200 × 34
Area (ha)	0.81	0.68
Number of trap stations	50	105
Installation effort (person hours)[a]	9.5	14
Effort/trap station	0.19	0.13
Effort/ha	11.7	20.5

[a]Averaged over the clearcut and older forest grids.

ecotone strata precluded the fitting of stratum-specific detection functions, so a global detection function was assumed.

Despite the large coefficients of variation in this study, a pattern was clear. Woodrat density was higher in the clearcut than it was in the ecotone and oldgrowth. There was also good evidence that deer mouse density was higher in the clearcut. The percentage of the variance in estimated density due to encounter rate was 71% and due to estimating detection probability was 29%. These percentages are typical of wild populations.

We noticed a number of logistic differences in establishing trapping line transects vs traditional trapping grids. A study using a 135 × 60 m trapping grid, with 50 trap stations spaced 15 m apart, had been conducted in 2001 in exactly the same locations; this allowed us to compare how much effort was required to set up the transect relative to the standard trapping grid normally used in small mammal studies. The transect required about 50% more effort (in person hours) to set up than the trapping grid even though it covered a smaller area (Table 9.8). Although almost twice as much effort per hectare was required for the transect, less effort was required per trap station setting up transects because of the close proximity of many trap stations to each other (Table 9.8). The transect also required more attention to detail in both setup and checking because trap stations were not established in a uniform and predictable manner, as was the grid. When checking trap stations for animals, trap stations were marked with stake flags and field personnel used a map of the transect and a check-off sheet to ensure that all trap stations were located and checked. Even when care was taken, some traps were missed resulting in trap mortalities. The complexity in the spatial distribution of trap stations in the transect requires more attention to detail than does the simpler design of the trapping grid.

9.3.5 *A critique of the trapping line transect*

The trapping line transect is a new method that needs refinement, but our example demonstrates that it can be a feasible field method. In a real study, more transects would be needed and it would be beneficial to run the transects over a short period of time (as was done in the example). Therefore, more effort is needed in terms of number of traps and person power, but the duration of trapping may be quite short depending on the birth and death rates of the species of interest.

We used unique individual tags to mark animals. This is not necessary for the analysis. To reduce cost, animals could be batch-marked on first capture. If a previously captured animal is caught again, the capture is noted but the animal does not need an additional mark.

In our example, a large number of traps are used. This is partly due to our goal of estimating density for two different species. Woodrats are large rodents that require Tomahawk traps while deer mice require smaller Sherman traps. If only one species was targeted, the number of traps could be reduced. If only Sherman traps were used, the total number of traps could be reduced from 170 to 105 per transect, given the same level of precision.

We chose to use data from the ends of the transects in our analysis. We do not think that large bias will occur from animals moving onto the strip from either end of the trapping transect, but if this were considered problematic, data from the ends of each transect could be truncated. More Monte Carlo simulation needs to be done to investigate this issue.

9.4 Discussion and summary

Passive distance sampling methods have a tight link to active distance methods and estimation theory is largely identical. GOF and model selection issues are equivalent for both methods. The main distinction between the two methods is that animals detect themselves by getting 'trapped'. This raises the issue of multiple sampling occasions, which are beneficial in the sense that they allow extra data to be obtained from the recaptures.

A critical step towards a high quality passive distance sampling study is careful attention to design. Valid inference relies on some form of random placement of trapping transects or webs within a defined study area. A sufficient sample size must be obtained to yield a reliable density estimate with the desired level of precision. WebSim is freely available to researchers to help answer questions about sample size, number of traps needed, allocation of traps to webs or trapping transects, and other design issues.

General theory and simulations suggest that the trapping line transect performs better than the trapping web. Fewer trapping occasions are required for the trapping line transect; 2–3 occasions rather than five. At low animal density, the trapping line transect often requires more traps and/or transects to trap enough animals to fit the detection function than does a trapping web. This appears to be due to the fact that the trapping line transect has a far larger area where trap density is high. Despite needing more traps for a trapping line transect, the decrease in the number of trapping occasions results in an overall low number of trap nights for a trapping line transect. Precision is generally better because more data exist near zero distance.

Many species show time variation in capture probability, behavioural response to capture, individual heterogeneity in capture probability, or a combination of these effects. When only first encounters are used to estimate density from trapping webs or trapping line transects, none of these issues affect the expected density estimate. Sampling is done over a short period, therefore time variation is not large. When recaptures are not used, behavioural response is irrelevant. As long as individuals at and near zero distance are caught with certainty, individual heterogeneity in capture probability is not important because P_a is estimated as an average quantity over a large and unknown number of factors.

Behavioural response becomes more problematic when recaptures are used in the analysis. If capture probability increases after first capture, density will be biased high. A negative trap response has little impact on the density estimate. Unfortunately, no GOF test currently exists to look for behavioural response in trapping web data. Therefore, use of recaptures is best with species that exhibit little behavioural response to capture.

Covariates can be used to better estimate the detection function with passively caught animals as well as actively detected animals (see Chapter 3). With passive distance methods, covariates are easy to measure because animals are in hand and trap locations are fixed. Potential covariates may include gender or age of the animal, distance from the trap to the nearest burrow, vegetation characteristics, etc.

Model selection and multimodel inference is another step towards reliable inference (Burnham and Anderson 2002). Distance 4 provides ΔAIC_c as an option in the output. Akaike weights can be computed from the ΔAIC_c values. Model-averaged estimates of density and the unconditional variance can be computed using the Akaike weights. These computations can be easily done in a spreadsheet.

Trapping webs and trapping line transects are just two possible passive distance sampling approaches. These approaches may be modified as innovative methods of encountering animals are developed. More work needs to be done to develop new passive approaches to distance sampling. Work is especially needed on field methods and technology for passive distance methods. Some field methods are quite inexpensive and can be used on a very large scale (e.g. insect sticky boards, hair snags, various types of pitfall traps). Animals that can be trapped with these sorts of methods lend themselves well to passive distance sampling methods.

10

Assessment of distance sampling estimators

R. M. Fewster and S. T. Buckland

10.1 Introduction

In this chapter we return to conventional line transect sampling, as outlined in Chapter 2, Sections 2.2–2.5. On a number of occasions in the literature, distance sampling methods have been tested by simulation studies. In some cases, the methods are reported to work well (e.g. Cassey and McArdle 1999), while in other cases the performance is criticized (Barry and Welsh 2001; Melville and Welsh 2001). The array of simulation options can seem bewildering: should the transect lines be regarded as fixed while object positions are allowed to vary, or should the object positions be fixed while the lines are allowed to vary? Indeed, should they both be random or both fixed? Should the data truncation distance be decided separately for every simulation, or kept constant?

Simulation studies are a valuable aid to understanding and intuition, and are recommended for all practitioners. However, results can be misleading unless the hypothesis under test is clear and relevant. This chapter aims to explore the inferential framework on which distance sampling is based, to clarify how simulations should be formulated to meet the assumptions underlying the conventional estimators. Investigators are free to test the methods in any framework they wish, but it is important to recognize when the simulations are departing from established protocol.

For simplicity, we focus on line transect sampling. The issues and recommendations can readily be extended to point transect sampling.

10.1.1 *Notation*

The following notation is used.

A the whole survey region of interest. (The area of the region is also called A, with the intended use clear from the context.)

a the area of the covered or sampled region

N	the unknown number of objects in the area A
$\widehat{N}$	the estimated number of objects in the area A
N_c	the unknown number of objects in the covered area a
$\widehat{N}_c$	the estimated number of objects in the covered area a
n	the number of objects detected in the covered region
w	the half-width of the search strips
$y_1, \ldots, y_n$	the perpendicular distances from the transect line of the n detected objects: $0 \leq y_i \leq w$ for all i
$g(y, \theta)$	detection function: the probability of detecting an object, given that it is at distance y from the transect line. θ is a scalar or vector parameter
$\pi(y)$	probability density function (pdf) of the perpendicular distances of all objects (detected and undetected) from the transect line
$P_a(\theta)$	$\int_0^w g(y, \theta)\pi(y)dy$: the average probability that an object is detected, given that it is in the covered region

10.2 Estimation framework

We consider the following estimation framework. This is the standard framework implemented by program Distance, except that polynomial or cosine adjustment terms are not considered here.

Step 1. Model step: to estimate parameters of the detection process using observations within the covered region.

(i) Assume detections are independent, given the object distances.

(ii) Assume the detection function belongs to a family $g(y, \theta)$ parameterized by θ, with $g(0, \theta) = 1$.

(iii) Assume that within-strip object distances are independent, with pdf $\pi(y) = w^{-1}$ (uniform), so $P_a(\theta) = w^{-1} \int_0^w g(y, \theta)dy$.

The parameter θ is estimated by maximizing the conditional likelihood of the distance data $y_1, \ldots, y_n$, given n. Hence $\hat{\theta}$ corresponds to the value of θ that maximizes

$$\prod_{i=1}^n \frac{g(y_i, \theta)}{wP_a(\theta)}.$$

Step 2. Design step: to extrapolate from the covered region to the whole area A.

(i) Ensure transect lines have been chosen according to an equal coverage design. This means that, before the design is realized (i.e. before the transect lines are chosen), every point in the region A has an equal probability a/A of being included in the covered region a.

(ii) Unconditional on the design realization, every object in the region A has an equal probability $aP_a(\theta)/A$ of being included in the sample of *detected* objects. Given $\hat{\theta}$ from Step 1, this probability is estimated by $aP_a(\hat{\theta})/A$.

The number of objects N in the area A is estimated by

$$\widehat{N} = \frac{n}{aP_a(\hat{\theta})/A} = \frac{nA}{aP_a(\hat{\theta})}.$$

10.3 Model and design

The estimation procedure above depends partly upon model-based reasoning (Step 1), and partly upon design-based reasoning (Step 2). We expand upon both of these ideas here. A discussion of design-based and model-based philosophy is given by Borchers *et al.* (2002: 46–53).

10.3.1 *Model-based inference*

A *model* is a set of assumptions. Models are used for quantities that are unknown. Model-based approaches use probability distributions to summarize the behaviour of unknowns. These probability distributions are linked to observable quantities to allow estimation using maximum likelihood or other methods.

In conventional line transect sampling, the observable quantities are the number of objects detected, n, and the sample of detected distances, $y_1, \ldots, y_n$. Assumptions, or models, are needed for the unknown detection process and the unknown positions of objects within search strips. These are assumptions (i), (ii), and (iii) in Step 1 of Section 10.2. No assumption is usually made about object positions across the whole survey area A; assumptions are confined to the search strips.

The assumptions enable us to find the probability distribution of the observable quantities $y_1, \ldots, y_n$. From this we use a conditional likelihood, conditional on n, to estimate the parameters of the detection function.

Every time we conduct a survey, we can expect a new value of n and a new set of detected distances $y_1, \ldots, y_n$. In the model-based framework, these quantities are considered to be drawn anew from their probability distributions each time. The variance in estimation is due to the randomness in drawing the quantities from their probability distributions. Conceptually, *if* all quantities are governed by the rules of the assumed distributions, then the estimated variance reports how much variability would be obtained from one occasion to the next if the exact same survey were repeated with new draws from the probability distributions.

Note that the variance of a model-based estimator is determined by the model. A model that assumes probability distributions with higher levels of variability will produce estimators with higher variance.

Clearly, the danger of model-based approaches is that the model assumptions might be wrong. As a rule, model-based estimators become more precise as the assumptions become stronger. If the assumptions are wrong, however, the estimators could be badly biased and estimated variances could be misleading. Model assumptions should be based on external knowledge about the biological population and physical survey conditions. Usually for biological populations, it is wise to make few assumptions about object positions, at the expense of high variability in the estimators. However, if realistic models can be constructed, they can improve estimation precision considerably (Chapter 4 and Section 10.8).

The following observation is central to model-based inference.

Every model applies only within the scope of its own assumptions. For example, in Step 1 of Section 10.2, assumptions about object positions are made only within the search strips. If we wish to test robustness of the estimators to failure of the assumptions, then we must restrict analyses to the search strips only. It would not be relevant to test the robustness of the estimators to changes in object positions beyond the limits of the search strips.

10.3.2 *Design-based inference*

A *design* is a sampling procedure. As such it is controlled by the investigator, and its specifications are known. In distance sampling, the design specifies the number of transect lines and a rule for generating their positions. An example of a design is to use 50 transects with fixed orientation and spacing but with a random start-point.

The chief requirement of a standard line transect survey design is that, a priori before the positions of the transect lines are determined, every point in the region A should have equal probability of being in the covered area a. This is known as *equal coverage probability*. For practical reasons, the exact requirement is often not met, especially for an irregularly shaped region A, but it should be satisfied to a good approximation if conventional analysis methods are to be used. Issues of coverage probability are discussed in Chapter 7, including methods for dealing with variable coverage probability.

Violation of the requirement of equal coverage probability generates bias in abundance estimates if object density is correlated with coverage probability, unless analysis methods specifically allow for variable coverage probability (Section 7.3). The issue is also important for variance estimation, and for evaluation of design-bias in a simulation setting. If design-based variance estimation assumes equal coverage probability, variance estimates may be biased when the assumption is violated.

In design-based inference, the objects are considered to be fixed. They could be surveyed using any one of the possible realizations of the chosen design. Every realization will give a different estimate. Randomness (and therefore variance) in estimation is due entirely to the randomness in the design realization. This equates to randomness due to the act of sampling. Replication of sampling units is necessary for design-based variance estimation. In distance sampling, this means the design should include many transect lines (preferably more than 20).

Design-based estimation uses only the known properties of the design, and does not involve assumptions about unknowns. To illustrate design-based reasoning, consider the problem of estimating the number of objects N in the region A, given observations only from the covered region a. Imagine that the detection function g and the detection parameter θ are known, so that assumptions about these quantities are not necessary. A model-based estimator for N would assume a probability distribution for object positions throughout the area A (typically uniform and independent), and construct a likelihood function (typically binomial) as in Section 2.4.2. A design-based estimator, on the other hand, reasons that the design has probability a/A of covering any point in A, and there is probability $P_a(\theta)$ (known) that any given object is detected given that it is in the covered region. On average, the number of objects that will be detected under this design is therefore $E(n) = N \times (aP_a(\theta)/A)$, giving the final estimator $\widehat{N} = n/(aP_a(\theta)/A)$.

As described, the design-based estimator does not involve any assumptions about the positions or independence of objects in the region A. Only the requirement of equal coverage probability is needed, which is a property of the design and therefore under the control of the investigators—and even this requirement can be relaxed using the methods of Section 7.3. The calculation of $P_a(\theta)$ involves $\pi(y)$, the pdf of the distances from objects to transect lines, but $\pi(y) = w^{-1}$ is a fact of the design if object positions are fixed and transect lines are equally likely to fall at any distance from them. Note that it is only possible to view $\pi(y)$ this way when θ is known. When θ is to be estimated, the requirement $\pi(y) = w^{-1}$ is used for estimating θ and is *conditional* on the placement of the search strip.

In practice, model-based and design-based estimators often provide similar point estimates of abundance. The variance of the estimators can be quite different, however, because they are measuring the variability from different sources. Model-based variance relates to the variance that would be obtained if the model is true, upon repeated draws from the same probability distributions. Design-based variance relates to the variance that would be obtained for a fixed set of objects if a new design realization was generated for every survey. If the model involves questionable assumptions such as the independence of object positions throughout A, the design-based variance is likely to be larger and more realistic.

The following observation is central to design-based inference.

Design-based properties of estimators and design-based variances depend upon the design. An estimator cannot be described as 'design-unbiased' without reference to the design involved. The same estimator may be unbiased under some designs and biased under others; 'bias' is a statement about the design as well as the estimator. Design-based estimators assume that coverage probabilities are known. If equal coverage probability is assumed, design-based estimators are likely to be biased under designs with variable coverage.

10.3.3   *Distance sampling: a composite approach*

From the above discussion, it is clear that it is not possible in distance sampling to adopt the purely design-based estimator $\widehat{N} = n/(aP_a(\theta)/A)$. The quantity $P_a(\theta)$ depends upon assumptions about the unknowns g, θ, and $\pi(y)$, and a model is therefore necessary. (The usual model $\pi(y) = w^{-1}$ does not involve any estimation, but it is nonetheless a model.)

On the other hand, a purely model-based approach is undesirable because it requires assumptions about the distribution of objects throughout the entire survey region A. Without using advanced model formulations and estimation techniques (Chapter 4), such assumptions are naive at best.

The conventional line transect estimator is therefore based on a composite philosophy. Modelling is kept to a minimum, to estimate the detection function parameter θ only. Once θ has been estimated, the quantity $P_a(\hat{\theta})$ can be computed. This estimate is plugged into the design-orientated formula $\widehat{N} = n/(aP_a(\theta)/A)$, to give the final estimator $\widehat{N} = n/(aP_a(\hat{\theta})/A)$ as in Step 2 of Section 10.2.

The philosophy behind this estimator cannot be neatly classified as 'design-based' or 'model-based', which may cause confusion for those undertaking simulation studies. However, the resulting estimator is far more defensible than a purely model-based estimator in the conventional framework.

A recommended approach for simulation studies is described in the next section.

10.4   Simulation framework

The usual purposes of simulation studies are to test properties of estimators such as bias, variance, and confidence interval coverage when all assumptions are fulfilled, and to test robustness of estimators to failures in assumptions or changes in survey design. Simulations should therefore be designed to highlight the question of interest. Some recommendations are provided below.

10.4.1 *Testing the design*

Testing the design by simulation is useful for two main purposes:

1. Testing the variance of $\widehat{N}$ under different designs. For example, investigating the impact of changing the number of transect lines, or changing the orientation of the transect lines with respect to known gradients in object density.

2. Investigating bias in the estimation of $\widehat{N}$ due to unequal coverage probabilities in the design.

Properties specific to the design are best tested with all other factors kept fixed. This ensures that results are not complicated by variance due to the estimation of g and θ, for example. We recommend the following procedure for testing properties specific to the design:

1. Select the area of interest, A.

2. Generate a fixed number N of object positions throughout A, according to some pre-determined spatial distribution.

3. Select a detection function family, $g(\cdot, \theta)$ (e.g. half-normal). Select the value of θ, and an appropriate distance w for the search strip half-width. Any combination of w and θ may be chosen, although combinations with $g(w, \theta)$ appreciably less than 0.15 are not consistent with recommended practice in real surveys (Buckland *et al.* 2001: 104–5).

4. Calculate $P_a(\theta) = w^{-1} \int_0^w g(y, \theta) dy$.

The object positions and specifications of g, θ, w, and $P_a(\theta)$ do not change for the remainder of the experiment.

5. Generate a single set of transect line placements from the chosen design, and establish search strips of width $2w$ around each line.

6. Generate artificial 'sightings' data, such that any object in the search strip at distance y from the transect line is detected with probability $g(y, \theta)$.

7. Let n be the total number of objects detected in the covered area a (where $a = 2wL$ if L is the total line length surveyed). Estimate $\widehat{N} = nA/(aP_a(\theta))$.

Steps 5–7 are repeated many times to obtain a sample of estimated values $\widehat{N}$. The design bias for the design, for the particular fixed object positions in step 2, is estimated by comparing the mean of the $\widehat{N}$ sample with the known value N from step 2. The design-based variance of the estimator $\widehat{N}$ for these object positions is estimated by the sample variance of the $\widehat{N}$ sample. Methods for calculating confidence intervals can also be tested, by calculating (say) a 95% confidence interval for N in each simulation. The proportion of simulations in which the confidence interval encloses the true value N should be close to the nominal coverage of 0.95.

The whole procedure may be repeated using object positions from step 2 generated according to different plans, bearing in mind any known biological trends in object density throughout the region. The variance of $\widehat{N}$ will often vary for different schemes of object placement.

The procedure above is fully design-based. No modelling is required, because g and θ are known. This is possible in simulations, although not in real life, and it allows the properties of the design to be tested in isolation. Note that $\pi(y) = w^{-1}$ is assumed, but in a fully design-based setting, this does not make any assumption about uniformity or independence of the object positions, even within the search strips. Instead, it indicates that the randomly placed transect lines are equally likely to fall at any distance from a fixed object. It is a statement about the design, not about the object positions.

Any design with equal coverage probability is design-unbiased for $\widehat{N}$. This is clear from the derivation in Section 10.3.2: we have $E(n) = N \times (aP_a(\theta)/A)$, so if $\widehat{N} = nA/(aP_a(\theta))$, then $E(\widehat{N}) = E(n)A/(aP_a(\theta)) = N$.

10.4.2   Testing the model

Modelling is applied within the search strips only, to estimate the detection parameter θ. Mistakes in the model assumptions are most likely to concern the form of the detection function g, and the assumption that $\pi(y) = w^{-1}$. We use the subscript T to denote the *true* specification of g, namely g_T, and M to denote the specification chosen in the model, g_M. Because simulations should be restricted to the search strips, let N_c be the true number of objects in the covered region.

Testing the model by simulation is particularly useful for the following two purposes.

(i) Testing the effect of mis-specifying the detection function g on estimating N_c.

(ii) Testing the impact of non-uniform object distances ($\pi(y) \neq w^{-1}$) on estimating θ, with correctly specified g.

It is also useful for testing other requirements, such as the effects of sample size n on estimation of θ. Again, we suggest that only one factor is varied at a time. Testing the effects of mis-specification of g requires estimation of θ, so it is usually easiest to employ simulation studies. Testing the impact of $\pi(y) \neq w^{-1}$, with g fixed, is most easily done mathematically. Both procedures are described below.

10.4.2.1 *Testing mis-specification of g*

The following simulation procedure is recommended. The simulations can usually be adequately conducted within a single search strip, depending upon the questions of interest. To test the effects of mis-specifying the detection function in isolation of other factors, we use $\pi(y) = w^{-1}$ for $0 < y < w$.

1. Set up a search strip with half-width w.
2. Select a true value for the number of objects in the strip, N_c.
3. Select a true detection function family, $g_T(\cdot, \theta)$, and select a true value for θ, θ_T. Now select a detection function family for the model, $g_M(\cdot, \theta)$, parameterized by a possibly different θ, where g_M differs from the true family g_T.

The specifications above remain fixed for the remainder of the experiment.

4. Generate N_c object distances $y_1, \ldots, y_{N_c}$ from the pdf $\pi(y) = w^{-1}$.
5. Generate artificial sightings data, such that object i at distance y_i is detected with probability $g_T(y_i, \theta_T)$.
6. Let n be the total number of objects detected, and let $y_1, \ldots, y_n$ be renamed as the distances of the *detected* objects.
7. Estimate $\hat{\theta}_M$ by finding the value of θ that maximizes $\{\prod_{i=1}^n g_M(y_i, \theta) / (w P_a(\theta))\}$, where $P_a(\theta) = w^{-1} \int_0^w g_M(y, \theta) dy$.
8. Estimate $\widehat{N}_c = n / P_a(\hat{\theta}_M)$.

Steps 4–8 are repeated many times to obtain a sample of estimated values $\widehat{N}_c$. The bias of $\widehat{N}_c$ can be estimated by comparing the sample mean with the true value N_c. The estimator variance is estimated from the sample variance of $\widehat{N}_c$.

Note that the estimation of g involves object *distances* from the transect line, not the absolute object *positions*. It is therefore ambiguous whether the above procedure involves randomness in line placement, or randomness in object placement. It can be thought of in both contexts.

10.4.2.2 *Testing non-uniformity of within-strip object positions: $\pi(y) \neq w^{-1}$*

As above, $\pi(y)$ refers to the distribution of *distances*, so it is ambiguous whether it arises from randomness in line placement or object placement. However, it is most useful to think of $\pi(y)$ as a description of object response to fixed survey lines. For example, $\pi(y)$ can be used to describe scenarios such as responsive movement towards or away from the transect lines, disturbance around permanently established lines, or use of fixtures such as roads or waterways as natural transects.

It is often easier to use mathematical methods to find the model-based expectation of $\widehat{N}_c$ when g is known. This also has the advantage that it removes simulation error and provides an exact result. Almost any functional form for $\pi(y)$ can provide an exact result for $E_y(\widehat{N}_c)$ using simple numerical integration software.

To isolate the impact of $\pi(y) \neq w^{-1}$, we first consider that $g(y, \theta)$ is known and fixed: so $g_M = g_T = g$ and θ is known. Asymptotic results (for large N_c) are also possible when θ is estimated: see Section 10.6.2. For smaller N_c, it may be easiest to revert to simulations to test the effects of $\pi(y) \neq w^{-1}$ on estimation of θ.

Suppose a spatial distribution $\pi(y)$ has been selected for $-w < y < w$. (If π is symmetric, that is, if $\pi(-y) = \pi(y)$, then $\pi(y)$ can be redefined for the interval $0 < y < w$.) We now derive the model-based expectation of $\widehat{N}_c$. This is the expectation of $\widehat{N}_c$ over the randomness in the distribution π. That is, if the survey were repeated many times using new y-values $y_1, \ldots, y_{N_c}$ drawn from $\pi(y)$, then $E_y(\widehat{N}_c)$ would be the average of the $\widehat{N}_c$ values obtained.

Consider that $\widehat{N}_c = n/P_a(\theta) = n/(w^{-1} \int_0^w g(r, \theta) dr)$, because $P_a(\theta)$ is calculated under the (possibly erroneous) assumption that $\pi(y) = w^{-1}$. Then

$$E_y(\widehat{N}_c) = \frac{E_y(n)}{w^{-1} \int_0^w g(r, \theta)\, dr} = \frac{E_y\left(\sum_{i=1}^{N_c} I\{\text{object } i \text{ is detected}\}\right)}{w^{-1} \int_0^w g(r, \theta)\, dr},$$

where $I\{\text{object } i \text{ is detected}\}$ is the usual indicator function, equal to 1 if object i is detected, and 0 otherwise. From basic probability theory,

$$E_y\left(\sum_{i=1}^{N_c} I\{\text{object } i \text{ is detected}\}\right) = N_c \int_{-w}^{w} g(|y|, \theta)\pi(y)\, dy.$$

Thus

$$E_y(\widehat{N}_c) = \frac{N_c \int_{-w}^{w} g(|y|, \theta)\pi(y)\, dy}{w^{-1} \int_0^w g(r, \theta)\, dr}. \tag{10.1}$$

Clearly, if $\pi(y)$ is uniform ($\pi(y) = (2w)^{-1}$ for $-w < y < w$, or equivalently, $\pi(y) = w^{-1}$ for $0 < y < w$), then $E_y(\widehat{N}_c) = N_c$ and the distance sampling estimator is model-unbiased for known θ. However, for any other formulation $\pi(y)$, eqn (10.1) allows the model-based bias of $\widehat{N}_c$ to be calculated exactly when θ is known.

10.4.3 *Testing the full line transect estimation procedure*

Up to this point, we have been careful to test only one factor at a time, keeping other factors fixed as far as possible. Frequently, however, practitioners will simply wish to know whether the whole survey will work, including design issues and the specification and estimation of g. In addition to bias and variance of $\widehat{N}$, the properties of many other estimators are of interest, such as the reciprocal $1/P_a(\hat{\theta})$, estimators for var$(\widehat{N})$, and confidence interval estimators. Note that the bias of an estimator is best understood in the context of its own variability. Bias is of little practical importance if it is small compared with the standard error of the estimator. A good rule of thumb is that, if the ratio given by the bias divided by the standard error has magnitude less than about 0.2, there is little cause for concern (Cochran 1977: 14–15).

The natural framework for testing the full methodology is to keep objects fixed and allow the design realization (transect lines) to vary. This follows the way that standard line transect variance estimation takes place. One recommended method of estimating var$(\widehat{N})$ is to bootstrap, using transect lines as sampling units (Buckland *et al.* 2001: 83). This provides an estimate of the variance in $\widehat{N}$ due to the design realization, where object positions are conceptually fixed. A second recommended estimator for var$(\widehat{N})$, which yields similar results, is the analytic variance estimator in program Distance. This incorporates design-based estimation of var(n) (Buckland *et al.* 2001: 78–80) with model-based estimation of var$[\hat{f}(0)] = $ var$[1/\{wP_a(\hat{\theta})\}]$ (Buckland *et al.* 2001: 62, 66). Once again, the objects can be considered fixed, and the model-based variance of $1/\{wP_a(\hat{\theta})\}$ can be thought of as the variability arising from random lines over fixed objects.

We therefore recommend that the full procedure is tested with fixed objects and random line placement, as in Section 10.4.1, but incorporating the extra step of estimating θ. This framework is also valid for testing robustness of the estimators to mis-specification of g, as in Section 10.4.2.1. Within-strip effects such as responsive movement are issues of field practice and are best tested separately, as in Section 10.4.2.2.

For a full test of the methods, we recommend that the analysis is performed using program Distance for estimation, and incorporating the full suite of modelling techniques, including polynomial or cosine adjustment terms, automated model selection via Akaike's information criterion (AIC), and data-driven truncation width w. The selected model and truncation distance may therefore be different for every simulation. Truncation tends to make analyses more robust, with little if any loss of precision. One possible rule for deciding truncation width is to select w so that approximately 5% (line transects) or 10% (point transects) of the detection distances are truncated (Buckland *et al.* 2001: 103–4, 151–3). An alternative is to select

w so that $g(w, \theta) \simeq 0.15$, although this rule is harder to automate. As with any real survey design, simulations should employ many transect lines, and transects should run parallel to any known gradients in object density.

An enhancement to program Distance is planned that will allow simulations such as the above to be conducted entirely within Distance. Until this is ready, data sets must be simulated externally, and read into Distance to be analysed. Computer code is then needed to extract the relevant quantities from the Distance output file.

10.5   Example: testing the design

In this section, we give some examples of testing the design, as described in Section 10.4.1. For simplicity, all simulations are conducted in a square region A, where $A = [0, 1] \times [0, 1]$. We use two sets of fixed object configurations, described as Uniform and Coastal. For each configuration, the total number of objects is fixed at $N = 1000$. The horizontal and vertical coordinates of the fixed object positions are named $(h_1, v_1), \ldots, (h_N, v_N)$. The object configurations are shown in Fig. 10.1.

1. Uniform configuration. Objects are independently and uniformly distributed throughout the survey area A. Object positions are (h, v), generated from $h \sim$ Uniform$[0, 1]$, and $v \sim$ Uniform$[0, 1]$.

2. Coastal configuration. Objects have a marked density gradient in both the h and v directions, but especially in the v direction. Positions (h, v) are generated from $h \sim$ Beta$(1, 1.5)$, $v \sim$ Beta$(5, 1)$. The distribution of (h, v) is artificial, but it is used to mimic an effect such as a coastline that can cause dramatic density gradients if, for example, the population is largely concentrated in inshore waters.

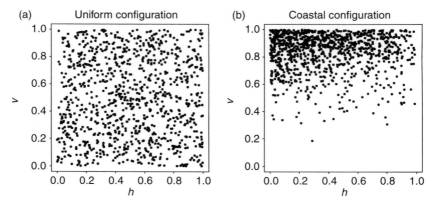

Fig. 10.1. Fixed object configurations for testing the design.

As described in Section 10.4.1, we use simulations to test the variance properties of $\widehat{N}$ under different design specifications such as the number and orientation of transects, and to test the effects of unequal coverage probability on the bias of $\widehat{N}$.

10.5.1 Testing equal coverage designs for var($\widehat{N}$)

We first examine properties of $\widehat{N}$ under different survey designs with equal coverage probability. For equal coverage designs, when $g(\cdot, \theta)$ is known, $\widehat{N}$ is an unbiased estimator of N. However, we would like the design-based variance of $\widehat{N}$ to be as small as possible, otherwise any single realization of the design could provide an estimate $\widehat{N}$ wholly remote from the true value N. In practice for field surveys, one single realization of the design is all that is available. This underlines the importance of striving for low design-based variance.

10.5.1.1 Establishing a design with equal coverage probability

The design specifies the number of transects, and a rule for generating their positions and orientation. We will consider simple designs that always orientate the transect lines horizontally or vertically in the region A. Considering first the case of horizontally orientated transects, let $u \in [0, 1]$ be the vertical v-coordinate of the transect line. To test the design as in Section 10.4.1, u should be randomly generated from some distribution. We need to devise a rule for the distribution of u that ensures equal coverage probability across the area A.

The obvious choice of $u \sim \text{Uniform}[0, 1]$ does not quite guarantee equal coverage, unless careful attention is paid to edge effects at the upper and lower vertical limits of A. Search strips centred at position u will extend beyond the edge of the survey area if $u < w$ or $u > 1 - w$. (Recall that w is the half-width of the strip, considered fixed in these simulations.) For exactly equal coverage probability, any search strips protruding from one edge of the survey area should be wrapped around to the other edge. We can think of the area A being glued onto the surface of a cylinder with the horizontal edges $v = 0$ and $v = 1$ joined. Any part of the search strip lying beyond one edge of A reappears at the other edge.

In this scheme, a transect position at u effectively establishes three non-overlapping search strips, with centrelines at $u - 1$, u, and $u + 1$. (This assumes that $w < 0.5$ for the present specification of A, which will invariably be the case in real surveys.) By design, we survey all parts of the search strips that overlap with A. The easy way to simulate surveys using the wraparound design is to concatenate the vector of object v-coordinates three times, namely $(v_1 - 1, \ldots, v_N - 1, v_1, \ldots, v_N, v_1 + 1, \ldots, v_N + 1)$, then proceed with the concatenated object positions but using a single centreline u. Practical implementations of the wraparound design, using buffer zones, are described in Chapter 7.

Without the wraparound design, coverage probability is not equal because the sampling probability is lower for points with v-coordinates in $[0, w]$ or in $[1 - w, 1]$ than for points in $[w, 1 - w]$. This is because points in the middle of the survey area can be covered by strips with centrelines above them or below them, whereas points at the edges can only be covered by strips coming from one direction. They are therefore less likely to be covered.

It is worth noting that the edge effect described above is negligible in almost all real surveys, where the strip widths w are a tiny fraction of the width of the survey area A. However, it does serve to illustrate the fact that even the most innocuous-looking designs can be subject to bias in sampling probability. If w extends to a large fraction of the survey area, the design bias of the non-wraparound design becomes severe. This explains the severely biased results reported in table 1 of Barry and Welsh (2001), who failed to ensure equal coverage probability in their simulations.

10.5.1.2 *Simulation experiments*
We have so far established the positions of $N = 1000$ objects and selected a basic design framework with equal coverage probability (the wraparound design). Following the procedure in Section 10.4.1, we must also select specifications for g and for w. We choose the half-normal detection family for g, so that $g(y, \theta) = \exp\{-y^2/(2\theta^2)\}$ for $0 \leq y \leq w$. In this section, we use $w = 0.02$ and $\theta = 0.01$. Fig. 10.2(a) shows the resulting detection function over the width of the strip. With these specifications, data truncation occurs at an approximate detection probability of 0.14.

Using the Uniform and Coastal object configurations, we test the effects of the following properties of the design on the variance of $\widehat{N}$:

(1) the number of transect lines (1 or 10);

(2) the orientation of transect lines (horizontal or vertical);

(3) with multiple transect lines, the choice between an evenly spaced grid of transects and a completely random selection of transects.

Figure 10.2(b) shows a design with one transect line and horizontal orientation. With one transect, the objects stay fixed while a new transect position u is drawn from $u \sim \text{Uniform}[0, 1]$ for each simulation. Figure 10.2(c) shows a design with 10 transect lines and grid spacing. For this plan, called the grid scheme, objects stay fixed while the first transect position u is drawn from $u \sim \text{Uniform}[0, 0.1]$ for every simulation. The remaining nine transects are evenly spaced at distance 0.1 apart. A further possibility, not shown, is a 10-line scheme with completely random transect spacing, in which the 10 transect positions are drawn randomly and independently from Uniform $[0, 1]$ for every simulation. This scheme is called the simple random sample scheme (SRS).

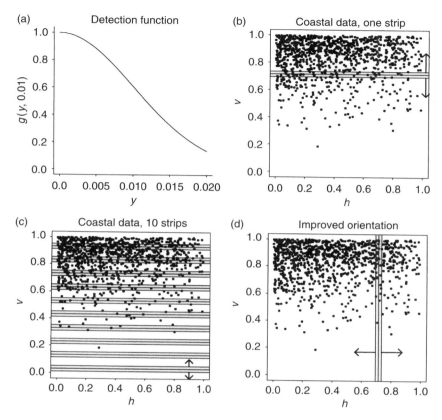

Fig. 10.2. Simulation framework for testing the design. The strip placements are random, indicated by arrows, while the objects are fixed.

Finally, Fig. 10.2(d) shows a vertical orientation of the transect for the Coastal configuration. Because the major gradient in object density is vertical for the Coastal configuration, vertical orientation of the transect is likely to be a great improvement over horizontal orientation. This is because a vertical transect captures the full range of object density within a single search strip. Using horizontal transects, object density will be extremely high on some transects, and extremely low on others. The design-based variance of $\widehat{N}$ is therefore expected to be much higher with the horizontal transect orientation than with the vertical orientation.

Aligning the transect lines parallel to any known gradients in object density is an important part of survey practice. It is discussed in Section 7.2.2.1 and by Buckland *et al.* (2001: 238–9).

For all the design formulations above, any transect at u whose search strip extends beyond the edge of the region A is wrapped around to the other edge. The wrapped fragment is surveyed as if the centreline were positioned at $u - 1$ or $u + 1$.

Results from the simulation studies are shown in Table 10.1. For each object configuration and survey design, results are presented from 10,000 simulations of $\widehat{N}$. As expected for equal coverage designs, the bias in $\widehat{N}$ is negligible. However, the variance in $\widehat{N}$ changes dramatically according to the number of transects, grid or random transect spacing, the object configuration, and the orientation of transects.

Increasing the number of transects reduces s.d.$(\widehat{N})$ because it increases the number of detected objects, n, as shown in Table 10.1. For constant transect lengths, as we have here, increasing the number of transects from one to ten also increases n by a factor of about ten, on average. If SRS is used for transect spacing, this decreases s.d.$(\widehat{N})$ by a factor of roughly $1/\sqrt{10}$. In general for this scheme, s.d.$(\widehat{N})$ behaves as $1/\sqrt{K}$, where K is the number of transects.

When the transects are placed according to the grid scheme, variance reduction can be much more spectacular. This depends upon the object configuration. For the Uniform configuration, all parts of the survey region are alike, and there is no gain in using the grid scheme rather than the SRS. With the Coastal configuration, however, the grid scheme achieves about a tenfold reduction in s.d.$(\widehat{N})$ for a tenfold increase in transect lines. As a rule, the grid scheme will produce significant improvements over SRS whenever there are marked gradients across the survey region.

In simulations, we have the luxury of increasing the number of transects with no extra cost. In reality, practitioners often have to choose between

Table 10.1. Results of testing an equal coverage design, with true value $N = 1000$. Every row represents results from 10,000 simulations from the wraparound design, with $w = 0.02$ and $\theta = 0.01$. For the 10-transect designs, transects are selected either as a simple random sample (SRS), or as an evenly spaced grid with a random start point (GRID)

Objects	Transects	Orientation	$\widehat{N}$ Mean	s.d.	n Mean	s.d.
Uniform	1	Horizontal	1000.3	184	24	4
Uniform	10: SRS	Horizontal	1000.3	58	239	14
Uniform	10: GRID	Horizontal	1000.5	61	239	15
Coastal	1	Horizontal	999.0	1287	24	31
Coastal	10: SRS	Horizontal	999.7	409	239	98
Coastal	10: GRID	Horizontal	999.8	118	239	28
Coastal	1	Vertical	1000.5	401	24	10
Coastal	10: SRS	Vertical	1001.0	126	240	30
Coastal	10: GRID	Vertical	1000.0	40	239	10

using many short transects, or fewer long transects. Using many short transects in these circumstances will not necessarily reduce the variance, but it will provide a better variance estimate.

As expected, Table 10.1 also shows that changing the transect orientation from horizontal to vertical for the Coastal configuration brings greatly reduced variance in $\widehat{N}$.

10.5.2 A design without equal coverage probability

We now test what happens when the designs above do not use the wraparound method to ensure equal coverage probability. This means that parts of the search strips that fall outside the survey area A are simply ignored, rather than wrapped around.

The first six rows of Table 10.2 show results for the same design specifications used in the wraparound design of Table 10.1, with horizontal transects. Neglecting to ensure equal coverage probability has clearly introduced bias, but the bias is only 1–2% and is of little practical importance, with the ratio of bias to s.d.($\widehat{N}$) well within the guideline magnitude of 0.2 mentioned in Section 10.4.3. The variance of $\widehat{N}$ appears slightly inflated when compared with the results in the first six rows of Table 10.1.

The edge effect due to the non-wrapped strips only really becomes problematic when the search strip half-width w extends to an appreciable proportion of the survey area. Using a wide strip ($w = 0.3$) with high detectability ($\theta = 0.5$) in the seventh row of Table 10.2 shows that design-bias of the estimator $\widehat{N}$ is now appreciable. The last row of Table 10.2

Table 10.2. Results from the non-wrapped design, with true value $N = 1000$. Every row represents results from 10,000 simulations with horizontal transect orientation. The 10-transect designs are generated either as a simple random sample (SRS), or as an evenly spaced grid with a random start point (GRID)

Wrap	(w, θ)	Objects	Transects	$\widehat{N}$ mean	s.d.	n mean	s.d.
No	(0.02, 0.01)	Uniform	1	988.0	185	24	4
No	(0.02, 0.01)	Uniform	10: SRS	990.0	59	237	14
No	(0.02, 0.01)	Uniform	10: GRID	990.5	70	237	17
No	(0.02, 0.01)	Coastal	1	979.1	1300	23	31
No	(0.02, 0.01)	Coastal	10: SRS	986.0	408	236	98
No	(0.02, 0.01)	Coastal	10: GRID	980.8	136	235	32
No	(0.3, 0.5)	Coastal	1	752.9	668	426	378
Yes	(0.3, 0.5)	Coastal	1	997.2	534	564	302

reassures us that the equal coverage design with the new specification $(w, \theta) = (0.3, 0.5)$ remains design-unbiased, within simulation variability.

For practical field surveys, the wraparound design is inconvenient to implement, as it means crossing from one edge of the survey area to the other just to sample a lone fragment of a search strip. Wrapping may even be physically impossible, or may alter detection properties. The results above help to evaluate the cost of omitting the wrapped strips in practice. For the majority of surveys, the search strip half-width w is far less than 2% of the width of the survey area ($w = 0.02$ in our example corresponds to 2%), and the design bias introduced by the unequal coverage design is not worth worrying about. In general, if the search strips are wide enough for the edge effects to cause appreciable bias, then the survey region is probably small enough for the wraparound design to be implemented without great cost.

For a full analysis of the design-bias in the non-wrapped design with wide search strips and unusual object configurations, see table 1 of Barry and Welsh (2001), but note that the authors overlook the fact that the design does not provide equal coverage probability.

10.6   Example: non-uniformity within the strip

10.6.1   *Estimation of N_c*

This example illustrates the mathematical method presented in Section 10.4.2. We have a fixed search strip, with true object abundance N_c, and known g and θ. We wish to test the robustness of $\widehat{N}_c$ to failure of the assumption that $\pi(y)$ is uniform. For simplicity, we take $w = 1$ and position the centreline of the strip at 0, so the strip ranges from -1 to $+1$. We use the half-normal detection function, $g(y, \theta) = \exp\{-y^2/(2\theta^2)\}$, and study a range of values for θ, from $\theta = 0.25$ (limited detection), to $\theta \gg 1$ (perfect detection).

We study four different shapes for $\pi(y)$ for $-1 < y < 1$:

1. Uniform: $\pi(y) = 0.5$.
2. Linear: $\pi(y) = 0.5 + 0.25y$.
3. $\cap$-Quadratic: $\pi(y) = 0.6 - 0.3y^2$.
4. $\cup$-Quadratic: $\pi(y) = \dfrac{1}{3} + 0.5y^2$.

The notation $\cap$-Quadratic indicates a peaked quadratic shape, which might occur in the field if there is a responsive movement of animals towards the transect. Similarly, $\cup$-Quadratic indicates a U-shaped quadratic that can represent a responsive movement away from the transect. The functions are selected for their shapes, with the exact specifications being arbitrary.

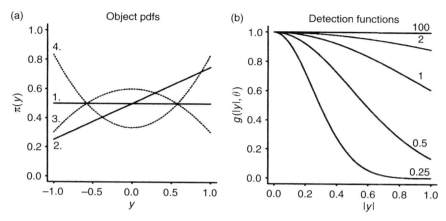

Fig. 10.3. Object pdfs, $\pi(y)$, and detection functions, $g(|y|, \theta)$, for testing the effects of non-uniform $\pi(y)$. The functions $\pi(y)$ are 1. Uniform; 2. Linear; 3. ∩–Quadratic; 4. ∪–Quadratic. The detection functions are half-normal with parameter θ marked on the curve.

Table 10.3. $E_y(\widehat{N}_c)/N_c$ when the half-normal detection function with parameter θ is known. Results are given for strip half-width $w = 1$ and the object distributions shown in Fig. 10.3

	$E_y(\widehat{N}_c)/N_c$ for the following object distributions:			
θ	Uniform	Linear	∩–Quadratic	∪–Quadratic
0.25	1	1	1.16	0.73
0.50	1	1	1.08	0.86
1	1	1	1.03	0.96
2	1	1	1.01	0.99
100	1	1	1.00	1.00

Figure 10.3 shows the four functions $\pi(y)$, and the range of detection functions considered. We apply eqn (10.1) to gain the exact model-based expectation $E_y(\widehat{N}_c)$ under the different schemes. The results for $E_y(\widehat{N}_c)/N_c$ are summarized in Table 10.3.

When detection is high, namely $\theta > w$, the ratio is approximately unity for each of the four π functions. This is because the estimated number of objects is roughly equal to the observed number, for high detection.

For low detection, bias appears for the quadratic object distributions, but not for the linear and uniform distributions. The bias arises for the quadratic distributions because the object density changes beyond the part of the strip where the change can be properly detected. To some extent this can be counteracted by truncating the observations to a smaller value of w, but not entirely. The bias is not negligible for these object distributions.

For linear and uniform object distributions, we can prove that $\widehat{N}_c$ will always be model-unbiased. In fact, a sufficient condition for model-unbiasedness is that $\pi(-y) + \pi(y)$ is constant, or equivalently that $\pi(y) - \pi(0)$ is an odd function of y. (An odd function h is one that satisfies $h(-y) \equiv -h(y)$.) The proof is straightforward.

First we note that if $\pi(-y) + \pi(y)$ is constant, then the condition that $\int_{-w}^{w} \pi(y)dy = 1$ forces the constant to be w^{-1}. Then by eqn (10.1),

$$\frac{E_y(\widehat{N}_c)}{N_c} = \frac{\int_{-w}^{w} g(|y|, \theta)\pi(y)\, dy}{w^{-1}\int_{0}^{w} g(r, \theta)\, dr} = \frac{\int_{0}^{w} g(y, \theta)\{\pi(-y) + \pi(y)\}\, dy}{w^{-1}\int_{0}^{w} g(r, \theta)\, dr} = 1.$$

Intuitively, when $\pi(-y) + \pi(y)$ is constant, an overabundance of objects on one side of the line is exactly compensated for by an underabundance on the other side, so the overall density estimate is correct.

This is an important result, as it means that the assumption that $\pi(y)$ is uniform is stronger than necessary. Any function $\pi(y)$ that is uniform, linear, or contains only constants and odd powers of y (e.g. $y^3, y^5, \ldots$) will ensure model-unbiased estimates of N_c when θ is known. The most significant of these results is the linear form, $\pi(y) = ay + b$ for $-w < y < w$. Because the width of the search strip is usually very small compared with that of the survey area (often as little as 0.1%), any large-scale trends in object density over the survey area are likely to appear approximately linear over the search strip. As long as there are no other factors such as responsive movement that influence the density gradient across the strip, we have high confidence in the estimator $\widehat{N}_c$. The large-scale changes in object density across the survey area are allowed for by surveying several strips, following an equal coverage design.

The poor results for the quadratic formulations for π highlight the importance of guarding against responsive movement of animals prior to detection. Responsive movement can cause the object distribution across the strip to become non-linear in such a way that $\pi(-y) + \pi(y)$ is not constant, and causes bias in the estimation of N_c even when g and θ are known. Issues of responsive movement are covered in Buckland *et al.* (2001: 31–4). Note that the bias is exacerbated if the responsive movement extends beyond the strip, so that animals beyond w move into the strip (attraction), or animals within the strip move outside (avoidance). In that case, even if N_c were estimated correctly after movement, design-based estimation of N from N_c would be biased high (attraction) or low (avoidance). Hence the overall bias in $\widehat{N}$ would be more severe than that in $\widehat{N}_c$ shown in Table 10.3.

10.6.2   *Asymptotic result when θ is estimated*

The analyses respecting non-uniform $\pi(y)$ in the previous section raise the question of what happens when the detection parameter θ is not known but has to be estimated. In the asymptotic case, as $N_c \to \infty$, we can answer this question exactly. It transpires that the estimator $\hat{\theta}$ is asymptotically unbiased for θ as $N \to \infty$, as long as $\pi(y) + \pi(-y) \equiv$ constant.

We consider a single strip of width w, centred at 0. There are N_c objects in the strip, at positions $y_1, \ldots, y_{N_c}$. The sightings data are the object positions y of detected objects. The true value of the detection parameter is θ_T, while the estimated value is $\hat{\theta}$.

We can write $\hat{\theta}$ (the value of θ that maximizes the conditional likelihood) as

$$
\hat{\theta} = \arg\max_\theta \left\{ \prod_{i=1}^{N_c} \left(\frac{g(|y_i|, \theta)}{wP_a(\theta)} \right)^{I\{\text{object } i \text{ is detected}\}} \right\}.
$$

Equivalently, taking logs, and using $wP_a(\theta) = \int_0^w g(r, \theta)dr$,

$$
\hat{\theta} = \arg\max_\theta \left\{ \sum_{i=1}^{N_c} \log \left(\frac{g(|y_i|, \theta)}{\int_0^w g(r, \theta)\, dr} \right) I\{\text{object } i \text{ is detected}\} \right\}.
$$

We define the function $\mathcal{L}(\theta)$ as

$$
\mathcal{L}(\theta) = \sum_{i=1}^{N_c} \log \left(\frac{g(|y_i|, \theta)}{\int_0^w g(r, \theta)\, dr} \right) I\{\text{object } i \text{ detected}\} = \sum_{i=1}^{N_c} \mathcal{L}_i(\theta),
$$

so that $\hat{\theta} = \arg\max_\theta \{\mathcal{L}(\theta)\}$. The argument θ denotes the detection parameter.

Now as $N_c \to \infty$, $\mathcal{L}(\theta)/N_c \to E_y(\mathcal{L}_i(\theta))$ by the strong law of large numbers, as each $\mathcal{L}_i(\theta)$ has the same expectation. The detection of object i involves the true detection parameter θ_T, so we obtain

$$
E_y(\mathcal{L}_i(\theta)) = \int_{-w}^w g(|y|, \theta_T) \log\{g(|y|, \theta)\} \pi(y)\, dy
$$
$$
- \left(\int_{-w}^w g(|y|, \theta_T)\pi(y)\, dy \right) \log \left(\int_0^w g(r, \theta)\, dr \right).
$$

If $\pi(-y) + \pi(y) \equiv \text{constant} = w^{-1}$, this becomes

$$E_y(\mathcal{L}_i(\theta)) = w^{-1} \int_0^w g(y, \theta_T) \log\{g(y, \theta)\} \, dy$$

$$- w^{-1} \left(\int_0^w g(y, \theta_T) \, dy \right) \log \left(\int_0^w g(r, \theta) \, dr \right).$$

Differentiating with respect to θ, we find that $\frac{\partial}{\partial \theta} E_y(\mathcal{L}_i(\theta))|_{\theta=\theta_T} = 0$. Thus $\hat{\theta}$, the maximum of the asymptotic likelihood function, is equal to the true parameter θ_T.

The above working holds for any detection function g such that $g(-y, \theta) \equiv g(y, \theta)$, and for any object distribution π such that $\pi(-y) + \pi(y) \equiv \text{constant}$: for example, when object distribution is uniform or linear. When these conditions are satisfied, $\hat{\theta}$ is asymptotically model-unbiased for θ_T.

10.7 Example: full estimation procedure

Here we give a brief example of testing the full line transect procedure, including estimation of θ, as in Section 10.4.3. We use the Uniform and Coastal configurations from Section 10.5, together with the 10-line wraparound designs with horizontal transect orientation and equal transect spacing. For each configuration, we collect a sample of 10,000 values of $\widehat{N}$. This follows exactly the same procedure as in Section 10.5, with the exception that the parameter θ is now estimated for each of the 10,000 simulations. As before, the true specifications are $w = 0.02$, $\theta_T = 0.01$, and $N = 1000$. The results are as follows.

1. Uniform configuration: mean($\widehat{N}$) = 1000.9; s.d.($\widehat{N}$) = 91; mean($\hat{\theta}$) = 0.010; s.d.($\hat{\theta}$) = 7.8×10^{-4}.

2. Coastal configuration: mean($\widehat{N}$) = 1002.9; s.d.($\widehat{N}$) = 165; mean($\hat{\theta}$) = 0.010; s.d.($\hat{\theta}$) = 1.1×10^{-3}.

Both estimators perform well. The higher estimator variances in the Coastal case are due to the spatial inhomogeneity of the data. Estimation of θ has caused some increase in the variance of $\widehat{N}$, compared with the corresponding values 61 and 118 from Table 10.1.

10.8 Trial by simulation: a completely model-based approach

Conventional distance sampling methods use a design step (Step 2 of Section 10.2) to extrapolate abundance estimates from the covered area a to the whole survey region A. In this section, we use simulation studies to investigate the effects of replacing this step with a model step. We demonstrate that a carefully constructed model can improve precision in $\widehat{N}$, without needing a randomized design.

We again consider the area $A = [0, 1] \times [0, 1]$. We use a non-randomized design in which 10 horizontal transects are fixed at positions $u = 0.05$, $0.15, \ldots, 0.95$. To mimic the situation in most real surveys, we set the strip half-width w to be a small proportion (0.5%) of the width of the survey area: $w = 0.005$. The total covered area a, over the 10 search strips, is 10% of the area A. The total number of objects in the region is set at $N = 10,000$.

The aim is to model explicitly any large-scale changes in object density across the area A, using the fixed transects. Suppose that the horizontal and vertical coordinates of the N object positions are $(h_1, v_1), \ldots, (h_N, v_N)$. Because we are using horizontal transects, results with fixed transects will only be biased if object density changes in the v-direction. Consequently, we imagine that the v-coordinates of objects are drawn from a pdf $\pi_V(v)$.

Let $D(v)$ be the instantaneous object density at vertical position v. $D(v)$ is directly related to the object pdf $\pi_V(v)$ via $D(v) = (N/A)\pi_V(v)$. Because $\pi_V(v)$ is a pdf on $0 < v < 1$, it integrates to unity: $\int_0^1 \pi_V(v)dv = 1$. It follows that $\int_0^1 D(v)dv = N/A$, so $N = A\int_0^1 D(v)dv$. Our aim is to generate an explicit model for $D(v)$. Once we have an estimated curve $\widehat{D}(v)$, we can estimate N using $\widehat{N} = A\int_0^1 \widehat{D}(v)dv$. Note that $A = 1$ in these simulations.

Let $u_1, \ldots, u_{10}$ be the v-coordinates of the 10 transect lines, and let $n_1, \ldots, n_{10}$ be the number of detected objects on the 10 lines. Each line has length one, so the area covered is $2w$. We estimate the instantaneous object density $D(u_1), \ldots, D(u_{10})$ for the 10 strips via $\widehat{D}(u_i) = n_i/(2wP_a(\hat{\theta}))$. The detection parameter θ is estimated by pooling sightings data from all 10 search strips, and $P_a(\hat{\theta}) = w^{-1}\int_0^w g(y, \hat{\theta})dy$. The full density curve $\widehat{D}(v)$ for $v \in [0, 1]$ is then obtained by fitting an interpolating cubic spline to the 10 point estimates, $\widehat{D}(u_1), \ldots, \widehat{D}(u_{10})$, extrapolating the output from $v = u_1$ to $v = 0$ and from $v = u_{10}$ to $v = 1$ by straight lines.

Note that the method requires w to be small enough for the density in the search strip $[u_i - w, u_i + w]$ to be a good estimate of the instantaneous density $D(u_i)$.

To test the method by simulation, we need each simulation to provide a new realization of objects from their assumed probability distributions. This means that every simulation should keep the same fixed lines, but draw a new set of object v-coordinates $v_1, \ldots, v_N$ from the probability distribution $\pi_V(v)$. We consider two distributions $\pi_V(v)$:

1. Uniform distribution $v \sim \text{Uniform}[0, 1]$.
2. Coastal distribution $v \sim \text{Beta}(5, 1)$.

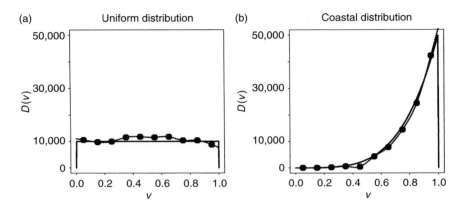

Fig. 10.4. Estimated density functions from a single simulation using the interpolating spline method. The solid lines represent the true function $D(v)$. The overlaid lines give the estimated function $\widehat{D}(v)$. The 10 transect positions are marked with dots. In each case, $N = 10{,}000$, $w = 0.005$, and $\theta = 0.002$.

The v-coordinates of objects in the distributions above can be seen in Fig. 10.1(a) and (b).

We test the method using the half-normal detection function $g(y, \theta) = \exp\{-y^2/(2\theta^2)\}$, with true values $\theta = 0.001$, 0.002, 0.004, and 0.006. This gives a range from low detection to high detection. Typical outputs for $\widehat{D}(v)$ are shown in Fig. 10.4 for the two object configurations and $\theta = 0.002$. The interpolating spline method gives estimated curves $\widehat{D}(v)$ that follow the true curves $D(v)$ closely.

To evaluate the performance of the new method, we compare it with the usual design-based estimator of $\widehat{N}$. The design-based estimator requires a randomized grid, so the 10 transect positions $u_1, \ldots, u_{10}$ must be newly generated for every simulation. As in Section 10.5, we use a grid of 10 equally spaced transect lines, distance 0.1 apart, with a random start-point for the first line. Parts of the search strips that extend outside of the area A are not wrapped around to the other side. This is intentional, because the wrapped design is not necessary for search strips with most realistic dimensions.

For the design-based estimator, we again generate new object positions $v_1, \ldots, v_N$ for each simulation, for consistency with the previous method. Using fixed object positions makes little difference to the results.

The design-based estimator is $\widehat{N} = \sum_{i=1}^{10} n_i / (10 \times 2wP_a(\hat{\theta}))$, where $\hat{\theta}$ is again derived by pooling observations across all 10 search strips.

Table 10.4 summarizes results from 1000 simulations for each data configuration, using the new model-based estimator and the design-based estimator for each value of θ. For the model-based approach, the table gives the mean and standard deviation of the 1000 estimates of $\hat{\theta}$ and $\widehat{N}/N$

Table 10.4. Results from the interpolating spline method ('Model') and the conventional design-based estimator ('Design') for $\widehat{N}$ and $\hat{\theta}$, using 10 transect lines with $w = 0.005$. The true value of N is 10,000. Each entry summarizes the results from 1000 different v vectors, with fixed grid for the Model method and randomized grid for the Design method

Objects	θ	Method	$\widehat{N}/N$ Mean	s.d.	$\hat{\theta}$ Mean	s.d.
Uniform	0.001	Model	1.00	0.078	0.0010	4.5×10^{-5}
$v \sim \mathrm{U}(0,1)$		Design	1.00	0.079	0.0010	4.4×10^{-5}
	0.002	Model	1.00	0.055	0.0020	7.6×10^{-5}
		Design	1.00	0.058	0.0020	7.9×10^{-5}
	0.004	Model	1.00	0.049	0.0040	3.6×10^{-4}
		Design	1.00	0.051	0.0040	3.5×10^{-4}
	0.006	Model	1.00	0.055	0.0143	2.0×10^{-1}
		Design	1.00	0.049	0.0208	1.7×10^{-1}
Coastal	0.001	Model	1.00	0.077	0.0010	4.5×10^{-5}
$v \sim \mathrm{Beta}(5,1)$		Design	1.00	0.164	0.0010	4.7×10^{-5}
	0.002	Model	1.00	0.054	0.0020	7.8×10^{-5}
		Design	1.00	0.152	0.0020	8.3×10^{-5}
	0.004	Model	1.00	0.050	0.0041	3.6×10^{-4}
		Design	1.00	0.149	0.0040	3.8×10^{-4}
	0.006	Model	1.00	0.060	0.0273	3.5×10^{-1}
		Design	1.00	0.150	0.0137	9.7×10^{-2}

after the interpolating spline has been fitted. The same quantities are given in the row beneath for the design-based approach.

The results suggest that both the model-based approach and the design-based approach give unbiased estimates of $\widehat{N}$, but the model-based results have much lower variance for the coastal object configurations. Both methods allow θ to be estimated with similar precision and no bias, except for $\theta = 0.006$ when high detectability can lead to non-identifiability of θ. The bias in $\hat{\theta}$ does not cause bias in $\widehat{N}$ when this occurs.

If the interpolating spline approach is not used, our fixed grid can be expected to produce biased results. The estimator for our fixed grid is $\widehat{N} = \sum_{i=1}^{10} n_i / (10 \times 2 w P_a(\hat{\theta}))$. From simulations in the model-based framework, however, the introduced bias appears to be very small ($\widehat{N}/N \simeq 0.99$ for the Coastal distribution).

We conclude that a completely model-based approach with fixed transects, even with our simplistic implementation, can lead to better precision than design-based placement of transects. A full discussion of spatial models in distance sampling is given in Chapter 4.

10.9   Summary

The following guidelines should always be followed for assessing distance sampling by simulation or other methods.

1. Formulate the question of interest. Is the aim to test *properties* of the estimators when all assumptions are met, or *robustness* of the estimators to failures of the assumptions?

2. For model-based analyses, first map out the scope of the relevant model. Only test models within the scope of their own assumptions.

3. For design-based analyses, distinguish between properties of the estimators and properties of the design. Always state results in the context of the design used.

Distance sampling estimators are guaranteed to perform well when all the underlying assumptions are met. If the assumptions are not met, the test becomes one of robustness rather than validity. Confusion over these issues has led some authors to criticize the distance sampling estimators, notably Barry and Welsh (2001). That work reports severe design-bias in the estimators, but it is conducted with respect to a design that severely violates the requirement of equal coverage probability. The results are a consequence of severe undersampling at the edges of the survey region, compounded by extreme object configurations that place large proportions of objects into the undersampled regions. For example, their Beta (5, 0.25) distribution for object positions places about half of all objects within a 1% strip of the region boundary.

No matter how extreme the object configuration, the distance sampling estimators are design-unbiased with respect to an equal coverage design, if the detection function is known. If the detection parameter θ must be estimated, the estimators will perform well as long as the pdf of signed object distances, conditional on the placement of the search strip, satisfies $\pi(-y) + \pi(y) \equiv$ constant. This condition will generally hold if search strips are narrow with respect to large-scale changes in object density, so that object density is approximately constant or linear across the strip. Problems such as responsive movement before detection, however, can interfere with this requirement and more advanced sampling techniques may be needed.

Practical survey designs should always include many transect lines, and transect orientation should be parallel to extreme gradients, where possible. Systematic spacing of transect lines (the grid method of Section 10.5) offers improvements in precision over a completely random selection. Explicit models for object density over the survey region (Section 10.8) may also increase precision, and research in these areas is ongoing.

11
Further topics in distance sampling

K. P. Burnham, S. T. Buckland, J. L. Laake,
D. L. Borchers, T. A. Marques,
J. R. B. Bishop, and L. Thomas

11.1 Distance sampling in three dimensions

Conceptually, line transects can be considered as one-dimensional distance sampling, because only distances perpendicular to the line of travel are used, even though objects are distributed in two dimensions. Point transects sample distances in those two dimensions because radial detection distances are taken at any angle in what could be represented as an $x-y$ coordinate system. In principle, distance sampling can be conducted in three dimensions, such as underwater for fish, where objects can be located anywhere in three dimensions relative to the observer. The observer might traverse a 'line', and record detection distance in two dimensions perpendicular to the line of travel, or remain at a point, recording data in three dimensions within the sphere centred at the point. Given the assumption of random line or point placement with respect to the three-dimensional distribution of objects, the mathematical theory is easily developed for the three-dimensional case. In practice, the third dimension may pose a problem: there may only be a thin layer in three dimensions, and in the vertical dimension, objects may exhibit strong density gradients (e.g. birds in a forest canopy, or fish near the sea surface). In sampling for fish, it might be possible to use vertical transects, perhaps with remotely controlled cameras. Variable density by depth would then not be problematic; indeed, because the transects would then be parallel to the density gradient as recommended in Chapter 7, such a scheme might be very efficient.

11.1.1 Three-dimensional line transect sampling

Assume that we follow a line randomly placed in three dimensions. Now we sample volume, not area, so $D =$ objects/unit volume; line length is still L.

Assume we record distances r for all objects detected out to perpendicular distance w. Counting takes place in a cylinder of volume $v = \pi w^2 L$, rather than a strip of area $a = 2wL$. The statistical theory at a fixed slice through the cylinder perpendicular to the line of travel is just like point transect theory. This sort of sampling (i.e. a 'tube transect') is like 'pushing' the point transect sampling point a distance L in three dimensions.

Aside from volume v replacing area a, we need little new notation: n is the sample size of objects detected in the sampled cylinder of radius w; P_v is the average probability of detecting an object in the sampled cylinder of volume v; $g(r)$ = probability of detecting an object that is at perpendicular distance r, $0 \leq r \leq w$; D is the true density of objects in the study space.

A basic starting point to develop theory is the formula $E(n) = DvP_v = D\pi w^2 LP_v$. The unconditional detection probability is easily written down because objects are, by assumption, uniformly distributed in space within the cylinder. Therefore, the probability density function (pdf) of the radial distance r for a randomly specified object (before the detection process) is $\pi(r) = (2\pi r)/(\pi w^2)$. The unconditional detection probability is $P_v = E[g(r)]$, where expectation is with respect to pdf $\pi(r)$. This is a weighted average of $g(r)$:

$$P_v = \int_0^w \pi(r)g(r)\,dr = \frac{1}{\pi w^2} \int_0^w 2\pi r g(r)\,dr. \qquad (11.1)$$

Notice that this P_v is identical to the unconditional detection probability in point transects.

A direct approach can be used to derive $E(n)$. Let v_ε be a small volume in the cylinder centred at distance r and position l along the line ($0 \leq l \leq L$). Thus Dv_ε = the expected number of objects in volume $v_\varepsilon = 2\pi r dr dl$, and the expected count of these objects is then $g(r)Dv_\varepsilon$. The expectation $E(n)$ can now be expressed as

$$E(n) = \int_0^L \int_0^w g(r)D2\pi r\,dr dl = LD \int_0^w 2\pi r g(r)\,dr = LD\pi w^2 P_v. \ (11.2)$$

An estimator of D is $\widehat{D} = n/(\pi w^2 L\widehat{P_v}) = n/L\hat{\mu}$ where $\mu = \pi w^2 P_v = \int_0^w 2\pi r g(r)dr$.

The sample of distances to detected objects is $r_1, \ldots, r_n$. The pdf of distance r to detected objects is

$$f(r) = \frac{2\pi r g(r)}{\mu} = \frac{r g(r)}{\int_0^w r g(r)\,dr}. \qquad (11.3)$$

This result is identical to that for point transects and can be proven using the same theory. In fact, slight modifications of point transect theory suffice as a complete theory for line transect sampling in three dimensions. In particular,

$$f'(r) = \frac{2\pi g(r)}{\mu} + \frac{2\pi r g'(r)}{\mu}, \tag{11.4}$$

so if $g'(0)$ is finite and $g(0) = 1$, then $f'(0) = 2\pi/\mu$. For consistency with point transects, we use $h(0) = f'(0) = 2\pi/\mu$ and hence we have $\widehat{D} = (n\hat{h}(0))/(2\pi L)$. Compare this with the point transect estimator, $\widehat{D} = (n\hat{h}(0))/(2\pi k)$. The only difference is that L replaces k.

In fact, all the theory for point transects applies to line transect sampling in three dimensions if we replace k by L. Thus, estimation of $h(0)$ or P_v could be done using program Distance and treating the detection distances, r_i, as point transect data. The case of objects as clusters poses no additional problems, giving $\widehat{D} = (n\hat{h}(0)\widehat{E}(s))/(2\pi L)$. For the clustered case, all the theory for point transects applies to line transects in three dimensions with k replaced by L. We do not know of any data for three-dimensional line transects as described here; however, if any such studies are ever done, we note that a complete theory for their analysis already exists.

11.1.2 *Three-dimensional point transect sampling*

Point transect sampling in two dimensions can be extended to three dimensions. (To people who use the term variable circular plots, such extension becomes a variable spherical plot.) Now the detection distances r are embedded in a three-dimensional coordinate system. There is no existing theory for this type of distance sampling, although theory derivation methods used for line and point transects are easily adapted to this new problem, and we present some results here.

In this case, the observer would be at a random point and record detections in a full (or partial) sphere around that point. For a sphere of radius w, the volume enclosed about the point is $v = (4\pi w^3)/3$. Given truncation of the data collection process at distance w, the expected sample size of detections at k random points is $E(n) = kDvP_v$. To derive P_v, we note that the pdf of radial distance for a randomly selected object in the sphere is $\pi(r) = (4\pi r^2)/(4\pi w^3/3)$ and $P_v = E[g(r)]$ with respect to $\pi(r)$:

$$P_v = \int_0^w \pi(r)g(r)\, dr = \frac{3}{4\pi w^3} \int_0^w 4\pi r^2 g(r)\, dr, \tag{11.5}$$

so that

$$E(n) = kD\frac{4}{3}\pi w^3 P_v = kD \int_0^w 4\pi r^2 g(r)\, dr. \qquad (11.6)$$

An alternative derivation is to consider that the volume, v_ε, of space in the shell at distances r to $r + dr$ is $4\pi r^2 dr$ (to a first-order approximation, which is all we need as we let $dr \to 0$). Thus,

$$E(n) = kD \int_0^w g(r)v_\varepsilon\, dr = kD \int_0^w 4\pi r^2 g(r)\, dr. \qquad (11.7)$$

Now define $\mu = \int_0^w 4\pi r^2 g(r)\, dr$ so that

$$\widehat{D} = \frac{n}{k\hat\mu} \equiv \frac{n}{k(4\pi w^3/3)\widehat{P}_v}. \qquad (11.8)$$

The pdf of detection distance r is

$$f(r) = \frac{4\pi r^2 g(r)}{\mu}, \quad 0 < r < w. \qquad (11.9)$$

Taking second derivatives, we get

$$f''(r) = \frac{8\pi g(r) + 16\pi r g'(r) + 4\pi r^2 g''(r)}{\mu}. \qquad (11.10)$$

Hence, if $g(0) = 1$ and both $g'(0)$ and $g''(0)$ are finite (preferably zero as then the estimators have better properties), then $f''(0) = 8\pi/\mu$. For simplicity of notation, we define $d(0) = f''(0)$, so that $\widehat{D} = (n\hat{d}(0))/(8\pi k)$.

The estimation problem reduces to fitting a pdf $f(r)$, as given above, to the detection distances $r_1, \ldots, r_n$ based on some model for the detection function, $g(r)$. This will lead to $\hat{d}(0)$ and $\widehat{\mathrm{var}}\{\hat{d}(0)\}$ by any of a variety of statistical methods. Because the variance of $\hat{d}(0)$ is conditional on n,

$$\widehat{\mathrm{var}}(\widehat{D}) = \widehat{D}^2 \left[\{\mathrm{cv}(n)\}^2 + \{\mathrm{cv}[\hat{d}(0)]\}^2 \right]. \qquad (11.11)$$

As with point transects in two dimensions, the theory for three dimensions can be transformed to look like line transect theory in one dimension. The transform is from radial distance r to the volume sampled, $\eta = \frac{4}{3}\pi r^3$, giving the pdf of η as

$$f(\eta) = \frac{g(\xi)}{\mu}, \quad 0 < \eta < v = \frac{4}{3}\pi w^3, \qquad (11.12)$$

where $\xi = (3v/4\pi)^{1/3}$. Then, if $g(0) = 1$, $f(0) = 1/\mu$. As for two-dimensional point transects, we do not recommend that analysis be based on such a transformation (cf Buckland 1987a).

Consider the case of objects as clusters with size-biased detection for three-dimensional point transect sampling. First, we would have a conditional detection function, $g(r \mid s)$, and a distribution of cluster sizes in the entire population, $\pi(s)$. The following result holds for each cluster size:

$$D(s) = \frac{E[n(s)]d(0 \mid s)}{8\pi k} \quad 1 \leq s. \tag{11.13}$$

The density of clusters irrespective of size is

$$D = \frac{E(n)\,d(0)}{8\pi k}. \tag{11.14}$$

Thus, dividing the first by the second of these two formulae, we get

$$\frac{D(s)}{D} = \pi(s) = \frac{E[n(s)]}{E(n)}\frac{d(0 \mid s)}{d(0)} = f_s(s)\frac{d(0 \mid s)}{d(0)}, \tag{11.15}$$

where $f_s(s)$ is the distribution of detected cluster sizes. Summing both sides of the above leads to $d(0) = \sum f_s(s)\,d(0 \mid s)$ whereas rearranging the formula and summing produces $d(0) = 1/\left(\sum \pi(s)/d(0 \mid s)\right)$ where all summations are over $s = 1, 2, 3, \ldots$.

Thus $\pi(s) = f_s(s)\,d(0 \mid s)/\sum f_s(s)\,d(0 \mid s)$ and $E(s) = \sum s f_s(s)\,d(0 \mid s)/\sum f_s(s)\,d(0 \mid s)$ from which expressions, estimators of $\pi(s)$ and $E(s)$ are evident.

Straightforward expressions for $d(0)$ and $d(0 \mid s)$ are $d(0) = 2/\left(\int_0^w r^2 g(r)dr\right)$ and $d(0 \mid s) = 2/\left(\int_0^w r^2 g(r \mid s)dr\right)$. Two more formulae are just stated here: $g(r) = \sum g(r \mid s)\pi(s)$ and $f(r \mid s) = r^2 g(r \mid s)/\left(\int_0^w r^2 g(r \mid s)dr\right)$. Also of interest are conditional distributions of cluster size given detection distance, r. These distributions are useful for exploring $E(s \mid r)$, where now the s is from the size-biased detected sample. The result is

$$f_s(s \mid r) = \frac{g(r \mid s)\pi(s)}{\sum g(r \mid s)\pi(s)} \equiv \frac{g(r \mid s)\pi(s)}{g(r)} \tag{11.16}$$

This is exactly the same as for either line or point transect sampling.

Perhaps some day three-dimensional point transect data will be taken in deep space (although, given likely object speeds, either a 'snapshot' approach will be required or the time dimension will also need to be modelled).

11.2 Conventional distance sampling: full likelihood examples

11.2.1 *Line transects: simple examples*

Note that we use x to represent perpendicular distance in this section (instead of y, which is used to represent generic distance, perpendicular or radial).

11.2.1.1 *Half-normal $g(x)$, Poisson n*

Assume that objects are spatially distributed as a homogeneous Poisson process, that the detection function is half-normal, that $w = \infty$, and that objects are single entities (i.e. we take $s = 1$). Then data from replicate lines may be collapsed into just the total count, n, for total line length L, and the perpendicular distances $x_1, \ldots, x_n$. The detection function $g(x)$ and pdf $f(x)$ are

$$g(x) = \exp\left\{-\frac{1}{2}\left(\frac{x}{\sigma}\right)^2\right\} \tag{11.17}$$

and

$$f(x) = \sqrt{\frac{2}{\pi\sigma^2}}\, \exp\left\{-\frac{1}{2}\left(\frac{x}{\sigma}\right)^2\right\} \quad 0 \le x < \infty,\ 0 < \sigma. \tag{11.18}$$

The probability distribution of n is

$$\Pr(n) = \frac{\exp\{(-2LD)/f(0)\}\,\{(2LD)/f(0)\}^n}{n!}. \tag{11.19}$$

We have

$$f(0) = \sqrt{\frac{2}{\pi\sigma^2}}, \tag{11.20}$$

so there are only two parameters, σ and D, although it is sometimes simpler to leave $f(0)$ in the formulae. The full likelihood is

$$\mathcal{L}(D, \sigma) = \Pr(n) \cdot \left[\prod_{i=1}^{n} f(x_i)\right]$$

$$= \frac{\exp\{(-2LD)/f(0)\}\,\{(2LD)/f(0)\}^n}{n}$$

$$\times \prod_{i=1}^{n}\left[\left(\sqrt{\frac{2}{\pi\sigma^2}}\right)\exp\left\{-\frac{1}{2}\left(\frac{x_i}{\sigma}\right)^2\right\}\right]. \tag{11.21}$$

We define

$$T = \sum_{i=1}^{n} x_i^2 \tag{11.22}$$

and simplify $\mathcal{L}(D, \sigma)$ by collapsing terms where possible, giving

$$\mathcal{L}(D, \sigma) \propto \exp\{-LD\sigma\sqrt{2\pi}\}\{LD\sigma\sqrt{2\pi}\}^n \left[\frac{1}{\sigma^n} \exp\left\{-\frac{T}{2\sigma^2}\right\} \right], \tag{11.23}$$

or simplified as much as possible,

$$\mathcal{L}(D, \sigma) \propto \exp\left\{-\left(LD\sigma\sqrt{2\pi} + \frac{T}{2\sigma^2}\right)\right\} D^n. \tag{11.24}$$

Note that we ignore multiplicative constants in the likelihood function.

The joint maximum likelihood estimators (MLEs) from eqn (11.24) are

$$\hat{\sigma} = \sqrt{\frac{T}{n}} \tag{11.25}$$

and

$$\widehat{D} = \frac{1}{L}\sqrt{\frac{n}{2\pi T}} \equiv \frac{n\hat{f}(0)}{2L}. \tag{11.26}$$

Standard likelihood theory can be used to derive the theoretical variance of $\widehat{D}$, which may be expressed in a variety of ways:

$$\begin{aligned}
\mathrm{var}(\widehat{D}) &= \frac{1.5 D f(0)}{2L} \\
&= D^2 \left[\frac{1.5}{E(n)} \right] \\
&= D^2 \left[\frac{1}{E(n)} + \frac{1}{2E(n)} \right] \\
&= D^2 \left[\{\mathrm{cv}(n)\}^2 + \{\mathrm{cv}(\hat{f}(0))\}^2 \right].
\end{aligned} \tag{11.27}$$

Thus, these results are all exactly the same as those derived by the 'hybrid' method of using $E(n) = 2LD/f(0)$, the Poisson variance of n, and a likelihood only for the distance data.

In general, the hybrid approach with empirical estimation of $\mathrm{var}(n)$ will be almost fully efficient for $\widehat{D}$ and is more robust as no distribution need be assumed for n. One advantage of a full likelihood approach is the possibility of using a profile likelihood interval for D. Such intervals can be expected to perform better than $\widehat{D} \pm 1.96\widehat{se}(\widehat{D})$ or log-based intervals because the likelihood function encodes information about the sampling distribution of $\widehat{D}$, thus allowing for non-normality of $\widehat{D}$ (or $\log_e\{\widehat{D}\}$). Below, we give some general explanation of profile likelihoods, then derive the profile log-likelihood function for D for the above example.

11.2.1.2   *Profile likelihood intervals*

Let $\mathcal{L}(D, \theta)$ be the full likelihood such as that given in eqn (11.24). The profile likelihood for D is

$$\mathcal{L}(D, \hat{\theta}(D)) = \text{ maximum value of } \mathcal{L}(D, \theta) \text{ over } \theta \text{ for each value of } D.$$

This is then a function of just the single parameter D. For computing profile likelihood intervals, it is convenient to use the following function:

$$W(D) = 2\Big\{\log_e \mathcal{L}(\widehat{D}, \hat{\theta}) - \log_e \mathcal{L}(D, \hat{\theta}(D))\Big\}. \tag{11.28}$$

In eqn (11.28), $\mathcal{L}(\widehat{D}, \hat{\theta})$ is equivalent to $\mathcal{L}(\widehat{D}, \hat{\theta}(\widehat{D}))$, where $\widehat{D}$ is the MLE of D. The function $W(D)$ is a pivotal quantity, asymptotically distributed as a single degree of freedom chi-square, χ_1^2. This approximate distribution of $W(D)$ holds better at small sample sizes than the assumed normality of $\widehat{D}$ underlying the use of $\widehat{D} \pm 1.96\widehat{se}(\widehat{D})$. A $100(1 - \alpha)\%$ profile confidence interval for D is given as the set of all values of D such that $W(D) \leq \chi_1^2(\alpha)$, where $\chi_1^2(\alpha)$ is the $1 - \alpha$ percentile point of the χ_1^2 distribution (3.84 for a 95% confidence interval). We only need the interval endpoints, which are the two solutions to the equation

$$W(D) = 2\Big\{\log_e \mathcal{L}(\widehat{D}, \hat{\theta}) - \log_e \mathcal{L}(D, \hat{\theta}(D))\Big\} = \chi_1^2(\alpha) \tag{11.29}$$

Barndorff-Nielsen (1986) described the theory underlying the method, including ways to improve on the χ_1^2 approximation for $W(D)$. Severini (2000) gives a detailed introduction to the modern theory of likelihood methods, including descriptions of profile likelihood intervals (pp. 126–9) and modified profile likelihood intervals (pp. 323–53).

11.2.1.3   *Profile formulae, half-normal $g(x)$, Poisson n*

Starting with the likelihood in eqn (11.24), we first must find the maximum for σ given any fixed value of D. The steps are summarized below:

$$\log_e \mathcal{L}(D, \sigma) = -LD\sigma\sqrt{2\pi} - \frac{T}{2\sigma^2} + n\log_e(D) \tag{11.30}$$

(apart from a constant) and

$$\frac{\partial \log_e \mathcal{L}(D, \sigma)}{\partial \sigma} = -LD\sqrt{2\pi} + \frac{T}{\sigma^3}. \tag{11.31}$$

Setting the above expression to zero, we get

$$\hat{\sigma}(D) = \left[\frac{T}{LD\sqrt{2\pi}}\right]^{1/3}. \tag{11.32}$$

The joint MLEs $\widehat{D}$ and $\hat{\sigma}$ are given above, hence finding the expression in eqn (11.28) is now merely algebraic manipulation:

$$\log_e \mathcal{L}(D, \hat{\sigma}(D)) = -\frac{3n}{2}\left[\frac{D}{\widehat{D}}\right]^{2/3} + n\log_e(D) \tag{11.33}$$

from which

$$\log_e \mathcal{L}(\widehat{D}, \hat{\sigma}) = -\frac{3n}{2} + n\log_e(\widehat{D}). \tag{11.34}$$

Reduced to a very simple form, we get for this example

$$W(D) = 3n\left[\left(\frac{D}{\widehat{D}}\right)^{2/3} - 1 - \log_e\left\{\left(\frac{D}{\widehat{D}}\right)^{2/3}\right\}\right]. \tag{11.35}$$

To get a profile likelihood interval for D, we substitute the values of $\widehat{D}$ and n in eqn (11.35), tabulate $W(D)$ for a range of D, and pick off the two solutions to eqn (11.29). It can be useful to plot $W(D)$, as is shown in another context by Morgan and Freeman (1989).

Below we look at some numerical examples comparing different confidence intervals. However, first we determine eqn (11.28) explicitly for a few more examples. These, and the above, are overly simplistic compared to real data, but only very simple cases lead to analytical, or even partially analytical, solutions for $W(D)$.

11.2.1.4   *Negative exponential g(x), Poisson n*
The negative exponential $g(x)$ is not a desirable detection function, but for $w = \infty$ and a Poisson n, we can derive closed form results for this case. Some formulae are

$$g(x) = \exp\left\{-\frac{x}{\lambda}\right\}, \tag{11.36}$$

$$f(x) = \frac{1}{\lambda} \exp\left\{-\frac{x}{\lambda}\right\} \quad 0 \le x < \infty, \ 0 < \lambda, \tag{11.37}$$

and $f(0) = 1/\lambda$. The full likelihood is

$$\mathcal{L}(D, \lambda) = \frac{\exp\{-2LD/f(0)\}\{2LD/f(0)\}^n}{n!} \prod_{i=1}^{n}\left[\frac{1}{\lambda} \exp\left\{-\frac{x_i}{\lambda}\right\}\right]. \tag{11.38}$$

Defining

$$T = \sum_{i=1}^{n} x_i, \tag{11.39}$$

we simplify $\mathcal{L}(D, \lambda)$ to

$$\mathcal{L}(D, \lambda) \propto \exp\left\{-\left(2LD\lambda + \frac{T}{\lambda}\right)\right\} D^n. \tag{11.40}$$

The joint MLEs are $\hat{\lambda} = T/n$ and $\widehat{D} = n^2/(2LT) \equiv (n\hat{f}(0))/(2L)$. From likelihood theory,

$$\text{var}(\widehat{D}) = \frac{D}{L\lambda} = D^2\left[\frac{2}{E(n)}\right] \equiv D^2\left[\{\text{cv}(n)\}^2 + \{\text{cv}(\hat{f}(0))\}^2\right]. \tag{11.41}$$

Fixing D in $\mathcal{L}(D, \lambda)$, we find $\hat{\lambda}(D)$ as follows:

$$\log_e \mathcal{L}(D, \lambda) = -2LD\lambda - \frac{T}{\lambda} + n\log_e(D) \tag{11.42}$$

and

$$\frac{\partial \log_e \mathcal{L}(D, \lambda)}{\partial \lambda} = -2LD + \frac{T}{\lambda^2} = 0, \tag{11.43}$$

so that

$$\hat{\lambda}(D) = \sqrt{\frac{T}{2LD}}. \tag{11.44}$$

Finally, we derive

$$W(D) = 4n\left[\left(\frac{D}{\widehat{D}}\right)^{1/2} - 1 - \log_e\left\{\left(\frac{D}{\widehat{D}}\right)^{1/2}\right\}\right]. \tag{11.45}$$

11.2.1.5   *Negative exponential g(x), Poisson n, Poisson s*

To the above example, we add the feature of varying cluster size, but with detection probability independent of cluster size, s. Let s be Poisson with mean κ. The parameter D is the density of individuals, not clusters. In the Poisson model, as given above, for counts of clusters, the density parameter is cluster density, D_s, not D. To parameterize this likelihood component in terms of density of individuals, we must replace D_s by D/κ. The full likelihood for this model is

$$\mathcal{L}(D, \lambda, \kappa)$$

$$= \frac{\exp\left\{(-2LD\lambda)/\kappa\right\}\left\{(2LD\lambda)/\kappa\right\}^n}{n!} \prod_{i=1}^{n} \left[\frac{1}{\lambda} \exp\left\{-\frac{x_i}{\lambda}\right\} \left\{ \frac{\exp(-\kappa)\kappa^{s_i}}{s_i!} \right\} \right]. \tag{11.46}$$

Using $\bar{s}$ to denote mean cluster size and $\bar{x} = T/n$ (from eqn (11.39)), this likelihood can be reduced to

$$\mathcal{L}(D, \lambda, \kappa) \propto \exp\left\{ -\left(\frac{2LD\lambda}{\kappa} + \frac{T}{\lambda} + n\kappa \right) \right\} D^n \kappa^{n(\bar{s}-1)} \tag{11.47}$$

and (apart from a constant)

$$\log_e \mathcal{L}(D, \lambda, \kappa) = -\frac{2LD\lambda}{\kappa} - \frac{T}{\lambda} - n\kappa + n\log_e(D) + n(\bar{s} - 1)\log_e(\kappa). \tag{11.48}$$

The hybrid and full likelihood results agree here; in particular,

$$\widehat{D} = \frac{n\bar{s}}{2L\bar{x}} \tag{11.49}$$

$$\mathrm{var}(\widehat{D}) = D^2 \left[\frac{3}{E(n)} \right]. \tag{11.50}$$

To get the profile likelihood, we need $\hat{\lambda}(D)$ and $\hat{\kappa}(D)$. Closed form results do not seem to exist. However, from the two partial derivatives set to zero,

$$\frac{\partial \log_e \mathcal{L}(D, \lambda, \kappa)}{\partial \lambda} \quad \text{and} \quad \frac{\partial \log_e \mathcal{L}(D, \lambda, \kappa)}{\partial \kappa} \tag{11.51}$$

we can derive the equations

$$\hat{\kappa}(D) = \frac{\bar{x}}{\hat{\lambda}(D)} + \bar{s} - 1 \tag{11.52}$$

and

$$\hat{\lambda}(D) = \sqrt{\frac{n\bar{x}}{2LD}\left[\frac{\bar{x}}{\hat{\lambda}(D)} + \bar{s} - 1\right]}. \tag{11.53}$$

The function $W(D)$ can be written as

$$W(D) = 6n\left[\hat{\kappa}(D) - \bar{s} - \log_e\left[\left(\frac{D}{\widehat{D}}\right)^{1/3}\right] - (\bar{s} - 1)\log_e\left[\left(\frac{\hat{\kappa}(D)}{\bar{s}}\right)^{1/3}\right]\right]. \tag{11.54}$$

To compute $W(D)$, choose a value of D, solve eqn (11.53) iteratively (easily done as $\hat{\lambda}(D)$ is a stable fixed point), compute $\hat{\kappa}(D)$, then compute eqn (11.54) (also using $\widehat{D}$, which is closed form).

11.2.2   Point transects: simple examples

Note that we use r to represent radial distance in this section (instead of y, which is used to represent generic distance, perpendicular or radial).

11.2.2.1   Negative exponential $g(r)$, Poisson n

Results for point transects can be obtained for a couple of simple cases. Here we assume a Poisson distribution for n, a negative exponential detection function, $g(r) = \exp(-r/\lambda)$ and k randomly placed points. Basic theory then gives $E(n) = 2\pi k D\lambda^2$ and the pdf of detection distance r is

$$f(r) = \frac{r\exp\{-r/\lambda\}}{\lambda^2} \quad 0 < r, \ 0 < \lambda. \tag{11.55}$$

The full likelihood is

$$\mathcal{L}(D, \lambda) = \frac{\exp\{-2\pi k D\lambda^2\}\{2\pi k D\lambda^2\}^n}{n!}\prod_{i=1}^{n}\left[\frac{r_i\exp\{-r_i/\lambda\}}{\lambda^2}\right], \quad (11.56)$$

which simplifies to

$$\mathcal{L}(D, \lambda) \propto \exp\left\{-\left(2\pi k D\lambda^2 + \frac{T}{\lambda}\right)\right\}D^n, \tag{11.57}$$

where $T = \sum_{i=1}^{n}r_i$.

Apart from a constant, the log likelihood is thus

$$\log_e \mathcal{L}(D, \lambda) = -2\pi k D \lambda^2 - \frac{T}{\lambda} + n \log_e(D). \qquad (11.58)$$

Standard likelihood theory now leads to

$$\hat{\lambda} = \frac{T}{2n} = \frac{\bar{r}}{2} \qquad (11.59)$$

$$\mathrm{var}(\hat{\lambda}) = \frac{\lambda^2}{2E(n)} \qquad (11.60)$$

$$\hat{D} = \frac{n}{2\pi k \hat{\lambda}^2} = \frac{n \hat{h}(0)}{2\pi k} \qquad (11.61)$$

$$\mathrm{var}(\hat{D}) = D^2 \left[\frac{3}{E(n)} \right]. \qquad (11.62)$$

In order to find the profile likelihood, we solve

$$\frac{\partial \log_e \mathcal{L}(D, \lambda)}{\partial \lambda} = -4\pi k D \lambda + \frac{T}{\lambda^2} = 0, \qquad (11.63)$$

getting

$$\hat{\lambda}(D) = \left[\frac{T}{4\pi k D} \right]^{1/3}. \qquad (11.64)$$

Using the above to form $\log_e \mathcal{L}(D, \hat{\lambda}(D))$ allows us to find the expression for $\log_e \mathcal{L}(\hat{D}, \hat{\lambda})$, from which we construct a simple representation of $W(D)$:

$$W(D) = 6n \left[(D/\hat{D})^{1/3} - 1 - \log_e \left\{ (D/\hat{D})^{1/3} \right\} \right]. \qquad (11.65)$$

11.2.2.2   Half-normal g(r), Poisson n

Instead of a negative exponential $g(r)$, let us assume $g(r)$ is half-normal; other assumptions are as in the above case. Now basic theory gives $E(n) = 2\pi k D \sigma^2$, and the pdf of detection distances is

$$f(r) = \frac{r \exp\left\{ -1/2 (r/\sigma)^2 \right\}}{\sigma^2} \qquad 0 < r,\ 0 < \sigma \qquad (11.66)$$

The full likelihood is

$$\mathcal{L}(D,\sigma) = \frac{\exp\{-2\pi kD\sigma^2\}\{2\pi kD\sigma^2\}^n}{n!} \prod_{i=1}^n \left[\frac{r_i \exp\{-1/2\left(r_i/\sigma\right)^2\}}{\sigma^2}\right], \tag{11.67}$$

which simplifies to

$$\mathcal{L}(D,\sigma) \propto \exp\left\{-\left(2\pi kD\sigma^2 + \frac{T}{2\sigma^2}\right)\right\} D^n \tag{11.68}$$

for T defined as the total, $T = \sum_{i=1}^n r_i^2$.

Standard likelihood theory now leads to

$$\hat{\sigma}^2 = \frac{T}{2n} \tag{11.69}$$

$$\mathrm{var}(\hat{\sigma}^2) = \frac{\sigma^4}{E(n)} \tag{11.70}$$

$$\widehat{D} = \frac{n}{2\pi k\hat{\sigma}^2} = \frac{n\hat{h}(0)}{2\pi k} \tag{11.71}$$

$$\mathrm{var}(\widehat{D}) = D^2 \left[\frac{2}{E(n)}\right]. \tag{11.72}$$

To find the profile likelihood, we solve

$$\frac{\partial \log_e \mathcal{L}(D,\sigma)}{\partial \sigma} = -4\pi kD\sigma + \frac{T}{\sigma^3} = 0 \tag{11.73}$$

getting

$$\hat{\sigma}(D) = \left[\frac{T}{4\pi kD}\right]^{1/4}. \tag{11.74}$$

Carrying through the algebra and simplifications, we have

$$W(D) = 4n \left[(D/\widehat{D})^{1/2} - 1 - \log_e\{(D/\widehat{D})^{1/2}\}\right]. \tag{11.75}$$

Notice that eqn (11.75) is identical to eqn (11.45); we expected this because there is a duality in the mathematics between the case of line transects with a negative exponential detection function and point transects with a half-normal detection function, both with n distributed as Poisson and with $w = \infty$.

11.2.3 *Some numerical confidence interval comparisons*

We used the above results on $W(D)$ and $\text{var}(\widehat{D})$ to compute a few illustrative numerical examples of profile, log-based, and standard confidence intervals (nominal 95% coverage). To facilitate comparisons, what is presented are the ratios, (interval bound)/$\widehat{D}$. Thus, the standard method yields relative bounds as $1 \pm 1.96\,\text{cv}(\widehat{D})$, and the log-based relative bounds are $1/C$ and C, where

$$C = \exp\left[1.96\sqrt{\log_e\{1 + [\text{cv}(\widehat{D})]^2\}}\right]. \qquad (11.76)$$

Some of our results are based on sample sizes that are smaller than those which would be justified for real data; our intent is to compare the three methods, and the differences are biggest at small n. The actual coverage of the intervals is not known to us; we take the profile likelihood intervals as the standard for comparison. Results are shown in Tables 11.1–11.4. (Note that in all these tables, the denominator $\widehat{D}$ is the same in all columns within a table.) One reason for the comparison is to provide evidence that the log-based intervals are generally closer to the profile intervals.

Table 11.1 corresponds to the line transect case in which $g(x)$ is half-normal and objects have a Poisson distribution. The relative interval end points in the table depend only upon sample size n, so these results are quite general under the assumed model. The invariance property of the ratios $\widehat{D}_{\text{lower}}/\widehat{D}$ and $\widehat{D}_{\text{upper}}/\widehat{D}$ applies also to Table 11.2 (line transect, negative exponential $g(x)$, and Poisson n) and Table 11.3 (point transect, negative exponential $g(x)$, and Poisson n). The log-based interval is slightly to be

Table 11.1. Some relative 95% confidence intervals, $\widehat{D}_{\text{lower}}/\widehat{D}$ and $\widehat{D}_{\text{upper}}/\widehat{D}$, for the profile, log-based, and standard method, for line transects with a half-normal detection function, $w = \infty$, and Poisson distributed sample size n. Equation (11.35) is the basis of the profile interval; results are invariant to the true D and σ. ($^+$ marks the method with the widest confidence interval in the row; $^-$ marks the method with the narrowest)

n	Profile interval	Log-based interval	Standard interval
5	0.296–2.610	0.366–2.729$^+$	−0.074–2.074$^-$
10	0.437–2.014	0.481–2.081$^+$	0.241–1.759$^-$
20	0.565–1.660	0.590–1.694$^+$	0.462–1.537$^-$
40	0.673–1.439	0.687–1.457$^+$	0.621–1.380$^-$
70	0.744–1.321	0.752–1.330$^+$	0.713–1.287$^-$
100	0.781–1.263	0.787–1.270$^+$	0.760–1.240$^-$

Table 11.2. Some relative 95% confidence intervals, $\widehat{D}_{\text{lower}}/\widehat{D}$ and $\widehat{D}_{\text{upper}}/\widehat{D}$, for the profile, log-based, and standard method, for line transects with a negative exponential detection function, $w = \infty$, and Poisson distributed sample size n. Equation (11.45) is the basis of the profile interval; results are invariant to the true D and λ. ($^{+}$ marks the method with the widest confidence interval in the row; $^{-}$ marks the method with the narrowest)

n	Profile interval	Log-based interval	Standard interval
5	0.251–3.076^{+}	0.321–3.117	-0.240–2.240^{-}
10	0.389–2.264	0.433–2.309^{+}	0.124–1.877^{-}
20	0.520–1.803	0.546–1.831^{+}	0.380–1.620^{-}
40	0.635–1.526	0.649–1.542^{+}	0.562–1.438^{-}
70	0.711–1.380	0.720–1.390^{+}	0.669–1.331^{-}
100	0.753–1.311	0.759–1.318^{+}	0.723–1.277^{-}

Table 11.3. Some relative 95% confidence intervals, $\widehat{D}_{\text{lower}}/\widehat{D}$ and $\widehat{D}_{\text{upper}}/\widehat{D}$, for the profile, log-based and standard method, for point transects with a negative exponential detection function, $w = \infty$, and Poisson distributed sample size n. Eqn (11.65) is the basis of the profile interval; results are invariant to the true D and λ. ($^{+}$ marks the method with the widest confidence interval in the row; $^{-}$ marks the method with the narrowest)

n	Profile interval	Log-based interval	Standard interval
5	0.191–4.056^{+}	0.261–3.833	-0.518–2.518^{-}
10	0.319–2.754^{+}	0.366–2.729	-0.074–2.074^{-}
20	0.453–2.072^{+}	0.481–2.081	0.241–1.759^{-}
40	0.575–1.684^{+}	0.590–1.694	0.463–1.537^{-}
70	0.660–1.487^{+}	0.669–1.494	0.594–1.406^{-}
100	0.708–1.395^{+}	0.714–1.401^{+}	0.661–1.339^{-}

preferred to the standard method in Table 11.1, and more strongly preferred for the cases of Tables 11.2 and 11.3. The choice in Table 11.4 (line transect, negative exponential $g(x)$, Poisson n, and Poisson s) is unclear. Note that the results in Table 11.2 for line transects with a negative exponential $g(x)$ are identical to results for the same values of n for point transects with a half-normal $g(r)$.

Table 11.4 reflects a case where the population of objects is clustered. The relative confidence intervals are for density of individuals. This is an interesting case because the log-based and standard relative confidence intervals do not depend upon D, λ, or κ (because the relative intervals do not depend upon the specific values of $\bar{x}$ or $\bar{s}$). The relative profile intervals do not depend upon D or λ (thus the results in Table 11.4 are

Table 11.4. Some relative 95% confidence intervals, $\widehat{D}_{\text{lower}}/\widehat{D}$ and $\widehat{D}_{\text{upper}}/\widehat{D}$, for the profile, log-based, and standard method, for line transects with a negative exponential detection function (parameter λ), $w = \infty$, Poisson distributed sample size n, and cluster size as Poisson, mean κ. Equation (11.54) is the basis of the profile interval; results are invariant to true D and λ, but depend weakly on true κ; $\hat{\kappa} = 3.0$ was used for these results. ($^+$ marks the method with the widest confidence interval in the row; $^-$ marks the method with the narrowest)

n	Profile interval	Log-based interval	Standard interval
5	0.230–3.440	0.261–3.833$^+$	−0.518–2.518$^-$
10	0.365–2.440$^-$	0.366–2.729$^+$	−0.074–2.074
20	0.497–1.900$^-$	0.481–2.081$^+$	0.241–1.759
40	0.614–1.583$^-$	0.590–1.694$^+$	0.463–1.537
70	0.693–1.419$^-$	0.669–1.494$^+$	0.594–1.406
100	0.737–1.341$^-$	0.714–1.401$^+$	0.661–1.339

Table 11.5. Relative 95% profile confidence intervals, $\widehat{D}_{\text{lower}}/\widehat{D}$ and $\widehat{D}_{\text{upper}}/\widehat{D}$, for line transects with a negative exponential detection function (parameter λ), $w = \infty$, Poisson distributed sample size n, and cluster size as Poisson, mean κ. Note the dependence of the intervals on $\hat{\kappa} = \bar{s}$

n	$\bar{s}$	Profile interval
20	1.25	0.465–2.023
20	3.00	0.497–1.900
20	30.00	0.518–1.813
20	300.00	0.520–1.804

independent of the choice of D and λ), but they do depend weakly upon κ because the specific value of $\bar{s}$ (three in this example) affects even the relative profile intervals. Heuristically, this seems to be because the sample size of number of individual animals detected increases as $\bar{s}$ increases and the likelihood function uses this information. To illustrate this point, we give in Table 11.5 the relative profile interval endpoints (based on eqn (11.54) and 95% nominal coverage) for a few values of $\bar{s}$ at $n = 20$. There is quite a noticeable effect here of the average cluster size and this is an effect that is not found in either standard or log-based methods. We speculate that in realistic likelihood models, the profile interval would be generally more sensitive to information in the data than simpler confidence interval methods.

11.2.3.1   *One more example*

Consider line transect sampling, in which $g(x)$ is half-normal, clusters have a homogeneous Poisson distribution (i.e. n is Poisson), and cluster size is a geometric random variable. Further assume that $g(0) < 1$, but that it can be estimated by an independent source of information, from which it is known that, of m clusters 'on' the line, z are detected. We assume that z is distributed as binomial$(m, g(0))$. The counts, n, will be Poisson with mean $E(n) = 2LDg(0)/\{\kappa f(0)\}$, where $\kappa = E(s)$ and D is density of individuals. The geometric distribution is used here in the form $\pi(s) = \phi^{s-1}(1 - \phi)$, hence $\kappa = 1/(1 - \phi)$. Also, $f(0) = \frac{1}{\sigma}\sqrt{2/\pi}$, so that we have

$$E(n) = \sqrt{2\pi}\sigma LDg(0)(1 - \phi). \qquad (11.77)$$

Maximum likelihood estimators are

$$\hat{\sigma} = \frac{\sum_{i=1}^{n} x_i^2}{n} = \frac{T}{n} \qquad (11.78)$$

$$\widehat{D} = \frac{n\hat{f}(0)\bar{s}}{2L\hat{g}(0)} \qquad (11.79)$$

$$\hat{\phi} = \frac{\bar{s} - 1}{\bar{s}} \qquad (11.80)$$

$$\text{and}\quad \hat{g}(0) = \frac{z}{m} \qquad (11.81)$$

and the asymptotic estimated var$(\widehat{D})$ is

$$\widehat{\text{var}}(\widehat{D}) = \widehat{D}^2 \left[\frac{1}{n} + \frac{1}{2n} + \frac{\hat{\phi}}{n} + \frac{1}{m}\left(\frac{1}{\hat{g}(0)} - 1\right) \right], \qquad (11.82)$$

where $1/n$ relates to the variance of n, $1/2n$ relates to the variance of $\hat{\sigma}^2$, $\hat{\phi}/n$ relates to the variance of $\hat{\phi}$, and $(1/\hat{g}(0) - 1)/m$ relates to the variance of $\hat{g}(0)$.

The full likelihood of the data entering into $\widehat{D}$ is given by the products of the likelihoods of the independent data components:

$$\mathcal{L}(D, \sigma, \phi, g(0)) = \frac{\exp\{-[\sqrt{2\pi}\sigma LDg(0)(1 - \phi)]\}\{\sqrt{2\pi}\sigma LDg(0)(1 - \phi)]^n}{n!}$$

$$\times \prod_{i=1}^{n}\left[\left(\sqrt{\frac{2}{\pi\sigma^2}}\right)\exp\left\{-\frac{1}{2}\left(\frac{x_i}{\sigma}\right)^2\right\}\right] \times \prod_{i=1}^{n}\left[\phi^{(s_i-1)}(1 - \phi)\right]$$

$$\times \left[\binom{m}{z}(g(0))^z(1 - g(0))^{m-z}\right]. \qquad (11.83)$$

Dropping constants and otherwise simplifying this likelihood gives

$$\mathcal{L}(D, \sigma, \phi, g(0)) \propto \left[\exp\left\{ -\left(\sqrt{2\pi}\sigma L D g(0)(1-\phi) + \frac{T}{2\sigma^2} \right) \right\} \right]$$
$$\times \left[D^n \phi^{n(\bar{s}-1)} (1-\phi)^{2n} (g(0))^z (1-g(0))^{m-z} \right]. \qquad (11.84)$$

Closed form expressions for $\hat{\sigma}(D)$, $\widehat{W}(D)$, and $\hat{g}(0, D)$ do not seem possible, but $W(D)$ can be computed using numerical optimization. A slight 'trick' simplifies the process of getting $W(D)$.

By setting the partial derivatives of L with respect to σ, ϕ, and $g(0)$ to zero, and with D arbitrary, we derived the following results:

$$\phi(\sigma) = \frac{\bar{s} - 1}{\bar{s} + 1 - (T/n\sigma^2)}, \qquad (11.85)$$

$$g(0, \sigma) = \frac{n + z - (T/n\sigma^2)}{n + m - (T/n\sigma^2)}, \qquad (11.86)$$

$$\text{and} \quad D = \frac{T}{\sqrt{2\pi}\sigma^3 L g(0, \sigma)(1 - \phi(\sigma))}. \qquad (11.87)$$

We cannot easily select D and compute $W(D)$. However, we can specify a value of σ and compute the unique associated $\phi(\sigma)$ and $g(0, \sigma)$ that apply for D, which is then computed. These are then the values of $\hat{\sigma}(D)$, $\widehat{W}(D)$, and $\hat{g}(0, D)$ to use in computing $W(D)$ for that computed value of D. All we need do is select a range of σ, which generates a range of D, and then we treat $\mathcal{L}(D, \sigma, \phi, g(0))$ as a function of D, not of σ. The MLEs are known, so the absolute maximum of L is known, thus normalizing L to ϕ is easy.

Table 11.6 gives a few numerical results for the model considered here. Sample sizes n and m are the dominant factors influencing the Table 11.6 results. In fact, these results do not depend on true D, L, or σ. However, they do depend on $E(s)$ and $g(0)$ too strongly to draw broad conclusions here. Inputs to the likelihood used for Table 11.6 were $T/n = 1$ (so MLE $\hat{\sigma} = 1$), $\bar{s} = 10$ ($\hat{\phi} = 0.9$), and $z = 16$, $m = 20$ ($\hat{g}(0) = 0.8$). The log-based intervals are closer to the profile intervals than are the standard intervals.

It is also worth noting that if n (i.e. the line transect sampling effort) is increased while m is fixed, the estimate of $g(0)$ is the weak link in the data. Studies that estimate $g(0)$ need to balance the effort for the two data types. It would be best to collect data on $g(0)$ during the actual distance sampling study to achieve both such balance of effort (with respect to n and m) and relevance of $\hat{g}(0)$ to the particular study. If in this example for

Table 11.6. Some relative 95% confidence intervals, $\widehat{D}_{\mathrm{lower}}/\widehat{D}$ and $\widehat{D}_{\mathrm{upper}}/\widehat{D}$, for the profile, log-based, and standard method, for line transects for the likelihood in eqn (11.84). Sample size n is Poisson, x is half-normal, s is geometric, and $g(0)$ is estimated from $z \sim \mathrm{binomial}(m, g(0))$. ($^+$ marks the method with the widest confidence interval in the row; $^-$ marks the method with the narrowest)

n	Profile interval	Log-based interval	Standard interval
5	0.238–4.009^+	0.289–3.457	-0.376–2.376^-
10	0.365–2.688^+	0.395–2.535	0.015–1.985^-
20	0.488–2.054^+	0.501–1.996	0.287–1.713^-
40	0.593–1.721^+	0.595–1.680	0.472–1.528^-
70	0.663–1.568^+	0.658–1.521	0.576–1.424^-
100	0.700–1.503^+	0.690–1.449	0.626–1.374^-

$n = 100$ we also put $z = 80$ and $m = 100$, then the three relative intervals are more similar, especially profile and log-based:

Profile interval	Log-based interval	Standard interval
0.726–1.377	0.728–1.373	0.681–1.319

In practice, it is unlikely that such a high proportion of detections (80%) could be considered as 'on' the line, necessitating the use of methods that utilize detections off the line.

11.2.3.2 *A general comment on precision*

The relative confidence intervals in Tables 11.1–11.4 and 11.6 have been computed in a variety of cases: line and point transects, some with clustered populations, different detection functions, and one case with an adjustment for $g(0) < 1$. A general conclusion is that sample size has the overwhelming effect on relative precision of $\widehat{D}$. Relative confidence intervals are quite wide at $n = 40$, being roughly $\pm 45\%$ of $\widehat{D}$. At $n = 70$ and 100, the intervals are roughly $\pm 35\%$ and $\pm 25\%$, respectively. This level of precision is under very idealized conditions that will not hold in practice for real data. With comparable sample sizes, we expect that relative interval widths will exceed the tabulated values. These results and our experience in distance sampling suggest strongly that reliable, precise abundance estimates from distance data require minimum sample sizes around 100. Coefficients of variation of around 20% are often adequate for management purposes; the results presented here indicate minimum sample sizes of 40–70 in this circumstance.

11.3   Line transect surveys with random line length

11.3.1   *Introduction*

Theory and application of line transect sampling has been almost exclusively in terms of fixed line lengths (and a fixed number of replicate lines). This approach means that line lengths $l_1, \ldots, l_K$ (and K itself), and of course L, are *a priori* fixed measures of sampling effort; that is, they are known before traversing the transects. It is then sample size n (overall and per line) that is random. In principle it is possible to do the reverse: fix the total sample size to be achieved and traverse the line(s) until that predetermined n is reached. This sampling scheme results in L being a random variable (Seber 1979).

The purpose of this section is to provide some results comparing the cases of random and fixed n, under simplistic but tractable assumptions, and to comment upon this alternative design. We conclude that the two schemes (fixed L and random n, or fixed n and random L), under some idealized conditions, are not importantly different in their statistical properties. Primarily, field (i.e. applied) considerations dictate the choice between sampling schemes.

A common example contrasting fixed and random effort sampling is provided by the (positive) binomial and negative binomial distributions. For the binomial distribution, sample size is fixed at n and we record the number of successes, $\tilde{y}$, in n independent binary trials. For the negative binomial, we fix the number of successful trials (y) and sample the binary events until y successes occur, so that the number of trials, $\tilde{n}$, is random. (The added notation, '$\sim$', is needed here to indicate which variable is random.) The corresponding probabilities, expectations, and variances are given below for the positive and negative binomial cases respectively:

$$\Pr\{\tilde{y} = i \,|\, n\} = \binom{n}{i} p^i (1-p)^{n-i}$$

$$E(\tilde{y}) = np$$

$$\mathrm{var}(\tilde{y}) = np(1-p)$$

$$\Pr\{\tilde{n} = i + y \,|\, y\} = \binom{i + y - 1}{y - 1} p^y (1-p)^i$$

$$E(\tilde{n}p) = y$$

$$\mathrm{var}(\tilde{n}p) = E(\tilde{n}p)(1-p).$$

Despite the differences in the two sampling schemes, the sampling variances are essentially the same. In particular, with reference to a fixed n under the direct (binomial) sampling approach, if we could select y for the inverse sampling such that $y = np$, then both sampling methods would have the same sampling variance.

Moreover, the respective MLEs and their variances are, for the positive binomial, $\hat{p} = \tilde{y}/n$ and $\mathrm{var}(\hat{p}) = p(1-p)/n$, and for the negative binomial, $\hat{p} = y/\tilde{n}$ and $\mathrm{var}(\hat{p}) = p(1-p)/E(\tilde{n})$. Thus, again, if we design the inverse sampling so that $E(\tilde{n}) = n$, there is no important large sample difference between the two approaches.

Another example more related to distance sampling is the use of randomly placed quadrats vs a sample of random points, with the data being distance to the nearest plant (e.g. Patil *et al.* 1979). In quadrats, the area is fixed and counts are random. In the nearest neighbour sampling, the plant count is fixed but the area sampled is random. Under an appropriate matching of the effort expended under the two schemes, the corresponding density estimates have almost equal large sample sampling variances when plants are randomly distributed (Holgate 1964).

We surmise that this relationship holds for most positive and negative sampling schemes, that is, there exist pairs of schemes such that the sampling variance of the parameter estimator is almost the same under either approach. In line transects, we have either L as fixed and n as random, or we fix n and traverse a random line length until n detections are made. To be consistent with the usual definitions of direct (positive) and indirect (negative, or inverse) sampling, we label these as below:

Positive case	n fixed	$\widetilde{L}$ random
Negative case	$\tilde{n}$ random	L fixed

Comparability of sampling variances requires that comparable effort be used in both schemes; this translates into the pair of relationships $E(\widetilde{L} \,|\, n) = L$ and $E(\tilde{n} \,|\, L) = n$.

11.3.2   *Line transect sampling with fixed n and random $\widetilde{L}$, under Poisson object distribution*

We examine here some properties of $\widehat{D}$ under such comparable schemes assuming a homogeneous Poisson distribution of objects, a constant detection function everywhere in the sampled area (spatially invariant $g(x)$), and independent detections. For the (usual) L-fixed case under these assumptions, $\tilde{n}$ has a Poisson distribution with mean $2LD/f(0)$. For $\widetilde{L}$ random, we assume a random starting point for the line and we move along it until n detections are made. Thus, the first segment is from the starting point to the point perpendicular to the first detection, and there are $n-1$ random inter-observational segments. We denote the lengths of these segments as $\tilde{l}_i$, which add to $\widetilde{L}$. In general, the ith segment of length $\tilde{l}_i$ is the distance travelled between points perpendicular to detections $i-1$ and i, where

$i = 0$ is defined to be the starting point. Assuming that the number of objects in any area of size a, including the total area $(a = A)$, is Poisson with mean aD, then it can be shown that $\tilde{l}$ is an exponentially distributed random variable with mean $E(\tilde{l}) = f(0)/(2D)$. The pdf of $\tilde{l}$ is

$$f_{\tilde{l}}(\tilde{l}) = \frac{2D}{f(0)} \exp\left[-\frac{2lD}{f(0)}\right] \tag{11.88}$$

By the assumptions we have made here, $l_1, \ldots, l_n$ are independent. Because $\tilde{L}$ is the sum of independent exponential random variables, it is known that $\tilde{L} = \sum_{i=1}^{n} \tilde{l}_i$ is distributed as a gamma$[n, f(0)/(2D)]$ distribution, so it has pdf

$$f_{\tilde{L}}(\tilde{L}) = \frac{[(2D/f(0))]^n \tilde{L}^{n-1} \exp[-(2\tilde{L}D/f(0))]}{(n-1)!}. \tag{11.89}$$

It is also easily established that $E(\tilde{L} \,|\, n) = nf(0)/(2D)$, which leads to the estimator

$$\tilde{D} = \frac{n\hat{f}(0)}{2\tilde{L}}, \tag{11.90}$$

$\hat{f}(0)$ is computed conditional on n exactly as in the case of fixed L and random $\tilde{n}$, so $\hat{f}(0)$ is the same estimator in either sampling scheme.

Compare the estimator in eqn (11.90) to that when $\tilde{n}$ is random:

$$\hat{D} = \frac{\tilde{n}\hat{f}(0)}{2L}. \tag{11.91}$$

Under a sampling theory approach, the two estimators have different expressions for small sample bias. For random $\tilde{L}$, from eqn (11.90),

$$E(\tilde{D}) = \frac{n}{2} E\left[\frac{\hat{f}(0)}{\tilde{L}} \,\middle|\, n\right]. \tag{11.92}$$

Given a Poisson distribution of objects and constant $g(x)$, it is reasonable to assume that $\tilde{l}$ and x are independent. Then the above becomes

$$E(\tilde{D}) = \frac{n}{2} E\left[\frac{1}{\tilde{L}} \,\middle|\, n\right] E[\hat{f}(0) \,|\, n]. \tag{11.93}$$

Under the gamma distribution of $\widetilde{L}$,

$$E\left[\frac{1}{\widetilde{L}}\,|\,n\right] = \left[\frac{1}{n-1}\right]\left[\frac{2D}{f(0)}\right], \tag{11.94}$$

which yields

$$E(\widetilde{D}) = \frac{n}{n-1}D\frac{E[\hat{f}(0)\,|\,n]}{f(0)}. \tag{11.95}$$

When L is fixed,

$$E(\widehat{D}) = D\frac{E[\hat{f}(0)\,|\,n]}{f(0)} \tag{11.96}$$

so there is little difference between the two sampling schemes for large n in this example. For $\widetilde{L}$ random, the bias associated with $1/\widetilde{L}$ could be eliminated by using

$$\widetilde{D} = \frac{(n-1)\hat{f}(0)}{2\widetilde{L}}. \tag{11.97}$$

This adjustment for bias when n is fixed and $\widetilde{L}$ is random seems generally appropriate.

An asymptotic formula for the variance of $\widetilde{D}$ in eqn (11.90) is

$$\mathrm{var}(\widetilde{D}) = \left[\frac{f(0)n}{2E(\widetilde{L}\,|\,n)}\right]^2\left[\frac{\mathrm{var}(\widetilde{L}\,|\,n)}{[E(\widetilde{L}\,|\,n)]^2}\right] + \left[\frac{n}{2E(\widetilde{L}\,|\,n)}\right]^2\mathrm{var}[\hat{f}(0)\,|\,n]$$

$$= D^2\left[[\mathrm{cv}(\widetilde{L})]^2 + [\mathrm{cv}\{\hat{f}(0\,|\,n)\}]^2\right]. \tag{11.98}$$

For the Poisson distribution of objects and a spatially invariant $g(x)$, so that $\tilde{l}$ is exponential, we have

$$\mathrm{var}(\tilde{l}) = [E(\tilde{l})]^2 = \left[\frac{f(0)}{2D}\right]^2 \tag{11.99}$$

and

$$\mathrm{var}(\widetilde{L}\,|\,n) = n\left[\frac{f(0)}{2D}\right]^2. \tag{11.100}$$

Using these results and eqn (11.98) gives

$$\text{var}(\widetilde{D}) = D^2 \left[\frac{1}{n} + [\text{cv}\{\hat{f}(0\,|\,n)\}]^2 \right].$$

(11.101)

The variance of $\widehat{D}$ in eqn (11.91) is

$$\text{var}(\widehat{D}) = D^2 \left[\frac{1}{E(\tilde{n}\,|\,L)} + [\text{cv}\{\hat{f}(0\,|\,\tilde{n})\}]^2 \right].$$

(11.102)

Under comparable effort, in which case $E(\widetilde{L}\,|\,n) = L$ and $E(\tilde{n}\,|\,L) = n$, it is clear that for large samples, $\text{var}(\widetilde{D}) \doteq \text{var}(\widehat{D})$.

The condition under which the two sampling schemes have almost the same variance for estimated density is that the coefficients of variation for $\tilde{n}$ and $\widetilde{L}$ are equal:

$$\frac{\text{var}(\tilde{n}\,|\,L)}{[E(\tilde{n}\,|\,L)]^2} = \frac{\text{var}(\widetilde{L}\,|\,n)}{[E(\widetilde{L}\,|\,n)]^2}.$$

(11.103)

This relationship holds for the above case.

11.3.3 *Technical comments*

Assuming independent detections, then a general formula for the cumulative distribution function (cdf) of $\tilde{l}$ is

$$F_{\tilde{l}}(\tilde{l}) = 1 - \sum_{i=0}^{\infty} \left[1 - \frac{1}{wf(0)} \right]^i \Pr\{\widetilde{N} = i\,|\,a = 2w\tilde{l}\}.$$

(11.104)

The probability of moving distance $\tilde{l}$ and detecting no objects is $1 - F_{\tilde{l}}(\tilde{l})$. Assume the area examined for detections is of width $\pm w$ about the line. Then the unconditional probability of detecting an object is $1/[w \cdot f(0)]$. The event that there are no detections in distance $\tilde{l}$ happens if there are no objects in the area of size $2w\tilde{l}$, or if there is one object but it is not detected, or there are two objects and both remain undetected, and so forth. The joint probability that there are i objects in the area of size $2w\tilde{l}$ and all are undetected is (under the assumptions made)

$$\left[1 - \frac{1}{w \cdot f(0)} \right]^i \cdot \Pr\{\widetilde{N} = i\,|\,a = 2w\tilde{l}\}.$$

(11.105)

For $i = 0, 1, 2, \ldots$, these events are mutually exclusive, hence we get eqn (11.104). For the Poisson case,

$$\Pr\{\widetilde{N} = i \,|\, a = 2w\tilde{l}\} = \frac{\exp(-2w\tilde{l}D)(2w\tilde{l}D)^i}{i!}. \tag{11.106}$$

Thus

$$
\begin{aligned}
F_{\tilde{l}}(\tilde{l}) &= 1 - \sum_{i=0}^{\infty} \left[1 - \frac{1}{wf(0)}\right]^i \frac{\exp(-2w\tilde{l}D)(2w\tilde{l}D)^i}{i!} \\
&= 1 - \exp[-2w\tilde{l}D]\exp\left\{\left[1 - \frac{1}{wf(0)}\right](2w\tilde{l}D)\right\} \\
&= 1 - \exp\left[\frac{-2\tilde{l}D}{f(0)}\right],
\end{aligned} \tag{11.107}
$$

which is the cdf of an exponential distribution. Notice also that w drops out of $F_{\tilde{l}}(\tilde{l})$ in this example.

Closed form results can also be derived if a negative binomial distribution is assumed for $\Pr\{\widetilde{N} = i \,|\, a = 2\tilde{l}w\}$ (Burnham *et al.* 1980: 197). However, we have not perceived a simple way to derive results for the random $\widetilde{L}$ case without making strong assumptions about $\Pr\{\widetilde{N} = i \,|\, a = 2w\tilde{l}\}$ and independence of detections.

Finally, we describe how to find $\Pr(\tilde{n} \,|\, L)$, as this is needed to compare the two schemes. Let $\widetilde{N}$ be the number of objects in the searched strip of area $2wL$. The event $\tilde{n} = i$ arises as the sum of mutually exclusive events: $\widetilde{N} = i$ and all i objects are detected, $\widetilde{N} = i + 1$, and only i are detected, $\widetilde{N} = i + 2$, and only i are detected, and so forth. The probability formula is

$$\Pr(\tilde{n} \,|\, L) = \sum_{\widetilde{N}=i}^{\infty} \Pr\{\tilde{n} = i \,|\, \widetilde{N}\} \cdot \Pr\{\widetilde{N} \,|\, a = 2wL\}. \tag{11.108}$$

For example, under the assumptions of a spatially constant detection function and independent detections, $\Pr\{\tilde{n} = i|\widetilde{N}\}$ is binomial:

$$\Pr\{\tilde{n} = i \,|\, \widetilde{N}\} = \binom{\widetilde{N}}{i}\left[\frac{1}{wf(0)}\right]^i\left[1 - \frac{1}{wf(0)}\right]^{\widetilde{N}-i}. \tag{11.109}$$

For the Poisson case,

$$\Pr\{\widetilde{N} \,|\, a = 2wL\} = \frac{\exp(-2wLD)(2wLD)^{\widetilde{N}}}{\widetilde{N}!}. \tag{11.110}$$

Applying eqn (11.108) with these distributions gives

$$\Pr(\tilde{n} \mid L) = \frac{\exp[-2LD/f(0)][2LD/f(0)]^i}{i!}, \tag{11.111}$$

which is a Poisson distribution.

11.3.4 *Discussion*

Having fixed n and random $\widetilde{L}$ is often not a practical design in line transect sampling. In particular, when planes or helicopters are used, you cannot set out to fly a random distance; $\widetilde{L}$ cannot exceed the fuel capacity of the plane. For most methods of traversing the line(s), the distance to travel must be specified in advance. This also allows an accurate cost estimate for a study, which is generally necessary. Further, in most studies, if effort ceased when the required sample size was reached, the observer would have to return to some point at the study boundary, and it would be wasteful not to seek and record detections for the maximum possible time in the study area.

Representative coverage of the area being sampled is also an important consideration. Lines (or points) are allocated so as to achieve a representative sampling of the area. This is critical to allow valid inference from the sample to the entire area. If $\widetilde{L}$ was random, one might finish before completing the *a priori* determined sample of lines, or finish the sample of lines and still need to sample more. With random line length, it is difficult to assure a representative sample over the area of interest; thus there is more danger of substantial bias in $\widehat{D}$ due to unbalanced spatial coverage.

Point transect sampling is potentially more amenable to a fixed n strategy. The fixed n can be set on a per point basis. Then a representative sample of K points can be selected and every point can be visited. The amount of time at each point will vary. An upper bound could be put on time, leading to a mixed strategy: stay at a point until n detections are made, or until the maximum time is reached. There is a potential to develop the theory for such a strategy. However, it is still not especially practical to ask a recorder to be aware of when n total detections are made, and then to stop effort. Also, should this n be a total for all species, or for one target species? A further difficulty occurs if there is object movement during the count at a given point; the longer the count, the greater the upward bias arising from movement. Variable time counts will therefore tend to generate greater upward bias in areas of lower object density.

We have not presented any theory here for point transects with fixed n and random time, as that theory is more difficult to conceptualize. For the typical application to birds, detections depend on cue generation, which would have their own temporal distribution. This adds another level of complexity to the case of a scheme with fixed n and random time.

Even for a fixed line length scheme, there is information in the inter-observational distances as defined here. For example, the $\tilde{l}_i$ may be used to assess the spatial distribution of objects (Burnham *et al.* 1980: 196–8). Under the (unlikely) hypothesis of a Poisson spatial distribution and constant $g(x)$, $\tilde{l}$ is an exponential random variable. There are many tests available for the null hypothesis that a random sample is from an exponential pdf. The distribution of $\tilde{l}$ under other object distributions can be determined by methods presented here. The information contained in the $\tilde{l}_i$ is reduced in practice, because they are likely to be serially correlated. However, if independence can be assumed, the information in the $\tilde{l}_i$ might be used to provide better estimates of the residual variation in the $\tilde{n}$; this subject may be worthy of study. The concept is that

$$\widehat{\mathrm{var}}(\tilde{n} \mid L) = \left(\frac{n}{L}\right)^2 \widehat{\mathrm{var}}(\tilde{L} \mid n) \qquad (11.112)$$

might hold true. Given independence of the interobservational distances,

$$\widehat{\mathrm{var}}(\tilde{L} \mid n) = \frac{\sum_{i=1}^{n}(\tilde{l}_i - \tilde{L}/n)^2}{n-1} \qquad (11.113)$$

thus giving an alternative estimator of $\mathrm{var}(\tilde{n} \mid L)$. The above variance estimator for $\tilde{L}$ also provides the basis for an empirical estimator of $\mathrm{var}(\tilde{D})$ for the fixed n scheme if such sampling is practical and valid.

A final important comment is required about the relative merits of random and fixed line lengths. Often, lines are *a priori* of fixed (frequently equal) length, in which case all the fixed length (and random n) theory holds. However, designs or field practice often result in unequal line lengths, for example, when lines are placed at random but then cross from one side to the other of an *a priori* defined area, or when bad weather causes effort termination during a ship survey, so that a transect is shorter than intended. Either of these instances gives the appearance of some stochasticity in line length, hence one might consider that the set of lines has a random component that should be accounted for in variances (and perhaps biases) of estimates (Seber 1979). We disagree with this thinking; it is entirely appropriate to condition on achieved line length in these and other cases, provided the stochastic variations in length are unrelated to the density of objects, or if it is not possible to fit a model that relates variation in line length to object density.

It does not follow that random line length theory applies, simply because the survey design or field protocol results in varying line lengths. The theory applies only if there is information about density D in the probability distribution of line lengths. Even in the case of randomly placed lines running across a predefined area, there is no information about object density in the probabilistic distribution of line length by itself. Moreover, in this case the line lengths are known before they are ever traversed, hence there is every reason in theory to consider line length as a fixed ancillary (i.e. it affects the precision of $\hat{D}$ but contains no direct information about D) in all the usual designs and field protocols.

Once we consider line lengths as known and fixed prior to data collection, or after the fact we condition on actual line lengths when appropriate, then some potential statistical methods are not relevant. For example, it is not relevant to apply finite population sampling ratio estimation theory to encounter rate; such ratio estimation theory leads to slightly different formulae for $\mathrm{var}(n)$ than we give in this book. The key point here is that line lengths l_i may often differ; this does not make them random in any sense that concerns us, especially if we know the actual line lengths before they are traversed. It is proper to take these line lengths as fixed unless there is a probability distribution on possible line lengths which depends on the density parameter of interest. The latter is only likely to be true under the scheme of fixed n, in which random linear effort continues until n detections are made, or under adaptive sampling schemes, in which sampling effort increases when areas of high density are found.

11.4 Models for the search process

There are many possible models for the detection function that may fit any given data set well. If all give similar estimates, then model selection may not be critical. However, when the observed data exhibit little or no 'shoulder', it is not uncommon that one model yields an estimated density around double that for another model. Although the development of robust, flexible models allows workers to obtain good fits to most of their data sets, it does not guarantee that the resulting density estimators have low bias. There is some value therefore in attempting to model the detection process, to provide both some insight to the likely form of the detection function and a parametric model that might be expected to fit real data well. Hazard-rate methods have proved particularly useful for this purpose, and have been developed by Schweder (1977), Butterworth (1982), Hayes and Buckland (1983), and Buckland (1985) for line transect sampling, and by Ramsey *et al.* (1979) and Buckland (1987a) for point transect sampling. There have been more recent developments in line transect sampling, summarized in Section 6.8.

11.4.1 *Continuous hazard-rate models*

11.4.1.1 *Line transect sampling*
At any one point in time, there is a 'hazard' that an object will be detected by the observer, which is a function of the distance r separating the object

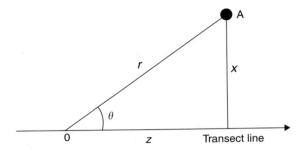

Fig. 11.1. The observer, at O, detects an object at A, with detection distance r and sighting angle θ. Perpendicular distance of the object from the line is x, and the observer is at distance z from the point of closest approach to the object.

and observer. If the object is on the line, the observer will be moving directly towards it, so that r decreases quite quickly. The farther the object is from the line, the slower the rate of decrease in distance r, so that the observer has more time to detect the object. Hazard-rate analysis models this effect, and also allows the hazard to depend on the angle of the object from the observer's direction of travel.

Suppose an object is at a perpendicular distance x from the transect line, and let the length of transect line between the observer and the point of closest approach to the object be z, so that r, the distance between the observer and the object, satisfies $r^2 = x^2 + z^2$ (Fig. 11.1). Suppose also that the observer approaches from a remote point on the transect so that z may be considered to decrease from ∞ to 0, and assume for simplicity that the object cannot be detected once the observer has passed his or her point of closest approach. Let

$$h(z,x)\,dz = \mathrm{pr}\{\text{object sighted while observer is in } (z, z - dz)\,|$$
$$\text{not sighted while observer is between } \infty \text{ and } z\}$$

and

$$G(z,x) = \mathrm{pr}\{\text{object not sighted while observer is between } \infty \text{ and } z\},$$

where both probabilities are conditional on the perpendicular distance x. Solving the forwards equations

$$G(z - dz, x) = G(z,x)\{1 - h(z,x)\,dz\} \tag{11.114}$$

for $G(z, x)$, and setting $G(\infty, x) = 1$, yields

$$G(z, x) = \exp\left\{ -\int_z^\infty h(v, x)\, dv \right\} \tag{11.115}$$

so that

$$g(x) = 1 - G(0, x)$$

$$= 1 - \exp\left\{ -\int_0^\infty h(z, x)\, dz \right\}, \quad 0 \le x < \infty. \tag{11.116}$$

Changing the variable of integration from z to r gives

$$g(x) = 1 - \exp\left\{ -\int_x^\infty \frac{r}{\sqrt{r^2 - x^2}} k(r, x)\, dr \right\}, \tag{11.117}$$

where $k(r, x) \equiv h\{\sqrt{r^2 - x^2}, x\}$.

Time could be incorporated in the model, but for a continuous hazard-rate process, there is little value in doing so provided that the speed of the observer is not highly variable. Otherwise the development has been general up to this point. To progress further, it is necessary to restrict the form of the hazard. A plausible hazard should satisfy the following conditions:

(1) $k(0, 0) = \infty$;

(2) $k(\infty, x) = 0$; and

(3) $k(r, x)$ is non-increasing in r for any fixed x.

For example, suppose that the hazard belongs to the family defined by:

$$\int_x^\infty \frac{r}{\sqrt{r^2 - x^2}} k(r, x)\, dr = \left(\frac{x}{\sigma}\right)^{-b} \tag{11.118}$$

for some σ and b.

Hayes and Buckland (1983) give two hazards from this family. In the first, the hazard of detection is a function of r alone:

$$k(r, x) = cr^{-d}, \quad r \ge x, \tag{11.119}$$

so that

$$b = d - 1 \text{ and } \sigma = \left\{ \frac{c\Gamma[(d-1)/2]\Gamma(0.5)}{2\Gamma(d/2)} \right\}^{1/(d-1)}.$$

The second hazard function allows the hazard of detection to be greater for objects directly ahead of the observer than for objects to the side. In practice this may arise if an object at distance r is more likely to flush when the observer moves towards it, or if the observer concentrates search effort in the forward direction. The functional form of the second hazard is:

$$k(r, x) = cr^{-d} \cos \theta, \quad \text{where} \quad \sin \theta = x/r \qquad (11.120)$$

so that

$$b = d - 1 \text{ and } \sigma = \left\{ \frac{c}{d-1} \right\}^{1/(d-1)}.$$

The family of hazards defined by eqn (11.118), to which the above two belong, yields the detection function

$$g(x) = 1 - \exp[-(x/\sigma)^{-b}]. \qquad (11.121)$$

This is the hazard-rate model derived by Hayes and Buckland (1983) and investigated by Buckland (1985), although the above parameterization is slightly different from theirs, and leads to better convergence properties. The parameter b is a shape parameter, whereas σ is a scale parameter. It may be shown that $g'(0) = 0$ for $b > 0$, which covers all parameter values for which the detection function is a decreasing function. Hence the above two hazards which are sharply 'spiked' (the derivative of the hazard with respect to r, evaluated at $r = 0$, is infinite) give rise to a detection function that always satisfies the shape criterion ($g'(0) = 0$; Buckland *et al.* 2001). For untruncated data, the detection function integrates to a finite value only if $b > 1$. For truncated data, the model has a long tail and a narrow shoulder if $b < 1$, and convergence problems may be encountered for extreme data sets. These problems may be avoided and analyses are more robust when the constraint $b \geq 1$ is imposed (Buckland 1987b).

Although in this book we describe the model of eqn (11.121) as 'the' hazard-rate model, any detection function may be described as a hazard-rate model in the sense that a (possibly implausible) hazard exists from which the detection function could be derived. Equation (11.121) is sometimes referred to as the complementary log–log model, a label which is both more accurate and more cumbersome.

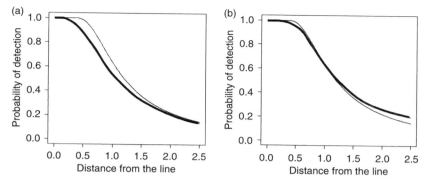

Fig. 11.2. Effect of heterogeneity in detectability on the pooled detection function. In both plots, the thin curve is the hazard-rate detection function $g(x) = 1 - \exp[-(x/\sigma)^{-b}]$ with $\sigma = 1$ and $b = 2$. In (a), the thick curve plots average detectability of objects subject to the same detection function model, still with $b = 2$, but with σ drawn from $\chi_{10}^2/10$ (so that $E(\sigma) = 1$). In (b), the thick curve plots average detectability of objects subject to the same detection function model, with $\sigma = 1$, but with b drawn from $\chi_{10}^2/5$ (so that $E(b) = 2$).

The above model should provide a good representation of the 'true' detection function when the hazard process is continuous, sighting (or auditory) conditions are homogeneous, and visibility (or sound) falls off with distance according to a power function. It appears to be fairly robust when these conditions are violated. However, strong heterogeneity in detectability changes the shape of the pooled detection function, as is shown in Fig. 11.2. A plausible model for heterogeneity is that the scale parameter σ varies from object to object, while the shape parameter b remains fixed. In Fig. 11.2(a), we show the average detection function obtained if the detection function is given by eqn (11.121), with $b = 2$ and with σ distributed as $\chi_{10}^2/10$. Also shown is the hazard-rate model with $b = 2$ and $\sigma = 1$ (the mean of $\chi_{10}^2/10$). It can be seen that the composite curve has a less flat shoulder than does the hazard-rate model. Indeed, apart from the thick tail, it looks more like a half-normal detection function. This possibly explains why the half-normal, Hermite polynomial or Fourier series models often in practice fit real data better than the hazard-rate model; they may better represent the shape of the detection function close to the line when strong heterogeneity in detectability is present. Under the less plausible scenario that σ is fixed and the shape parameter b varies from object to object, a similar effect is observed, though it is less marked (Fig. 11.2(b)).

11.4.1.2   *Point transect sampling*

For point transect sampling, the hazard-rate formulation is simpler, since there is only one distance, the sighting distance r, to model. The probability of detection is no longer a function of distance moved along the transect, but of time spent at the point. Define the hazard function $k(r, t)$ to be such that

$$k(r, t) \, dt = \text{pr}\{\text{an object at distance } r \text{ is detected during } (t + dt) \mid$$
$$\text{it is not detected during } (0, t)\}.$$

Then the detection function becomes:

$$g(r) = 1 - \exp\left[-\int_0^T k(r, t) \, dt\right], \qquad (11.122)$$

where T is the recording time at each point. If the observer is assumed to search with constant effort during the recording period, then $k(r, t) = k(r)$, independent of t, so that

$$g(r) = 1 - \exp[-k(r)T]. \qquad (11.123)$$

If the hazard is assumed to be of the form $k(r) = cr^{-d}$, then

$$g(r) = 1 - \exp[-(r/\sigma)^{-b}], \qquad (11.124)$$

where $b = d$ and $\sigma = (cT)^{1/b}$. The effect of increasing the time spent at each point is therefore to increase the scale parameter. This widens the shoulder on the detection function, making it easier to fit. Scott and Ramsey (1981) plotted the changing shape of a detection function as time spent at the point increases from four to 32 min. The disadvantages of choosing T large are that assumptions are more likely to be violated (Buckland *et al.* 2001), and after a few minutes, the number of new detections per minute will become small.

The parametric form of the above detection function is identical to that derived for line transects (eqn (11.121)). Moreover, if sightings are squared prior to analysis, the parametric form remains unaltered, so that maximum likelihood estimation is invariant to the transformation to squared distances. This property is not shared by other widely used models for the detection function. Burnham *et al.* (1980: 195) suggested squaring distances, to allow standard line transect software to be used for analysing point transect data, but we now advise against this strategy (Buckland 1987a).

11.4.2 *Discrete hazard-rate models*

The hazard-rate development above assumes that objects are continuously available for detection. Often this is not the case. For example, whales that travel singly or in small groups may surface at regular intervals, with periods varying from a few seconds to over an hour, depending on species and behaviour, when the animals cannot be detected. Discrete hazard-rate models have been developed for this case.

11.4.2.1 Line transect sampling

Schweder (1977) formulated both continuous and discrete sighting models for line transect sampling, although he did not use these to develop specific forms for the detection function.

Let $q(z,x)$ =pr{seeing the object for the first time | sighting cue at (z,x)}, where z and x are defined in Fig. 11.1. Then if the ith detection is recorded as (t_i, z_i, x_i), where t_i is the time of the ith detection, the set of detections comprises a stochastic point process on time and space. The first-time sighting probability depends on the speed s of the observer so that

$$q(z, x \mid s) = Q(z,x)E\left[\prod_{i>1}[1 - Q(z_i, x_i)]\right], \qquad (11.125)$$

where $Q(z,x)$ is the conditional probability of sighting a cue at (z,x) given that the object is not previously seen; $Q(z,x)$ is thus the discrete hazard. Assuming that detections behind the observer cannot occur, then

$$g(x \mid s) = \int_0^\infty \pi(z,x)q(z, x \mid s)\, dz, \qquad (11.126)$$

where $\pi(z,x)\, dz$ is the probability that a sighting cue occurs between (z,x) and $(z+dz, x)$, unconditional on whether it is detected; $\pi(z,x)$ is a function of both object density and cue rate.

More details were given by Schweder (1977, 1990), who used the approach to build understanding of the detection process in line transect sampling surveys. In a subsequent paper (Schweder et al. 1991), three specific models for the discrete hazard were proposed, and the corresponding detection function for the hazard they found to be most appropriate for north Atlantic minke whale data is:

$$g(x) = 1 - \exp[-\exp\{a' + b'x + c'\log_e(x)\}]. \qquad (11.127)$$

If we impose the constraints $b' \leq 0$ and $c' < 0$, this may be considered a more general form of the hazard-rate model of eqn (11.121), derived assuming a continuous sighting hazard. When $b' = 0$, eqn (11.127) reduces to eqn (11.121) with $a' = b\log_e(\sigma)$ and $c' = -b$. Thus a possible strategy for analysis is to use eqn (11.121) (the standard hazard-rate model) unless a likelihood ratio test indicates that the fit of eqn (11.127) is superior. Both models are examples of a complementary log–log model (Schweder 1990).

More recent developments of discrete hazard-rate models are described in Sections 6.8.3 (animal-based availability) and 6.8.4 (cue-based availability).

11.4.2.2 *Point transect sampling*

Point transects are commonly used to estimate songbird densities. In many studies, most cues are aural, and therefore occur at discrete points in time. Ramsey *et al.* (1979) defined both an 'audio-detectability function' $g_A(r)$ and a 'visual-detectability function' $g_V(r)$. (Both are also functions of T, time spent at the point seeking detections.)

Let $G(r, t) = \text{pr}\{\text{object at distance } r \text{ is not detected within time } t\}$. Then

$$g_A(r) = 1 - G(r, T) = 1 - \sum_{j=1}^{\infty} [1 - \gamma(r)]^j \text{pr}(j), \qquad (11.128)$$

where j is the number of aural cues the object makes in time T, $\text{pr}(j)$ is the probability distribution of j, and $\gamma(r)$ is the probability that a single aural cue at distance r is detected. This assumes that the probability of detection of an aural cue is independent of time, the number of cues is independent of distance from the observer, and the chance of detecting the jth cue, having missed the first $j - 1$, is equal to the chance of detecting the first cue. Hence the audio-detectability function is of the form

$$g_A(r) = 1 - \psi[1 - \gamma(r)], \qquad (11.129)$$

where

$$\psi(s) = \sum_{j=0}^{\infty} s^j \text{pr}(j) \qquad (11.130)$$

is the probability generating function of j.

The visual detectability function, $g_V(r)$, is modelled in a continuous framework, and yields eqn (11.123): $g(r) = 1 - \exp[-k(r)T]$. Ramsey *et al.* (1979) then combined these results to give

$$g(r) = 1 - \psi(1 - \gamma(r)) \exp[-k(r)T]. \qquad (11.131)$$

A detectability function may be derived by specifying (1) a distribution for the number of calls, (2) a function describing the observer's ability to detect a single call, and (3) the function $k(r)$ of visual detection intensities. Ramsey *et al.* considered possibilities for these, and plotted resulting composite detection functions. One of their plots shows an audio detection function in which detection at the point is close to 0.6 but which falls off slowly with distance and a visual function where detection is perfect at the point, but falls off sharply. The composite detection function is markedly 'spiked' and would be difficult to model satisfactorily. This circumstance could arise for songbird species in which females are generally silent and retiring. The spiked detection function can be avoided by estimating the male population alone, or singing birds alone, if adults cannot easily be sexed. If data are adequate, the female population could be estimated in a separate analysis.

11.4.3 *Further modelling of the detection process*

The detection of objects in distance sampling requires some type of active search effort. This will often be visual, so that observers must have some visual search pattern. Koopman (1980) discusses ideas on the search and detection process. We suggest that it is useful to consider some concept of search effort, and we pursue this suggestion here for line transects.

Conceptually, searching effort has its own distribution about the centreline for line transects. Can we separate this concept of search effort from some concept of 'innate' detectability? To a limited (but useful) extent, we think the answer is 'yes'. Let $e(x)$ be relative searching effort at perpendicular distance x, and let E be total absolute effort over all perpendicular distances. Then the perpendicular distance distribution of total effort is $E(x) = Ee(x)$. Total absolute effort, E, is conceptual because we do not precisely know what constitutes total effort, given that there are subjective aspects to the detection process; we do not know how to measure E on a meaningful scale. However, relative effort in the distance interval $(x, x + dx)$ could be the relative time spent searching in that interval. This measure of $e(x)$ is sensible and could be measured, in principle. Usually, we require that total effort E is sufficient to ensure $g(0) = 1$. Therefore, we will use the norm $e(0) = 1$ to scale $e(x)$. We maintain, and assume, that $e(x)$ should be non-increasing in perpendicular distance x.

We consider here some useful heuristic thinking, while recognizing that this is not the best mathematical approach. Use of a hazard-rate approach is coherent, but is not required for the points we wish to make. For the detection function, write

$$g(x) = d(x)e(x), \qquad (11.132)$$

where $d(x)$ is some innate, or standard, detection function, as for example, for some optimal effort, $e_0(x)$. By assumption, both $d(0)$ and $e(0)$ are unity and both functions are non-increasing in x. Based on eqn (11.132),

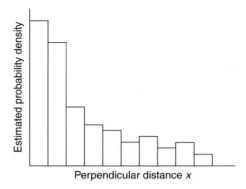

Fig. 11.3. Line transect data exhibiting a shape that is encountered too often; the idealized histogram estimator of the density function $f(x)$ suggests a narrow shoulder followed by an abrupt drop in detection probability, and a long, heavy tail.

$g'(x) = d'(x)e(x) + d(x)e'(x)$, so that $g'(0) = d'(0) + e'(0)$. It follows that if both effort and innate detectability have a shoulder, then so does $g(x)$. However, if effort is poorly allocated perpendicular to the line, then we can get $g'(0) < 0$, that is, no shoulder in the distance data, even when $d'(0) = 0$.

For line transect surveys in which there is visual searching for objects, especially aerial and ship surveys, histograms of the detection distances all too commonly have the shape of Fig. 11.3. It seems unlikely that the innate detectability would drop off this sharply; it is more likely that the data reflect an inadequate distribution of search effort or another field problem, such as heaping at zero distance or attraction of animals to the vessel before detection. We focus on effort here.

In order to pursue this idea mathematically, we need to be able to conceptualize innate detectability, $d(x)$. Although we may want to think of $d(x)$ as detectability under some optimal searching pattern $e_0(x)$, it is not possible to define an actual detection function, $g(x)$, free of some implicit or explicit underlying detection effort distribution. For $x \leq w$, we might allow $e(x)$ to be distributed as uniform $(0, w)$, but this becomes unreasonable for large w, and impossible as $w \to \infty$. Still, for small to moderate w, we could define $e_0(x)$ to be uniform. Then for a survey with this effort function, $g(x)$ would reflect the innate detectability of the object at perpendicular distances $x \leq w$.

Our intuition that detectability should not fall off sharply with increasing distance should be applied to $d(x)$. In most line transect sampling with which we have experience, the assumptions that $d(0) = 1$ and that $d(x)$ has a shoulder, that is, $d'(0) = 0$, seem reasonable to us. (Many marine

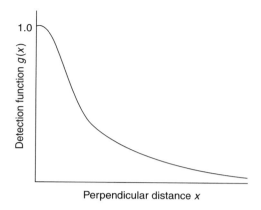

Fig. 11.4. An undesirable relative effort function $e(x)$ can give rise to the detection function shown here, and hence to data that exhibit the features of Fig. 11.3. Relative effort should be expended to ensure that the detection function has a wider shoulder relative to the tail than is shown here.

mammal surveys are exceptions to the first assumption (Chapter 6), and potentially to the second.) When the observed data appear not to exhibit a shoulder, we should bear in mind that the data really came from the detection function $g(x) = d(x)e(x)$ and hence the probability density function (pdf) of the observed perpendicular distance data is

$$f(x) = \frac{d(x)e(x)}{\int_0^w d(x)e(x)\,dx}. \tag{11.133}$$

If $g(x)$ is as shown in Fig. 11.4, the data may primarily reflect effort $e(x)$, not innate detectability $d(x)$. For any data set, we would like to know the general nature of the effort distribution $e(x)$ to assess our faith in the assumptions that $g(x)$ has a shoulder and satisfies $g(0) = 1$.

Desirable patterns for search effort should be addressed at the design stage, and observers should be trained to follow them; Fig. 11.4 suggests that the search pattern was poor. We suspect that in aerial and ship surveys, there are often two distinct search modes occurring simultaneously: (1) intense scanning of the area near the centreline for much of the time, and (2) occasional scans at greater distances and over large areas, with more lateral effort. This may occur because one observer 'guards' the centreline, searching with the naked eye, while another scans a wider area with binoculars, or a single observer may search mostly with the naked eye, with occasional scans using binoculars. Data then come from the composite pdf as indicated in Fig. 11.5. In this case, most choices of histogram interval will obscure the shoulder.

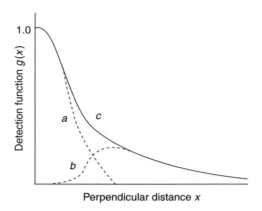

Fig. 11.5. The detection function c, which is the same as in Fig. 11.4, can arise from a mixture of curves a and b corresponding perhaps to two observers, one (a) 'guarding' the transect line and the other (b) scanning laterally; such minimal overlap of effort is undesirable.

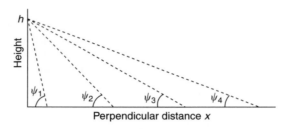

Fig. 11.6. The distance h represents (eye) height of an observer, and detections at various perpendicular distances x, indicated by the dashed lines, occur for angles of declination ψ. For some types of visual cue, the cue strength depends critically upon ψ.

For the innate detectability, $d(x)$, a shoulder should exist. Assume that an object on the centreline is moved just off the line. In an aerial transect survey (Fig. 11.6), the maximum angle of declination to the object would change from $90°$ to perhaps $89°$ or $88°$. Assuming the observer's view is not obstructed, the perceived properties of the object and detection cues will barely change. There is continuity operating, so that $g(x)$ will be almost the same at $x = 0$ as at a small increment from zero. Given continuity, we maintain that it is not reasonable for $d(x)$ to be spiked (i.e. $d'(x) < 0$).

We turn our attention now to considering what an optimal $e_0(x)$ might be. We consider an aerial survey, although the theoretical approach applies more generally. In Fig. 11.6, the angle of declination ψ is also the angle of incidence of vision, with $\pi/2 \geq \psi \geq 0$. If objects are assumed to be essentially flat and detection probability is proportional to the perceived area of an object, then the same object when moved further away shows less area and so is less detectable. The best you could achieve in this case is an innate detectability $d(x)$ proportional to

$$\cos(\psi) = \cos\left[\tan^{-1}\left(\frac{x}{h}\right)\right]. \tag{11.134}$$

This only allows for the loss in perceived object area due to the oblique view of the object as ψ decreases. Using heuristic arguments, if we add the effect of perpendicular distance off the centreline and generalize the result, we get as a plausible innate detection function the form

$$d(x) = \left[1 - e^{-\lambda O}\right] \frac{\cos\left[\tan^{-1}(x/\sigma)\right]}{1 + (x/\sigma)^2}, \quad 0 < x \tag{11.135}$$

for some scale factor σ. Here, O is the true area of the object. We would want λO such that $d(0) \simeq 1$, in which case the tail behaviour of $d(x)$ (i.e. as x gets large) is

$$d(x) \simeq \frac{1}{1 + (x/\sigma)}. \tag{11.136}$$

This is a very slow drop-off in detection probability. In fact, it cannot be used as a basis for general theory because it corresponds to $\int d(x)\,dx \to \infty$. This $d(x)$ does, however, give some sort of plausible upper bound on innate detectability, hence on possible effort. That is, we have reason to expect for any effort $e(x)$, properly scaled on x,

$$e(x) \leq \frac{\cos\left[\tan^{-1}(x)\right]}{1 + x^2}. \tag{11.137}$$

Also, note that in this simple situation, innate detectability would have a definite shoulder.

Motivated by an aerial line transect mode of thinking, we could express our effort in terms of a distribution on the angle ψ (Fig. 11.6). To make derivations easier, we focus on u, $0 \leq u \leq 1$, where

$$u = \frac{2}{\pi}\psi = \frac{2}{\pi}\tan^{-1}\left(\frac{x}{h}\right) \tag{11.138}$$

and define $q(u)$ as the pdf of u. For greater generality, we use $u = (2/\pi)\tan^{-1}(x/\sigma)$; now the distribution of effort is proportional to

$$\frac{q[u(x)]}{dx/du},\tag{11.139}$$

where dx/du is evaluated at $u = (2/\pi)\tan^{-1}(x/\sigma)$, giving

$$e(x) \propto \frac{(2/\pi)q\left[(2/\pi)\tan^{-1}(x/\sigma)\right]}{\sigma[1 + (x/\sigma)^2]}, \quad 0 \le x.\tag{11.140}$$

The proportionality constant is determined by the convention that $e(0) = 1$.

Let the effort be uniformly distributed over ψ, so that $q(u) \equiv 1$ for all u and

$$e(x) = \frac{1}{[1 + (x/\sigma)^2]}, \quad 0 \le x.\tag{11.141}$$

If effort is uniform on $\cos(\psi)$, then we spend more time looking away from the centreline, and the result is

$$e(x) = \frac{\cos\left[\tan^{-1}(x/\sigma)\right]}{1 + (x/\sigma)^2}.\tag{11.142}$$

It is interesting that if either ψ or $\cos(\psi)$ is uniform, the tail behaviour of the induced effort distribution is

$$e(x) \to \frac{1}{[1 + (x/\sigma)^2]} \to \left(\frac{\sigma}{x}\right)^2.\tag{11.143}$$

Note that for the hazard-rate model of $g(x)$ for large x

$$g(x) \to \left(\frac{\sigma}{x}\right)^b,\tag{11.144}$$

which of course includes the case of $b = 2$. Because effort decreases at large perpendicular distances, we would expect the applicable b to be ≥ 2.

It is also useful to consider total effort, E, and its likely influence on $g(x)$ near $x = 0$. Now eqn (11.132) must be replaced by the more coherent form (justified by a hazard-rate argument)

$$g(x) = 1 - e^{-Ee(x)}\tag{11.145}$$

with $e(0) = 1$, but where $e(x)$ is not mathematically identical to the $e(x)$ function considered above. Consider what happens at $x = 0$ as a function of total effort, E: $g(0) = 1 - \exp(-E)$. Some values of $g(0)$ as E increases are as follows:

E	$g(0)$
1	0.6321
2	0.8647
4	0.9817
8	0.9997
10	1.0000

It is obvious upon reflection, as the above illustrates, that if effort is inadequate, detection probability on the line can be less than unity even if innate detection probability at $x = 0$ is one. More interesting is what might happen to a shoulder under inadequate detection effort. Analytically from eqn (11.145), we have $g'(x) = \{1 - g(x)\}Ee'(x)$ and with E finite, we can write $g'(0) = (1 - g(0))Ee'(0)$. If total effort is large enough to achieve $g(0) = 1$, we are virtually sure that $g'(0) = 0$, regardless of the shape of the relative effort, $e(x)$ (provided $e(x)$ is not pathologically spiked at $x = 0$, with $e'(0) = -\infty$). Also if $e(x)$ has a shoulder, then $g'(0) = 0$. This could occur with insufficient total effort, E, to ensure $g(0) = 1$, hence the presence of a shoulder in the data is no guarantee that $g(0) = 1$.

The case in which relative effort has no shoulder is interesting. As noted above, it is possible that $e'(0) < 0$ and yet $g'(0) = 0$. As an example, consider the spiked relative effort $e(x) = e^{-x}$ for $E = 15$, so that $g(x) = 1 - e^{-15e^{-x}}$ and $g'(x) = (-e^{-15e^{-x}})(15e^{-x})$. A few values are given below:

x	$g(x)$	$g'(x)$
0.0	1.0000	−0.00001
0.1	1.0000	−0.00002
0.5	1.0000	−0.00102
1.0	0.9960	−0.02215
1.5	0.9648	−0.11778

Even though effort is spiked at $x = 0$, $g(x)$ has a distinct shoulder. However, if total effort is decreased, this shoulder will vanish and $g(0) = 1$ will fail.

The result illustrated above is due to a threshold effect. Once effort is large enough to achieve $g(0) = 1$, more effort cannot push $g(0)$ higher, but it can increase $g(x)$ for values of $x > 0$. We conclude that if there is sufficient total effort expended, then a shoulder is expected to be present even with a spiked relative effort function. The converse is disturbing: if total effort is too little, we can expect $g(0) < 1$, and there may be no shoulder. We emphasize the implications of guarding the centreline; if this is done, then as total effort decreases, more of the relative effort is likely to go near the centreline. This forces $e(x)$ to decrease more quickly, ultimately becoming spiked. The end result might be that we would have $g(0) < 1$, and $g(x)$ might be spiked (no shoulder). The data analysis implications are that if $\hat{g}(x)$ is, or is believed to be, spiked, then there is a basis to suspect that $g(0)$ is less than one. Conversely, if there is a shoulder, then there is a greater chance that $g(0) = 1$.

11.5   Combining mark-recapture and distance sampling surveys

Chapter 6 deals with what are essentially mark-recapture models used in distance sampling to estimate $p(0)$ (detection probability on the line given full availability). These applications require some form of dual observer data on the same animals. That is, a given animal may be detected, or not, by more than one observer. However, no mark is applied to the animal as part of this distance sampling. If animals can be caught and marked in the course of distance sampling, for example, the desert tortoise (Freilich *et al.* 2000), then classical mark-recapture data can be collected under a distance sampling design. Alternatively, if the animals have unique natural markings, as is the case for some whales, it may be possible for a study to collect distance sampling data and mark-recapture data, using photos, in the same study (Hammond 1990; Smith *et al.* 1999). The use of 'photo' mark-recapture stems from the early 1970s. More recently, the use of an animal's own DNA recoverable from faeces (Mills *et al.* 2000; Eggert *et al.* 2003) allows the possibility of combining distance sampling of dung with mark-recapture data collection. For any such combined surveys, we envision the primary application as being the monitoring of long-lived animals.

The envisioned study involves two or more complete passes (hence, capture occasions) through the study area and produces both distance sampling data sets (one at each occasion) and a classical mark-recapture data set. All these data are collected within the context of one design. This concept has been given, and developed for the two-pass design, by Alpizar-Jara and Pollock (1999). We are suggesting routine consideration of combining, where possible, line transect surveys and collection of mark-recapture data as part of research or monitoring programmes. The two types of data generally would be analysed separately to provide abundance and survival information. We note some applicable situations below.

For years demographic studies of the desert tortoise have been based on mark-recapture data (Krzysik *et al.* 1995; Freilich *et al.* 2000). Given an established study area, often a square mile $(2.59\,\text{km}^2)$ above ground, tortoises are found simply by walking a series of straight lines within the area and visually searching for tortoises. A sighted tortoise is easily approached, whereupon it withdraws into its shell and stays still. It is easy to put a unique mark on the tortoise such that marked animals are neither more nor less detectable than unmarked ones. Hence, with no basic design change, it is possible to collect distance data to detected, and hence captured animals, and to thus have a combined distance sampling and mark-recapture study within one survey.

At any given time, some tortoises are underground in burrows. Therefore, for the distance sampling study, not all animals are available, which we denote by $g_0 < 1$, where g_0 is the proportion of animals on the line that is potentially detectable. However with proper protocols $g(0) = 1$ will hold (i.e. available animals are certain to be detected if they are on the line; see, for example, Anderson *et al.* 2001) so that in distance sampling theory we can here equate g_0 to $p(0)$, where $p(0)$ is the probability of detection of an animal on the line, unconditional on whether it is available. The mark-recapture data typically arise from two complete passes over the study area in a two-month time interval each year. Also the studies last for many years. Therefore, any given tortoise has at some capture occasion a positive capture probability, but this comes at the expense of using much effort to obtain data required to estimate abundance. Moreover, the estimated abundance parameter is the number (N) of tortoises at risk of capture for an unknown area somewhat larger than the nominal study area (Freilich *et al.* 2000).

Distance sampling will directly produce an estimate of local density (D) with less time spent on a study site, given that we augment the design with radio-tagged animals to estimate g_0. This is quite feasible, as shown by McLuckie *et al.* (2002). The radio-tagged animals also provide information on movement and behaviour, so that such data are sometimes collected anyway. It is more efficient and informative to meld all of this research activity into one coherent distance sampling study design for locating animals, marking and recapturing them, and using radio-tagged animals to obtain $\hat{g}_0$. The estimated local density $\widehat{D}$ may be more useful than the mark-recapture $\widehat{N}$ $(= \widehat{D}A)$, but combined we can get $\widehat{A}$ to compare with other methods of estimating A. The main advantage of combining the two types of study is that distance data do not allow estimation of survival probability, whereas the mark-recapture data do.

Both line transect sampling and photo-based mark-recapture data have been collected for years on some whales. It is sensible to combine the two where possible. It is not required to close on all whales to take photos, as long as the probability of obtaining a photo is independent of the animal's mark (all animals are implicitly marked). Adaptive sampling might be especially useful when such mark-recapture photo data are also being collected, because more animals are encountered in adaptive surveys.

There is substantial potential for combining DNA-based mark-recapture data from dung in line transect sampling. This seems especially true for surveys on forest elephant abundance, for which it is usually dung that is surveyed (Eggert *et al.* 2003). In this case the DNA data are an add-on to the distance sampling design, which need not be altered to accommodate the mark-recapture component because elephants are highly mobile. The added cost for the DNA collection and analysis may well be modest.

Scientists should be alert for other possible applications of combined distance sampling and mark-recapture. The main theoretical constraint is the independence of the two data types. That independence requires only that detection, which is also capture, probability is independent of an animal's marked status. A second constraint may be the need for a more intense coverage of the study area in order to minimize a certain 'edge' effect. We should put lines close enough together that all animals in the defined study area have positive and substantial capture (detection) probability. The spacing needed depends on the mobility of the subject animals. The optimal situation occurs with highly mobile animals and a well-defined area to sample such that no animals permanently avoid the area where sampling occurs.

11.6   Combining removal methods and distance sampling

11.6.1   *Introduction*

Farnsworth *et al.* (2002) used removal models to analyse point count data. They split counts of songbirds into three time periods, recording only numbers of new birds detected in the second and third periods. Effectively, this ensures that birds detected in one time period are removed from counts in later time periods, allowing three-sample removal models to be used. Nichols *et al.* (2000) also developed a removal-type estimator, in which a primary observer 'removes' birds and a secondary observer records birds not removed by the primary observer, allowing the use of two-sample removal estimators.

The motivation for the work of Nichols *et al.* (2000) and of Farnsworth *et al.* (2002) is that for most point count surveys, detection distances are not recorded, and the counts are assumed to be a valid estimator of relative abundance. Farnsworth *et al.* (2002) and Buckland *et al.* (2001: 299) discuss the various problems associated with this assumption. Use of removal estimators is an attempt to estimate detection probabilities, so that counts can be corrected for detectability. As noted by Borchers *et al.* (2002: 83–8), individual heterogeneity causes substantial bias in removal estimators. Farnsworth *et al.* (2002) address this by introducing a heterogeneity model, under which birds belong to one of two groups: an easily detectable group, all of which are assumed to be detected in the first time period; and a less detectable group.

For a given species, and assuming only song is used as a cue, then much of the heterogeneity in detectability between individuals is due to distance of the individual from the point. (Most of the remaining heterogeneity is likely to be due to variation in cue frequency between individuals.) It seems natural therefore to model this source of heterogeneity in the removal analysis, and thus combining distance sampling with removal methods. Before we consider this option, we discuss two issues that most bird survey groups seem remarkably reluctant to address.

The first issue is whether it is feasible or practical to collect distance data during point counts. The reluctance to record such data is what motivated the work of Nichols *et al.* (2000) and Farnsworth *et al.* (2002); the methods of this section cannot be applied without detection distances. In surveys conducted in open habitat, there is usually little excuse for not estimating distances. Laser binoculars or rangefinders should be regarded as an essential item of equipment (Buckland *et al.* 2001: 300); they are now available at low cost, and allow distances of up to several hundred metres to be estimated to the nearest metre. They are quick and easy to use, so that there is little likelihood in open habitats at least that the observer will be swamped, and unable to cope with the quantity of data. If necessary, some of the distances can be measured after the end of the recording period at a given point, provided the observer has noted sufficient information to identify the location of a bird detected during the recording period. Closed habitats are more problematic, as most detections are likely to be aural only. If the habitat is not too dense, it may be possible to identify the tree from which a bird is singing, and then to use a laser binocular to measure the distance of the tree (the tree trunk if possible, to avoid systematic bias in distances, which would occur if the distance to the nearest part of the tree were measured). In denser habitats, it is usually possible to get reasonable distance estimates, by use of appropriate field methods coupled with careful observer training, and possibly more than one observer. This will often be at the expense of sample size, but the added reliability of estimates, coupled with the availability of estimates of absolute density, is likely to outweigh the loss for most surveys.

The second issue commonly ignored by bird survey groups is the effect of movement on point count methods. The point counts considered by Farnsworth *et al.* (2002) were of 10-min duration. Many songbirds move around within a territory that is similar in size to the plot surveyed around a point within a 10-min period. Many others do not stay strictly within a territory. This leads to substantial overestimation of densities using

standard point transect methods, and the removal estimates of Farnsworth
et al. (2002), and those considered below, will also be very sensitive to such
movement. The problem may be less serious for point counts used only as
relative measures of abundance, although variability in degree of movement
will cause non-comparability of counts across species, habitats, months, and
times of day. Point transect sampling is essentially a 'snapshot' method—
conceptually, birds are considered to be frozen at their location—so that
field methods need to attempt to record locations at a snapshot moment
for species that move around a lot (Buckland *et al.* 2001: 173). Alternatively, analyses based on cues (songs) rather than on individuals avoid
the need to assume that birds do not move during the recording period
(Section 11.7).

11.6.2 *Combining removal methods with distance sampling*

Borchers *et al.* (2002: 220–2) list three approaches for modelling heterogeneity in removal experiments. The first of these is animal-level stratification.
This is the solution adopted by Farnsworth *et al.* (2002), but is not ideal,
as we expect detectability to vary smoothly with distance. The second
approach is a conditional likelihood approach, in which we specify the likelihood of the n observed capture histories only. This approach is useful
when we do not know the distribution of the covariate in the population.
Given random point placement, however, we can ensure by design that
the distribution of distances r from the point is given by $\pi(r) = 2r/w^2$,
$r \leq w$, allowing us to use the third, full likelihood approach. Note that, as
birds are detected and so 'removed' from the population, the distribution of
remaining distances changes; the most detectable birds tend to be removed
first, and these tend to be at smaller distances.

   In the following, the sampling occasions of the removal experiment
might be successive observers, in which the primary observer records
everything she or he detects, the secondary observer records additional
birds not detected by the primary observer (as in Nichols *et al.* 2000), and so
on; or they might be successive time periods as in Farnsworth *et al.* (2002).
In the latter case, it might also be possible to develop a continuous-time
model; we do not pursue that here.

   Let

K = number of points; a circle of radius w is surveyed at each point
$a = K\pi w^2$ = covered area
n = number of birds detected
r_i = distance from the point of detection i, $i = 1, \ldots, n$
S = number of sampling occasions in the removal experiment (typically
     two or three)
n_s = number of new birds detected on occasion s, $s = 1, \ldots, S$
N_s = number of undetected birds present in a at the start of occasion s

Then the contribution to the likelihood of new detections on occasion s is

$$\mathcal{L}_s = \binom{N_s}{n_s} E[P(\underline{0}_{s+1} \,|\, \underline{0}_s, r)]^{N_s - n_s} \prod_i p(r_i, \underline{0}_s, l_s) \pi(r_i \,|\, \underline{0}_s), \qquad (11.146)$$

where $\underline{0}_s$ is the capture history corresponding to no capture up to and including occasion $s - 1$; l_s is the sampling effort on occasion s; $P(\underline{0}_{s+1} \,|\, \underline{0}_s, r)$ is the probability that a bird at distance r has not been detected by the end of occasion s, given that it has not been detected by the start of occasion s; $p(r_i, \underline{0}_s, l_s)$ is the probability that a bird is detected on occasion s, given that it is at distance r_i and previously undetected, and given effort l_s on occasion s; $\pi(r_i \,|\, \underline{0}_s)$ is the probability that a bird is at distance r_i, given that it is within a surveyed circle, and undetected prior to occasion s.

The full likelihood is obtained as $\prod_{s=1}^{S} \mathcal{L}_s$.

The following results follow from Borchers et al. (2002: 294–5).

$$\pi(r \,|\, \underline{0}_s) = \frac{P(\underline{0}_s \,|\, r)\pi(r)}{E[P(\underline{0}_s \,|\, r)]} = \frac{2r\, P(\underline{0}_s \,|\, r)}{w^2 E[P(\underline{0}_s \,|\, r)]}, \qquad (11.147)$$

where $P(\underline{0}_s \,|\, r)$ is the probability that a bird at distance r is undetected by the start of occasion s (given efforts $l_1, \ldots, l_{s-1}$), and $E[P(\underline{0}_s \,|\, r)]$ is the expected value of $P(\underline{0}_s \,|\, r)$ with respect to r:

$$P(\underline{0}_s \,|\, r) = \prod_{s^*=1}^{s-1} (1 - p(r, \underline{0}_{s^*}, l_{s^*})), \qquad (11.148)$$

$$E[P(\underline{0}_s \,|\, r)] = \int_0^w P(\underline{0}_s \,|\, r)\pi(r)\, dr$$

$$= \int_0^w \frac{2r}{w^2} \prod_{s^*=1}^{s-1} (1 - p(r, \underline{0}_{s^*}, l_{s^*}))\, dr. \qquad (11.149)$$

Noting that $P(\underline{0}_{s+1} \,|\, \underline{0}_s, r) = 1 - p(r, \underline{0}_s, l_s)$, we can now write

$$E[P(\underline{0}_{s+1} \,|\, \underline{0}_s, r)] = \int_0^w [1 - p(r, \underline{0}_s, l_s)]\pi(r \,|\, \underline{0}_s)\, dr$$

$$= \frac{E[P(\underline{0}_{s+1} \,|\, r)]}{E[P(\underline{0}_s \,|\, r)]} \qquad (11.150)$$

using the above results.

It remains to choose a suitable model for $p(r, \underline{0}_s, l_s)$, defining probability of detection of a previously undetected bird at distance r on occasion s with effort l_s. The likelihood is then maximized with respect to the parameters of this model. This allows the methods of Nichols *et al.* (2000) and Farnsworth *et al.* (2002) to be extended, to model the fall-off in detectability with distance, at the expense of having to record detection distances.

11.7 Point transect sampling of cues

11.7.1 *Introduction*

Good survey design for point transect sampling of songbirds is difficult due to two conflicting requirements. To give unbiased estimates of density in the presence of movement, the method must be regarded as a 'snapshot' method (Buckland *et al.* 2001: 173), whereas aural cues (songs or calls) are discrete, making it impossible to detect all birds near the point at the snapshot moment. Standard methods are therefore a compromise.

As noted by Buckland *et al.* (2001: 191), a possible solution is to use standard point transect sampling methods to estimate number of cues per unit area per unit time, and separately to estimate cue rate. By dividing the first estimate by the second, an estimate of bird density is obtained, without having to assume that birds do not move.

We consider this approach in more detail here. This section was motivated by a query on how to analyse data collected by audio recording, for which recording periods are substantially greater than would be normal for a point count (Hobson *et al.* 2002), so that the problem of movement cannot be ignored. However, it is potentially equally useful for standard point transect survey data conducted in woodland where most detections are aural. Although we assume that the objects of interest are birds, they could equally well be vocal mammals such as some monkey species or amphibians such as frogs.

Line transect sampling may also be conducted on cues. However, to apply the same approach, we would need to be able to estimate the effective area being searched at any moment in time, which cannot be done using standard line transect analysis. The cue count method for surveying whale blows (Hiby 1985; Buckland *et al.* 2001: 191–8) does allow this; it is like surveying a point transect from a ship or aircraft, so that the point moves along the line. Thus the design is that of a line transect survey, but the analysis is based on standard point transect analysis methods. Exactly the same approach could be adopted for songbird surveys. Instead of standing at a point for a fixed time, the observer could walk along a transect, estimating or measuring the distance to each cue detected. Effort would then be time taken to walk the transect. Note that the distances would be observer-to-cue distances, not the usual line transect perpendicular distances from the line. Provided the cues are instantaneous, or nearly so, movement of birds or observer does not bias the method. By contrast, conventional line transect methods based on observer-to-object (radial) distances are subject to bias as a consequence of the continuous availability of the object to be detected as the observer approaches (Hayes and Buckland 1983). In cue counting of whales, just a sector ahead of the ship or aircraft is surveyed. Since bird song is detected aurally, the full circle could be surveyed, as for point transect sampling. However, if reaction to the observer is suspected, just the forward half of the circle might be surveyed.

More complex methods for line transect sampling of cues, which allow uncertain detection of cues on the line, have been developed (Schweder 1974, 1977, 1999; Schweder and Høst 1992; Schweder *et al.* 1999).

11.7.2 *Estimation*

We assume that detections are of distinct, well-defined cues. For breeding season surveys of songbirds, the obvious cue is song. Calls might be included also, but for many species, calls may sometimes be much less detectable than song, in which case it may be preferable to restrict the definition of cue to song. In addition, by restricting recording to birds that are singing, we avoid the extreme heterogeneity in detectability that often occurs when both easily detected males and less detectable females are recorded (Section 11.4.2.2). It does not matter if the song lasts several seconds or several minutes even, provided time spent at each point is substantially longer than mean song length (which might be achieved using recording equipment). In fact, this restriction can be removed if, for example, the cue is defined to be the start of a song burst. Then birds that are already singing at the onset of recording are excluded.

The method we propose would be unsuitable for species for which it is difficult to define when a single burst of song starts and ends. Non-singing birds that are detected visually would not be recorded, due to the difficulty in defining an appropriate cue rate.

The cues are assumed to obey the usual assumptions of point transect sampling. For example, it is assumed that, if a bird located at the point sings, its song is certain to be detected. It is also assumed that distance of the cue from the point is recorded accurately. Rounding error in distance estimates can generate substantial bias (Buckland *et al.* 2001: 195). If the cues are aural, and especially if they are recorded on audio equipment, this could be problematic. In this case, it may be necessary to conduct experiments where distances to singing birds are estimated and then measured accurately. This allows correction of bias arising from inaccurate distance estimation. A better solution would be to use an acoustic array that allows location of cues to be estimated by triangulation.

In standard point transect sampling, effort at a point is equal to the number of times that point is visited. For point transect sampling of cues, the effort is the total time spent recording at that point (which may be from one or more visits).

Estimated bird density is given by

$$\widehat{D} = \frac{n}{\hat{\nu} T \hat{\eta}}, \tag{11.151}$$

where n is the number of cues detected, $\hat{\nu}$ is the estimated effective area of detection about a single point (estimated using conventional point transect analysis of detection distances), $\hat{\eta}$ is the estimated mean number of cues per unit time per bird, and $T = \sum_{i=1}^{k} t_i$, with t_i equal to the recording time at point i. Typically, t_i would be chosen to be the same for every point. If it is not, care should be taken to ensure that t_i is not correlated with bird density, which would generate bias, even if points were randomly located through the survey region.

Variance of estimated bird density can be approximated using the delta method:

$$\widehat{\mathrm{var}}(\widehat{D}) = \widehat{D}^2 \left\{ \frac{\widehat{\mathrm{var}}(n)}{n^2} + \frac{\widehat{\mathrm{var}}(\hat{\nu})}{\hat{\nu}^2} + \frac{\widehat{\mathrm{var}}(\hat{\eta})}{\hat{\eta}^2} \right\}. \tag{11.152}$$

The same time units must be used for both T and estimated cue rate. Note that $n/\hat{\nu}T$ is the estimated number of cues per unit time per unit area; to convert this to estimated bird density, we divide by the estimated average number of cues produced per unit time per bird.

Estimation of cue rate can be tackled by either a design-based approach or a model-based approach. In a design-based approach, it is important that a representative sample of rates is monitored. This might be done, for example, by gathering data on cue rates at a random or systematic sub-sample of points. For single-species surveys, or surveys of a small number of species, it may be possible to estimate the cue rate concurrently with the point transect survey work. If detection is near-certain within say u metres of a point, then whenever a singing bird is detected within this range, the number of cues detected within the timed recording period could be noted. If some birds have cue rates such that they might not sing at all during the timed period, there are two options. First, if all males sing, and are distinguishable visually from females, then any male detected visually within u of the point should be included in the calculation of cue rate; if it does not sing at all, its estimated cue rate is zero. In this case, u should be chosen so that detection of a non-singing male at distance u from the point is high, say around 0.8 or higher. The second option is to estimate cue rate at a random or systematic subset of points, to estimate cue rates of birds near those points. The period over which cue rates are monitored should be substantially (perhaps five times) longer than the typical period between cues for the species of interest, as otherwise estimates may be biased towards birds with higher cue rates. The monitoring should be conducted concurrently with the point transect survey, to ensure that the rates are those that apply at the time of the survey. This might be achieved using additional observers. Given this approach, there is no need to record covariates that affect cue rate.

A model-based strategy would be to record relevant covariates, both during the point transect survey and during the cue rate experiment, then to construct a model using the cue rate data, relating cue rate to variables such as time of day, date, topography, and habitat (Buckland *et al.* 2001: 191). For example, generalized additive models (Hastie and Tibshirani 1990) or generalized linear mixed models (Diggle *et al.* 1998) might be useful for this purpose. The model can then be used to estimate the cue rate at each surveyed point in the point transect sample, and hence the mean cue rate for the survey.

11.8 Migration counts

11.8.1 *Background*

Migration counts are used to estimate the size of populations of whales that file past convenient coastal watch points when migrating between feeding and breeding areas. Stocks of humpback, right, bowhead, and gray whales are assessed in this way, and if carefully designed, migration counts yield abundance estimates with low bias and high precision. The basic data are counts of whales, or more usually pods (clusters), passing during watch periods. Various covariates are recorded, including the distance of detected pods from shore. It is the modelling of these distance data that falls within the realm of distance sampling.

Migration counts therefore represent a non-standard application of distance sampling in which time as well as distances must be modelled. Furthermore, the underlying distribution of distances in the population is not uniform ($\pi(y) \neq 1/w$). The watch point is chosen such that most animals pass within the view of the observers, so that whale density typically increases with distance offshore, to a maximum, before rapidly falling off to nearly zero a few kilometres from shore. Double-platform methods are therefore required to model the detection function. One approach, described below, is to have two observation platforms at the watch point and to record data that allow identification of which pods were detected from both platforms. Two-sample mark-recapture methods then allow detection probabilities to be estimated as a function of distance from shore (Chapter 6). Either we must assume that all animals are potentially detectable, or an estimate is needed of the number of whales passing beyond the range of visibility of the shore-based platforms. Alternative approaches to estimate detection probability are described in Section 11.8.3 below.

Typically there will be regular, perhaps daily, counts of numbers of pods passing a watch point. The basic data are start and end times of watch periods and number of pods passing during each watch period. Thus the data are in frequency form, being grouped by watch period. There will be gaps between watch periods, corresponding to night or to poor visibility. For the basic method, animals are assumed to migrate at the same rate during unwatched periods as during watches. If no migration occurs at night, then time should be defined to end at dusk and start again at dawn. If migration occurs at night, but possibly at a different rate, this rate should be estimated, for example, by use of sonar (active or passive), radar, or thermal imagers (Perryman *et al.* 1999), or by radio-tagging animals (Swartz *et al.* 1987).

11.8.2 *Modelling migration rate*

To model migration rate as a function of date, we can use essentially the same methods for fitting a pdf to the data as are used in line transect sampling for modelling distances. In line transect sampling, the density function is assumed to be symmetric about the line, so only even functions (cosines for the Fourier series model and even powers for polynomial models) are used. For migration counts, odd functions are also needed. Related to this, the key function requires both a location and a scale parameter, whereas only a scale parameter is necessary for line transect sampling, because if sightings to the left of the line are recorded as negative and distances to the right as positive, the expected distance from the transect is zero under the assumption that the density function is symmetric about the line. Also, allowance must be made for a large number of counts, equal to the number of separate watch periods, whereas in a grouped analysis of line transect data, the number of groups for perpendicular distances seldom exceeds 20. Having fitted the density to migration times, abundance is estimated by taking the ratio of the area under the entire density function to the combined area corresponding to watch periods alone. The total number of animals estimated to have passed during watches is then multiplied by this ratio.

11.8.3 *Modelling detection probabilities*

We assume here that two shore-based observation platforms are used, and model the double-count data using the approach of Huggins (1989, 1991), which incorporates covariates to allow for heterogeneity in mark-recapture experiments. Essentially the same method was developed independently by Alho (1990). We assume that pods detected by both platforms are identified without error.

Assuming that the probability of detection of a pod from one platform is independent of whether it is detected from the other, and independent of whether other pods are detected by either platform, the full likelihood for all pods passing during watch periods is

$$\mathcal{L}^* = C \prod_{i=1}^{M} \prod_{j=1}^{2} p_{ij}^{\delta_{ij}} (1 - p_{ij})^{1-\delta_{ij}}, \tag{11.153}$$

where M = total number of pods passing during count periods,
p_{ij} = probability that pod i is detected by platform j,
$$i = 1, \ldots, M, \quad j = 1, 2,$$
$$\delta_{ij} = \begin{cases} 1, & \text{pod } i \text{ is detected by platform } j, \\ 0 & \text{otherwise,} \end{cases}$$

and C depends on M but not on the parameters that define p_{ij}.

Huggins (1989) showed that inference can be based on the conditional likelihood,

$$\mathcal{L} = \prod_{i=1}^{n} \frac{\prod_{j=1}^{2} p_{ij}^{\delta_{ij}} (1 - p_{ij})^{1-\delta_{ij}}}{p_i}, \tag{11.154}$$

where n = number of pods detected by at least one platform, and $p_i = 1 - \prod_{j=1}^{2}(1 - p_{ij})$ = probability that pod i is detected by at least one platform.

This likelihood is equivalent to $\mathcal{L}_\Omega$ as specified by eqn (6.3) in Chapter 6. Both Huggins (1989, 1991) and Alho (1990) maximized the likelihood directly, which is perfectly acceptable. However, as an alternative we present a method described by Buckland et al. (1993a,b) that uses standard logistic regression software to maximize the likelihood iteratively. Note that Buckland et al. (1993a,b) incorrectly implied that eqn (11.155) was the conditional likelihood instead of eqn (11.154), but this did not affect the validity of their fitting algorithm which we describe below.

Suppose we carry out a naive logistic regression, ignoring the fact that pods seen by neither platform are missing from our data. We would do that by having n trials for platform 1 of which n_1 are seen and n trials for platform 2 of which n_2 are seen. The total number of trials is $n = n_1 + n_2 - n_3 = \sum_{i=1}^{n} \delta_{i1} + \sum_{i=1}^{n} \delta_{i2} - \sum_{i=1}^{n} \delta_{i1}\delta_{i2}$, where n_3 is the number of pods that were detected by both platforms. The naive logistic regression maximizes the following likelihood

$$\mathcal{L} = \prod_{i=1}^{n} \prod_{j=1}^{2} \pi_{ij}^{\delta_{ij}} (1 - \pi_{ij})^{1-\delta_{ij}}, \tag{11.155}$$

where $\pi_{ij} = p_{ij}/p_i = \Pr(\text{pod seen by platform } j \mid \text{it is seen by at least one platform})$.

Note that the notation π_{ij} is not related to $\pi(y)$, which is used to indicate the distribution of distances of animals (whether detected or not) from the line or point in distance sampling. In this naive logistic regression, the linear predictor corresponding to observation ij is

$$\gamma_{ij} = \log \left[\frac{\pi_{ij}}{1 - \pi_{ij}} \right]. \tag{11.156}$$

To maximize the correct conditional likelihood of eqn (11.154), the linear predictor should be:

$$\theta_{ij} = \log \left[\frac{p_{ij}}{1 - p_{ij}} \right]. \tag{11.157}$$

Algebra yields:

$$\gamma_{ij} = \theta_{ij} - \log p_{ij'}, \quad \text{where } j' = 3 - j. \tag{11.158}$$

This relationship suggests the following iterative algorithm proposed by Buckland *et al.* (1993a) which uses $-\log p_{ij'}$ for trial i of platform j as an offset for the naive logistic regression:

1. Initially set all $p_{ij'} = 1$ which corresponds to a zero offset.
2. Fit the logistic regression using the offsets to get estimates of π_{ij}. For the initial fitting, $\theta_{ij} = \gamma_{ij}$.
3. Compute new offset values using the new estimates of $p_{ij} = \exp\{\theta_{ij}\}/(1 + \exp\{\theta_{ij}\})$.
4. Repeat steps 2 and 3 until estimates of θ_{ij} converge. The estimates of γ_{ij} and π_{ij} are non-informative because they always remain the same.

Let $p_{ij}^{(0)}, p_{ij}^{(1)}, p_{ij}^{(2)} \ldots$ be the sequence of estimates from iteration. With initial zero offsets, $p_{ij}^{(0)} = \pi_{ij}$. The offsets $(-\log p_{ij'})$ are always ≥ 0 because $p_{ij'}^{(k)} \leq 1$, which means $\theta_{ij}^{(k+1)} \leq \theta_{ij}^{(k)}$ and thus $p_{ij}^{(k+1)} \leq p_{ij}^{(k)}$. Because the sequences are monotone decreasing and bounded below, they will converge.

To demonstrate that the algorithm converges to the correct estimates, we can substitute $p_{ij} = \exp\{\theta_{ij}\}/(1 + \exp\{\theta_{ij}\})$ and solve the resulting simultaneous equations for θ_{i1} and θ_{i2}, and we get

$$\theta_{ij} = \log\left\{\frac{\exp(\gamma_{i1} + \gamma_{i2}) - 1}{1 + \exp(\gamma_{ij'})}\right\}, \quad \text{with } j = 1, 2 \quad \text{and} \quad j' = 3 - j.$$

$$(11.159)$$

However, the relationship is only valid if $(\gamma_{i1} + \gamma_{i2}) > 0$. If we consider the simple case where $p_{ij} = p_j$, the estimators of π_1 and π_2 are the intuitive ones, n_1/n and n_2/n, respectively. For this simple model, $(\gamma_{i1} + \gamma_{i2}) > 0$ and iteration is not necessary. Using eqn (11.159), we can derive with a little algebra the correct estimators $\hat{p}_j = n_3/n_{j'}$ from the initial naive estimators $\hat{\pi}_j = n_j/n$. Unfortunately, $(\gamma_{i1} + \gamma_{i2}) > 0$ will not hold in general, so an iterative approach is necessary.

From the estimates of p_{ij}, we can estimate M as

$$\widehat{M} = \sum_{i=1}^{n} \frac{1}{\hat{p}_i} \tag{11.160}$$

with

$$\widehat{\text{var}}(\widehat{M}) = \sum_{i=1}^{n} \frac{1}{\hat{p}_i^2}(1 - \hat{p}_i). \tag{11.161}$$

This variance ignores a component due to estimation of the detection probabilities. This component is small when the probabilities are near one, as typically occurs in migration counts. Huggins (1989) shows how this component may be estimated.

Thus a correction factor for pods missed by both platforms is given by

$$\hat{f}_m = \frac{\widehat{M}}{n}, \tag{11.162}$$

with

$$\hat{se}(\hat{f}_m \mid n) = \frac{\sqrt{\widehat{\text{var}}(\widehat{M})}}{n}. \tag{11.163}$$

This estimation scheme is only one of several possible methods of estimating detection probability for shore-based migration counts, and it does have some disadvantages. The double-platform scheme described above is equivalent to the 'full independence' method described in Chapter 6 in which estimation is based solely on $\mathcal{L}_\omega$. If there is any unmodelled heterogeneity in the detection probabilities (i.e. there are important missing covariates), the estimator will be biased; typically the bias in the abundance estimator will be negative. There are alternative approaches, but each requires an aircraft or equivalent means of sampling the offshore distribution. We list three alternative double-platform methods in order of increasing complexity:

1. Use a single shore-based observer and an aircraft that surveys independently perpendicular to the shore. No attempt is made to identify duplicate detections of the two platforms.

2. Use a single shore-based observer and an aircraft that surveys concurrently perpendicular to the shore. Duplicate detections of the two platforms are identified.

3. Use two independent shore-based observers and an aircraft that surveys independently perpendicular to the shore. Duplicate identifications are identified for shore-based observations but not with the aircraft.

For method 1 above, the aircraft is used to estimate $\pi(y)$, but we must assume that detection probability for the shore observer is unity ($p(y) = 1$) at some distance y. This is an extension of conventional distance sampling with an estimated non-uniform $\pi(y)$. Method 2 above is equivalent to the method described in this section except that an aircraft is used as one of the platforms. However, because the platforms have completely different vantage points, it is less likely that heterogeneity will be problematic. It has the advantage that we do not need to assume that $p(y) = 1$, but we must match detections between the aircraft and the shore station. The third method is probably the most costly but avoids the problems with matching shore and aircraft observations. It is also less susceptible to heterogeneity. Using method 3, we could assume 'point independence' to avoid unmodelled heterogeneity associated with increasing distance as described in Chapter 6. This would be possible because the aircraft observations could be used to estimate $\pi(y)$. A fourth method would be to match the two shore observers and aircraft, which would be equivalent to mark-recapture with three sampling occasions.

11.8.4 An example: gray whales

We use here as an example the gray whale census data collected near Monterey, California. The eastern North Pacific stock of gray whales migrates from feeding grounds in the Bering and Chukchi Seas to spend the winter in warm Mexican waters, returning north in spring. Aerial surveys confirm that almost the entire population passes close to shore near the counting station. Counts at this station can therefore be used to estimate population size. These counts were annual from 1967–8 through to 1979–80, and further surveys were carried out in 1984–5, 1985–6, 1987–8, 1992–3, 1993–4, 1995–6, 1997–8, 2000–01, and 2001–02. Reilly *et al.* (1980, 1983) give more information on the earlier surveys, and Buckland and Breiwick (2002) provide abundance estimates corresponding to surveys until 1995–6. We use analyses of the gray whale count data for 1987–8, extracted from Breiwick *et al.* (unpublished) and Buckland *et al.* (1993b), to illustrate analysis of migration count data. In that year, counts were made from two stations (north and south) a few yards apart, to allow estimation of numbers of pods missed during watch periods. The data analysed were numbers of pods passing, within each count period, so that the data are grouped, the group endpoints being the start and end of each watch period. Information on duplicate detections was used to reduce the data sets from both stations to a single set of counts of the number of pods detected by at least one station in each watch period. Pods detected travelling north were excluded from the analyses.

The key function selected for modelling migration rates was, apart from a scaling factor, the normal density:

$$\alpha(y) = \exp\left\{-\left(\frac{y-\mu}{\sigma}\right)^2\right\}, \tag{11.164}$$

where y corresponds to time, measured in days from a predetermined date. Adjustments to the fit of the key were made by adding Hermite polynomial terms sequentially, adjusting the fit first for skewness, then kurtosis, and so on. Likelihood ratio tests selected a three-term (i.e. five-parameter) fit, and this fit is shown in Fig. 11.7. To convert the fitted density to an estimate of population size, it is necessary to evaluate the proportion of the entire untruncated density that corresponds to watch periods. To ensure that the Hermite polynomial fits were sensible in the tails of the migration, zero counts were added for 1 December 1987, before the migration started, and 29 February 1988, after it ended. This had little effect for 1987–88, when counts took place throughout the main migration period (Fig. 11.7).

In total, and excluding pods travelling north, $n = 3593$ pods were seen from at least one station. The χ^2 goodness of fit statistic corresponding to Fig. 11.7 was $\chi^2_{125} = 334.55$. This value is more indicative of overdispersion of counts than of intrinsic lack of fit of the Hermite polynomial model; in other words, counts in successive watches show greater than Poisson variation. The overdispersion was compensated for

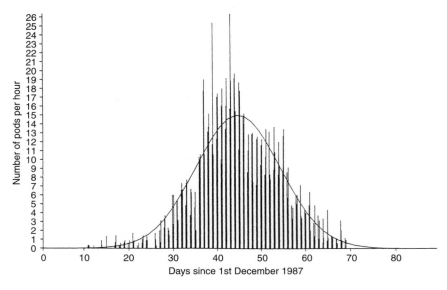

Fig. 11.7. Histogram of number of California gray whale pods sighted, adjusted for watch length, by date, 1987–8 survey. Also shown is the Hermite polynomial fit to the histogram.

by multiplying the Poisson variance on the total count by the dispersion parameter, estimated as the χ^2 statistic divided by its degrees of freedom; this multiplicative correction is sometimes termed a variance inflation factor (Cox and Snell 1989). Thus the dispersion parameter estimate is $334.55/125 = 2.676$, giving $\widehat{se}(n) = \sqrt{3593 \cdot 2.676} = 98.1$ pods. The fit of the Hermite polynomial model to the counts yields a multiplicative correction for animals passing outside watch periods of $\hat{f}_t = 2.4178$ with standard error 0.0068.

Swartz *et al.* (1987) reported on experiments in which whales were radio-tagged in 1985 and 1986. Of these, 15 were recorded both at night and in daylight. An unpaired t-test on the difference in log day and night speeds revealed no significant difference between Monterey and the Channel Islands ($t_{11} = -1.495$; $p > 0.1$). After pooling the data from both locations, a paired t-test revealed a significant difference in log speeds between day and night ($t_{14} = 2.284$; $p < 0.05$). A back-transformation with bias correction gave a multiplicative correction factor for hours of darkness of 1.100 ($\widehat{se} = 0.045$); thus it is estimated that rate of passage is 10% higher at night, and thus night counts, if they were feasible, would generate counts 10% higher. However, counts were limited to ten hours per day. Assuming whales treat twilight as daylight, we added an hour to each end of the day, giving roughly 12 h of daylight per 24 h. Thus the multiplicative correction applies approximately to one half of the total number of whales estimated, giving a multiplicative correction factor of $\hat{f}_n = 1.050\,(\hat{se} = 0.023)$. Swartz (personal communication) notes that the behaviour of the animals off the Channel Islands is very different from their behaviour near Monterey. If a correction factor is calculated as above from the nine radio-tagged whales off Monterey that were recorded both during the day and at night, we obtain $\hat{f}_n = 1.020\,(\hat{se} = 0.023)$. Although this does not differ significantly from one, we apply it, so that the variance of the abundance estimate correctly reflects the uncertainty in the information on this potentially important parameter.

During the 1987–8 season, counts were carried out independently by observers in identical sheds, 5 m apart. We analyse the double-count data using the methods of Section 11.8.3. The procedures for matching detections from the two stations are described by Rugh *et al.* (1993).

Potential covariates were date, Beaufort sea state, components of wind direction parallel and perpendicular to the coast, visibility code, distance offshore, pod size, and rate of passage (pods per hour); observer, station, and watch period were entered as factors. Probability of detection of a pod was adequately modelled as a function of five covariates: pod size; rate of passage; migration date; visibility code; and the component of wind direction parallel to the coast. None of the factors (observer, watch period, station) explained a significant amount of variation. Probability of detection increased with pod size ($p < 0.001$), with rate of passage ($p < 0.001$), and with migration date ($p < 0.05$), and decreased with visibility code ($p < 0.05$). It was also greater when the wind was parallel to the coast from 330° (slightly west of north), and smaller when from 150° (east of south). The correction factor f_m was estimated by $\hat{f}_m = 1.0632$, with standard error 0.00447.

The number of whales passing Monterey is equal to the number of pods multiplied by the average pod size, which was estimated by the average size of pods detected (excluding those moving north). This gave $\bar{s} = 1.959\,(\hat{se} = 0.020)$. A correction factor for mean pod size was calculated using data from Reilly *et al.* (1980), comparing recorded pod sizes with actual pod sizes, determined by observers in an aircraft. For pods whose recorded size was one, an additive correction of 0.350, with standard error $0.6812/\sqrt{225} = 0.0454$, was used. The correction for pods of size two was 0.178 ($\hat{se} = 0.9316/\sqrt{101} = 0.0927$), for pods of size three, 0.035 ($\hat{se} = 1.290/\sqrt{28} = 0.244$), and for pods of size four or greater, the correction was 0.333 ($\hat{se} = 0.7825/\sqrt{27} = 0.151$). A multiplicative correction factor for mean pod size was then found as:

$$\hat{f}_s = 1 + \frac{0.350n_1 + 0.178n_2 + 0.035n_3 + 0.333n_{4+}}{n\bar{s}} = 1.131$$

with

$$\widehat{se}(\hat{f}_s \mid n) \simeq \sqrt{[}(0.0454n_1)^2 + (0.0927n_2)^2 + (0.2438n_3)^2 + (0.1506n_{4+})^2$$
$$+ 0.6812^2 n_1 + 0.9316^2 n_2 + 1.290^2 n_3 + 0.7825^2 n_{4+}]/(n\,\bar{s})$$
$$= 0.026.$$

where n is the total number of pods recorded, n_i is number of pods of size i, $i = 1, 2, 3$, and n_{4+} is number of pods of size four or more.

The revised abundance estimate was thus found as follows. Counts of numbers of pods by watch period were combined across the two stations, so that each pod detected by at least one station contributed a frequency of one. The Hermite polynomial model was applied to these counts, to obtain a multiplicative correction factor $\hat{f}_t$ to the number of pods detected for whales passing at night or during poor weather. The correction for different rate of passage at night $\hat{f}_n$ was then made. Next, the multiplicative correction $\hat{f}_m$ was applied, to allow for pods passing undetected during watch periods. The estimated number of pods was then multiplied by the mean pod size, and by the correction factor $\hat{f}_s$ for underestimation of pod size, to obtain the estimate of the number of whales passing Monterey during the 1987–8

Table 11.7. Estimates of abundance and of intermediate parameters, California gray whales, 1987–8

Parameter	Estimate	SE	% contribution to $\widehat{\mathrm{var}}(\widehat{N})$	95% confidence interval
E(Number of pods seen by at least one station) $= E(n)$	3593	98	39	(3406, 3790)
Correction for pods passing outside watch periods, f_t	2.418	0.007	0	(2.405, 2.431)
Correction for night passage rate, f_n	1.020	0.023	27	(0.976, 1.066)
Correction for pods missed during watch periods, f_m	1.063	0.004	1	(1.054, 1.072)
Total number of pods passing Monterey	9419	337		(8781, 10,104)
Mean recorded pod size	1.959	0.020	5	(1.920, 1.999)
Correction for bias in recorded pod size, f_s	1.131	0.026	28	(1.081, 1.183)
Total number of whales passing Monterey	20,869	913		(19,156, 22,736)

migration. Thus the abundance estimate for 1987–8 is given by

$$\widehat{N} = n\,\hat{f}_t\,\hat{f}_n\,\hat{f}_m\,\bar{s}\,\hat{f}_s \qquad (11.165)$$

with

$$\mathrm{cv}(\widehat{N}) \simeq \sqrt{\begin{array}{c}[\mathrm{cv}(n)]^2 + [\mathrm{cv}(\hat{f}_t)]^2 + [\mathrm{cv}(\hat{f}_n)]^2 \\ +[\mathrm{cv}(\hat{f}_m)]^2 + [\mathrm{cv}(\bar{s})]^2 + [\mathrm{cv}(\hat{f}_s)]^2\end{array}} \qquad (11.166)$$

Table 11.7 shows the different components of the estimate $\widehat{N}$. Combining them, estimated abundance is 20,869 whales, with $\mathrm{cv}(\widehat{N}) = 0.0437$ and approximate 95% confidence interval (19,200, 22,700).

Buckland and Breiwick (2002) scaled their relative abundance estimates for the period 1967–8 to 1987–8 to pass through an absolute abundance estimate for 1987–8. Rescaling them to pass through the revised estimate above, and adding in the abundance estimates for 1992–3 and 1993–4 from Laake *et al.* (unpublished) and for 1995–6 from Hobbs *et al.* (unpublished) yields the estimates of Table 11.8. Figure 11.8 plots the absolute abundance estimates and shows the estimated increase in abundance assuming an exponential model with non-zero asymptote. The estimated mean annual rate of increase is 2.5% per annum ($\hat{se} = 0.3\%$). Buckland and Breiwick (2002) also fitted a logistic curve to these estimates, which gave an estimated carrying capacity of 26,000 whales ($\hat{se} = 6300$).

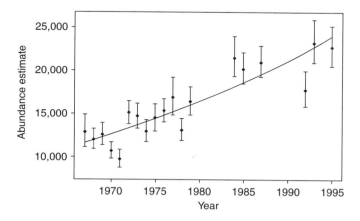

Fig. 11.8. Estimates of abundance by year of California gray whales, and predicted abundance from a weighted exponential regression of abundance estimates on year. Year 1967 signifies winter 1967–8, etc.

Table 11.8. Estimated number of pods, pod size, and number of whales by year. (Standard errors in parentheses.) For any given fit, the number of parameters is two greater than the number of terms, corresponding to the two parameters of the normal key

Year	No. of terms	χ^2 [df]	Sample size (pods)	Estimated no. of pods	Estimated average pod size	Relative abundance estimate	Absolute abundance estimate
1967–8	4	83.0 [45]	903	4051 (253)	2.438 (0.063)	9878 (667)	12,921 (964)
1968–9	0	70.6 [61]	1079	4321 (134)	2.135 (0.046)	9227 (348)	12,070 (594)
1969–70	1	104.5 [67]	1245	4526 (155)	2.128 (0.043)	9630 (383)	12,597 (640)
1970–1	2	116.2 [90]	1458	4051 (115)	2.021 (0.033)	8185 (267)	10,707 (487)
1971–2	0	71.3 [56]	857	3403 (127)	2.193 (0.048)	7461 (323)	9760 (524)
1972–3	4	91.5 [71]	1539	5279 (152)	2.187 (0.034)	11,543 (378)	15,099 (688)
1973–4	4	133.7 [66]	1496	5356 (186)	2.098 (0.034)	11,235 (431)	14,696 (731)
1974–5	0	159.2 [74]	1508	4868 (174)	2.034 (0.035)	9904 (394)	12,955 (659)
1975–6	2	101.1 [47]	1187	5354 (218)	2.073 (0.039)	11,100 (497)	14,520 (796)
1976–7	0	139.7 [87]	1991	5701 (153)	2.052 (0.028)	11,700 (353)	15,304 (669)
1977–8	0	50.2 [31]	657	7001 (356)	1.843 (0.046)	12,904 (731)	16,879 (1095)
1978–9	4	152.9 [84]	1730	4970 (159)	2.016 (0.034)	10,018 (361)	13,104 (629)
1979–80	4	109.3 [55]	1451	6051 (220)	2.068 (0.033)	12,510 (498)	16,364 (832)
1984–5	3	105.2 [49]	1756	7159 (301)	2.290 (0.038)	16,393 (740)	21,443 (1182)
1985–6	1	141.4 [104]	1796	6873 (191)	2.237 (0.042)	15,376 (515)	20,113 (927)
1987–8N	3	205.9 [92]	2426	7756 (221)	2.040 (0.027)	15,825 (497)	
1987–8S	3	152.8 [91]	2404	7642 (194)	2.104 (0.029)	16,082 (464)	
1987–8 (average)						15,954 (481)	20,869 (913)
1992–3						17,674 (1029)	
1993–4							23,109 (1262)
1995–6							22,263 (1078)

11.9 Estimation with distance measurement errors

Thus far, all the methods presented in this book assume that the distance data are recorded with no error. The effects when this assumption fails have received less attention than those for the assumption that $g(0) = 1$ or that there is no responsive movement. While it has been a fairly common practice to correct bias in distance estimation when results of estimated distance experiments are available, much less attention has been given to the effects of random (possibly unbiased) errors in distance estimation. Hiby *et al.* (1989) were probably the first to address the problem. They used maximum likelihood methods based on a multiplicative normal measurement error model to accommodate measurement error in cue counting surveys of minke whales. Others have addressed the problem since. Alpizar-Jara (1997) proposed a procedure based on the SIMEX algorithm (Cook and Stefansky 1994) to correct for the random component of additive measurement error in line transect surveys. Schweder (e.g. 1996, 1997) considered error models to remove error-induced bias in the analysis of line transect surveys of minke whales. Chen (1998) described the effect of additive errors in line transect surveys, and derived a corrected estimator based on the method of moments and assuming some knowledge of the distribution of the errors. Chen and Cowling (2001) generalized this approach for the case in which errors are present in both distances and other covariates. (They considered cluster size as well as distance.)

The following development of general distance sampling likelihood functions in the presence of measurement errors is taken from Marques (in press) and Borchers *et al.* (in preparation *b*). For simplicity of presentation, we neglect all explanatory variables other than distance. See Borchers *et al.* (in preparation *b*) for the development with other explanatory variables.

11.9.1 *Conventional distance sampling: $g(0) = 1$*

Let y_i be the true distance to the ith object (animal or cluster) on a distance sampling survey and let $g(y_i)$ be the probability that the object is detected, where $g(0) = 1$. Suppose that the pdf of the distances in the population (detected and undetected) is $\pi(y)$. Then the pdf of the distance to the ith detected object is

$$f(y_i) = \frac{g(y_i)\pi(y_i)}{\int g(y)\pi(y)\,dy} = \frac{g(y_i)\pi(y_i)}{P_a}, \qquad (11.167)$$

where $P_a = \int g(y)\pi(y)dy$.

If the detection function $g(y_i)$ has an unknown parameter vector $\underline{\theta}$ associated with it, the likelihood for $\underline{\theta}$ given the n observed distances

$y_1, \ldots, y_n$ is

$$\mathcal{L}(\underline{\theta}) = \prod_{i=1}^{n} f(y_i). \tag{11.168}$$

Equation (11.168) is the likelihood function that is maximized to estimate the detection function parameters $\underline{\theta}$ with conventional distance sampling methods.

Now suppose that distances are observed with error. Denote the observed distances (those with measurement error) $y_{01}, \ldots, y_{0n}$ and let $\pi(y_0 \,|\, y)$ be the conditional pdf of observed distance, given true distance. We discuss below how this pdf might be specified; for the moment suppose that we have specified it in some sensible way. This pdf has some parameter vector $\underline{\phi}$ associated with it; thus $\underline{\phi}$ is the parameter vector of the measurement error model. In this case, we can derive the likelihood for $\underline{\theta}$ and $\underline{\phi}$ given the observed data $y_{01}, \ldots, y_{0n}$, as follows.

The joint density of true distance y and observed distance y_0 for detected animals is $f(y, y_0) = f(y)\pi(y_0 \,|\, y)$. True distance y is not observed but the pdf of observed distance is

$$f_{y_0}(y_0) = \int f(y)\pi(y_0 \,|\, y) \, dy \tag{11.169}$$

and the likelihood for $\underline{\theta}$ and $\underline{\phi}$ given $y_{01}, \ldots, y_{0n}$ is

$$\mathcal{L}_1(\underline{\theta}, \underline{\phi}) = \prod_{i=1}^{n} f_{y_0}(y_{0i}). \tag{11.170}$$

The data from a survey are inadequate for estimation of the error model parameter vector $\underline{\phi}$ from this likelihood because true distances are not observed. However, given an estimate of $\underline{\phi}$, this likelihood provides us with a method for estimating $\underline{\theta}$ (and hence $g(y)$) when distances are observed with errors. This can be converted to an estimate of abundance using a Horvitz–Thompson-like estimator.

Given additional data to estimate $\underline{\phi}$, we can write the joint likelihood for $\underline{\theta}$ and $\underline{\phi}$ as

$$\mathcal{L}(\underline{\theta}, \underline{\phi}) = \mathcal{L}_1(\underline{\theta}, \underline{\phi})\mathcal{L}_2(\underline{\phi}), \tag{11.171}$$

where the additional data are utilized in $\mathcal{L}_2(\underline{\phi})$, while $\mathcal{L}_1(\underline{\theta}, \underline{\phi})$ is as specified in eqn (11.170).

For the likelihood of eqn (11.170) to be useful, we need to specify a model for $\pi(y_0 \,|\, y)$ and estimate $\underline{\phi}$.

A useful special case is that in which

(1) detection at distance zero is certain ($g(0) = 1$);
(2) measurement errors are multiplicative[1] (see below for details); and
(3) measurement errors are independent of true distance.

Marques (in press) developed parametric and nonparametric estimators of $1/P_a$ and abundance for this case with line transect surveys. Below, we outline his methods and extend them to point transect surveys.

11.9.2 *Independent multiplicative measurement errors*

The multiplicative error model we consider in this section is such that

$$y_0 = y\epsilon, \tag{11.172}$$

where ϵ is the measurement error and ϵ and y are independent. Notice that this error model implies that when $y = 0$, there is no measurement error. It does not require that the measurement errors are unbiased: $E(\epsilon)$ need not be 1.

Transformation of variable from y_0 to ϵ gives the result $\pi(y_0 \mid y) = \pi_\epsilon(\epsilon) \times \epsilon/y_0$. Hence

$$f_{y_0}(y_0) = \int_0^\infty f_y(y)\pi(y_0 \mid y)\, dy = \int_0^\infty f_y(y)\pi_\epsilon(\epsilon)\frac{\epsilon}{y_0}\, dy. \tag{11.173}$$

If we now change the variable of integration from y to ϵ, we obtain

$$f_{y_0}(y_0) = \int_0^\infty f_y\left(\frac{y_0}{\epsilon}\right)\pi_\epsilon(\epsilon)\frac{1}{\epsilon}\, d\epsilon. \tag{11.174}$$

11.9.2.1 *Line transects*
From eqn (11.174), the intercept of the pdf of observed distances (those with error) can be written as

$$f_{y_0}(0) = f_y(0)\int_0^\infty \pi_\epsilon(\epsilon)\frac{1}{\epsilon}\, d\epsilon = f_y(0) \times E\left[\frac{1}{\epsilon}\right]. \tag{11.175}$$

When measurement errors are unbiased ($E(\epsilon) = 1$), measurement errors of the kind considered in this section cause positive bias in the density or abundance estimator because (by Jensen's inequality) $E(1/\epsilon) > 1/E(\epsilon)$.

---

[1] The term 'multiplicative error' is somewhat ambiguous because any multiplicative error model can be written as an additive error model with a different variance structure.

If $g(0) = 1$, then $f_y(0) = 1/P_a$, and an estimate of $f_y(0)$ is needed for line transect estimators of density and abundance. Marques (in press) proposed estimators of the form

$$\hat{f}_y(0) = \hat{f}_{y_0}(0)\widehat{E}\left[\epsilon^{-1}\right]^{-1},\tag{11.176}$$

where $\hat{f}_{y_0}(0)$ is obtained using conventional line transect methods with the observed distances (those with measurement error) and $\widehat{E}[\epsilon^{-1}]^{-1}$ can be considered a correction for measurement error.

For ease of reference below, we refer to this method as the 'pdf correction approach'. The key to this method is the estimation of $E[\epsilon^{-1}]^{-1}$. Marques (in press) proposed both parametric and nonparametric estimators (see below).

11.9.2.2 *Point transects*
Borchers *et al.* (in preparation *b*) extended the 'pdf correction approach' of Marques (in press) for point transects as follows. Whereas conventional line transect estimators require an estimate of $f_y(0)$, conventional point transect estimators require an estimate of $h_y(0) = df_y(y)/dy|_{y=0}$.

Differentiating eqn (11.174) with respect to y_0, we get

$$
\begin{aligned}
h_{y_0}(y_0) = \frac{df_{y_0}(y_0)}{dy_0} &= \frac{d}{dy_0}\left(\int_0^\infty f_y\left(\frac{y_0}{\epsilon}\right)\frac{1}{\epsilon}\pi_\epsilon(\epsilon)\,d\epsilon\right)\\
&= \int_0^\infty \frac{df_y\left(y_0/\epsilon\right)}{dy_0}\frac{1}{\epsilon}\pi_\epsilon(\epsilon)\,d\epsilon\\
&= \int_0^\infty \frac{df_y\left(y_0/\epsilon\right)}{d\left(y_0/\epsilon\right)}\frac{1}{\epsilon^2}\pi_\epsilon(\epsilon)\,d\epsilon
\end{aligned}\tag{11.177}
$$

from which we get

$$h_{y_0}(0) = h_y(0) \times E\left[\epsilon^{-2}\right],\tag{11.178}$$

and $h_y(0)$ can be estimated as follows

$$\hat{h}_y(0) = \hat{h}_{y_0}(0)\widehat{E}\left[\epsilon^{-2}\right]^{-1}.\tag{11.179}$$

Here $\hat{h}_{y_0}(0)$ can be obtained using conventional point transect methods with the observed distances (those with measurement error) and $\widehat{E}[\epsilon^{-2}]^{-1}$ can be considered a correction for measurement error. The key to this method is estimation of $E[\epsilon^{-2}]^{-1}$. We outline parametric and nonparametric estimators below.

With error models of the sort covered in this section, if distance estimates are unbiased, the same error model will cause greater bias on a point transect survey than a line transect survey because $E(1/\epsilon^2) > E(1/\epsilon)$.

11.9.2.3 *Error model parameter estimation*

One convenient way to specify the measurement error model is to specify the conditional distribution of the error given the true distance, $\pi(\epsilon \,|\, y)$, together with an additive or multiplicative relationship between ϵ, y and y_0. This implicitly defines $\pi(y_{0k} \,|\, y_k)$. Marques (in press) used a beta distributional form for $\pi(\epsilon)$, assumed y_0 and y to be independent, and $y_0 = y\epsilon$; he also noted that a gamma form is a convenient alternative. Another alternative is to specify a functional form for $\pi(y_0 \,|\, y)$, the conditional distribution of observed distance given the true distance, directly. Borchers *et al.* (in preparation *b*) considered normal and gamma distributional forms for this distribution.

Estimated distance experiment data can be used to estimate the parameters ϕ of the measurement error model and/or $E[\epsilon^{-1}]^{-1}$ (line transects) or $E[\epsilon^{-2}]^{-1}$ (point transects). These data consist of pairs (y_{0k}, y_k), for $k = 1, \ldots, K$. However the measurement error model is specified, its parameters can be estimated by maximizing the likelihood

$$\mathcal{L}(\underline{\phi}) = \prod_{k=1}^{K} \pi(y_{0k} \,|\, y_k). \tag{11.180}$$

The parameter estimates and measurement error model can be used to evaluate $E[\epsilon^{-1}]^{-1}$ or $E[\epsilon^{-2}]^{-1}$. Alternatively, they can be estimated nonparametrically as follows:

$$\widehat{E}[\epsilon^{-1}]^{-1} = \frac{1}{(1/K) \sum_{k=1}^{K} y_k/y_{0k}}, \tag{11.181}$$

or

$$\widehat{E}[\epsilon^{-2}]^{-1} = \frac{1}{(1/K) \sum_{k=1}^{K} (y_k/y_{0k})^2}. \tag{11.182}$$

11.9.3 *Mark-recapture distance sampling: $p(0) < 1$*

When detection of animals at distance zero is not certain and double-platform methods are used, one distance is observed if the animal is detected by only one observer but two are observed in the case of 'duplicate' detections (those detected by both observers). It is convenient to define a vector $\underline{y}_{0i}$ which consists of the observed distance(s): $\underline{y}_{0i} = y_{0ij}$ if only observer j ($j = 1$ or 2) detected the animal and $\underline{y}_{0i} = (y_{0i1}, y_{0i2})^T$ if both

observers detected the animal. We assume that measurement errors by
the two observers are independent, so in the case of duplicate detections,
$\pi(\underline{y}_{0i} \mid y) = \pi(y_{0i1}, y_{0i2} \mid y) = \pi(y_{0i1} \mid y)\pi(y_{0i2} \mid y)$.

In the case of double-platform surveys, we define $p_.(y) = p_1(y) + p_2(y) - p_1(y)p_2(y)$ to be the probability that at least one of the observers detects
an animal at y, where the detection probability function for observer j is
$p_j(y)$. Also,

$$P_a = \int [p_1(y) + p_2(y) - p_1(y)p_2(y)]\pi(y)\,dy. \qquad (11.183)$$

For any given detection there are three possible capture histories $\underline{\omega}_i = (\omega_{i1}, \omega_{i2})$: $(1,0)$ for detections by observer 1 only; $(0,1)$ for detections by
observer 2 only; and $(1,1)$ for detections by both observers. The probability
of observing capture history $\underline{\omega}_i$, given true distance y_i, is

$$Pr\{\underline{\omega}_i \mid y_i\} = \prod_{j=1}^{2} p_j(y_i)^{\omega_{ij}}[1 - p_j(y_i)]^{1-\omega_{ij}}. \qquad (11.184)$$

The conditional probability of observing capture history $\underline{\omega}_i$ and the
distance(s) $\underline{y}_{0i}$, given that animal i was detected by at least one observer, is

$$Pr\{\underline{\omega}_i \cap \underline{y}_{0i} \mid \omega_{i1} + \omega_{i2} > 0\} = \frac{\int Pr\{\underline{\omega}_i \mid y\}\pi(\underline{y}_{0i} \mid y)\pi(y)\,dy}{P_a}, \qquad (11.185)$$

where $P_a = \int p_.(y)\pi(y)\,dy$ as before.

Maximum likelihood estimates of the measurement error model parameters (ϕ) and detection function parameters for each observer $(\underline{\theta}_1, \underline{\theta}_2)$ can
be obtained by maximizing the conditional likelihood

$$\mathcal{L}(\underline{\theta}_1, \underline{\theta}_2, \underline{\phi}) = \prod_{i=1}^{n} Pr\{\underline{\omega}_i \cap \underline{y}_{0i} \mid \omega_{i1} + \omega_{i2} > 0\}. \qquad (11.186)$$

Note that no separate distance experiment is required to estimate the parameters ϕ of the measurement error model in this case, provided distance
estimation is unbiased, animals do not move, and we know the relative size of errors from each platform; the paired observations (y_{0i1}, y_{0i2})
$(i = 1, \ldots, n)$ provide data from which ϕ can be estimated simultaneously
with $\underline{\theta}$. If data from a distance experiment are available, these can be
incorporated by multiplying the likelihood of eqn (11.186) by the experiment likelihood of eqn (11.180) and maximizing the product with respect
to $\underline{\theta}$ and $\underline{\phi}$.

11.9.4 *Maximum likelihood vs pdf correction approach*

Although comparisons between maximizing the likelihood of eqn (11.170) and the pdf correction approach are not entirely clear because the methods are new, the following considerations are relevant.

1. The pdf correction approach is easy to apply and allows the full power of existing distance software to be used to estimate $f_{y_0}(0)$ or $h_{y_0}(0)$.

2. Because some forms of measurement error can result in $f_{y_0}(y_0)$ having a 'spike' at distance zero in the case of line transect surveys, it may be difficult to fit $f_{y_0}(y_0)$ reliably even when $f_y(y)$ does not have a 'spike'. Because the likelihood approach models $g(y)$ (or $p(y)$), not $f_{y_0}(y_0)$, it is less likely to suffer from this problem.

3. Unlike the pdf correction approach, the likelihood approach is not restricted to any particular error structure and can cope with situations in which detection at distance zero is not certain.

11.10 Relating object abundance to population abundance for indirect sampling

11.10.1 *Introduction*

One use of distance sampling is indirect inference about animal abundance by surveying, for example, dung (Barnes *et al.* 1997; Barnes 2001; Marques *et al.* 2001; Buckland *et al.* 2001: 183–9). The underlying idea is that the animals of interest frequently and reliably produce objects, typically dung or nests, that can be surveyed. Buckland *et al.* (2001: 182–3) use the term 'indirect methods' to include both methods where the objects are detectable and survive for a time period before vanishing (such as dung), and methods based on (nearly) instantaneous cues (such as whale blows). We consider here only the former case, for which observers typically encounter the object days or weeks after the animal is gone.

The relevant literature seems not to have spelled out general mathematics of the relationships among animal (N) and object abundance (M) via object creation and survival probabilities. In general, all four quantities are time varying. The purpose of this section is to show a way to interrelate and explain these quantities under the assumption that doing so will help with insights into indirect distance sampling. Deterministic mathematics is used; a stochastic version is possible but is not needed to convey these basic ideas.

We assume that the study area is well-defined and of size A. At time instant t, there are $N(t)$ individual animals in the study area; density is $D(t) = N(t)/A$. Also at time t there are $M(t)$ objects extant, such as dung piles or nests. Let $B(t)$ denote the instantaneous object creation rate based on all $N(t)$ animals. Hence, the total number of objects created in any time interval t_1 to t_2 is $\int_{t_1}^{t_2} B(t)dt$; this total is not of interest to us because objects decay over time. In a stochastic version of this modelling, $B(t)$ is the intensity function of a non-homogeneous Poisson process. Finally, once created at time t, an object has a probability $S(t, x)$ of survival until any future time x.

11.10.2 Discrete-time modelling

If we consider everything in terms of discrete days, the number of objects extant on day x is the sum of objects created on day i that survive to day x, for all days i up to day x. Thus the individual terms of this sum can be expressed as

$$\left[\int_i^{i+1} B(t)\, dt\right] S(i, x), \qquad (11.187)$$

which gives the expected number of objects extant ('alive') of age $x - i$ at day x. Summing such terms gives

$$M(x) = \sum_{i \leq x} \left[\int_i^{i+1} B(t)\, dt\right] S(i, x). \qquad (11.188)$$

11.10.3 Continuous-time modelling

The above representation is an approximation because the survival term assumes that objects are created at the start of the day in which they appear. It is also an awkward expression from which to derive theory. The solution to both these problems is to use a continuous-time formulation, obtainable as a limit as we discretize time into shorter and shorter intervals. The result is

$$M(x) = \int_{-\infty}^{x} B(t)S(t, x)\, dt. \qquad (11.189)$$

Note that $S(t, x)$ implicitly refers to objects of age $x - t$ at time x since they were created at time t.

It is convenient to use a hazard (i.e. mortality) rate function $h(\cdot)$ to express

$$S(t, x) = \exp\left\{-\int_t^x h(\tau)\, d\tau\right\}. \qquad (11.190)$$

Models for $S(t, x)$ can be time-specific, or age-specific, or depend on both age and time. We specify such models by modelling $h(\cdot)$ and its integral in

$S(t, x)$. Substituting eqn (11.190) in eqn (11.189), we obtain

$$M(x) = \int_{-\infty}^{x} B(t) \exp\left\{-\int_{t}^{x} h(\tau)d\tau\right\} dt. \qquad (11.191)$$

It is important to understand that there is no concept of absolute time here, so that how we index time is irrelevant. That is, we can have time run from either 0 or $-\infty$ (mathematically, either has advantages), but there is not actually negative time. All that intrinsically matters is the lengths of intervals of time, such as $x - t$, or age-related time intervals such as object age at disappearance ('death'), hence expected survival time units, or an object's age at its detection. Also, time is not causal; rather, $h(\tau)$ merely reflects hazards occurring in the environment and/or the 'effects' of object age.

We could interpret the formula for $M(x)$ as applying to one animal, with animal-specific rate functions, and get the total $M(x)$ as a sum over individuals. However, for inference purposes from data, a model using animal-specific covariates is useless here because we cannot expect to know what animal created what object. Rather, we need to use the model as $B(t) = N(t)b(t) = AD(t)b(t)$ and interpret $b(t)$ as the average per-animal instantaneous rate of object creation. Hence, the basic model for understanding indirect distance sampling of objects is

$$M(x) = \int_{-\infty}^{x} N(t)b(t) \exp\left\{-\int_{t}^{x} h(\tau)\,d\tau\right\} dt \equiv \int_{-\infty}^{x} N(t)b(t)S(t, x)\,dt,$$
$$(11.192)$$

where $M(x)$ is total objects extant in area A at time x. We envision the time unit as 1 day and all continuous functions are considered smoothed over the full 24-h day so that daily periodic effects are removed. A discrete version of eqn (11.192) eliminates that smoothing step; however, continuous functions are simpler to use in theory development.

We can estimate $M(x)$ by distance sampling, but we want inference about $N(x)$. The practical context is distance sampling in some short time interval t_1 to t_2 during when we expect $M(x) \approx \bar{M}$ to be quite stable. Both pragmatically and by the mean value theorem of calculus, there exists at least one reference time point t_*, in $[t_1, t_2]$, such that $M(t_*) = \bar{M}$. Thus we can say it is $M(t_*)$ we are estimating, even though we may not know t_*. (This issue arises in direct distance sampling as well; if we need a single time point to reference $\widehat{N}$ to, we just pick an end point or the mid-point.)

The animals creating the objects were almost all present after some time t_0 such that $S(t_0, t_*) = \epsilon$; we will take ϵ to be small, such as 0.01. The value of ϵ gives us an interval of time t_0 to t_2 that has practical relevance,

in that our estimate $\widehat{N}$ refers to an average number of animals per day
within this not quite precise time interval.

The application of indirect distance sampling is best if $N(t) = N$ is
constant in t_0 to t_2. Then both $b(t)$ and $S(t, x)$ are interpretable as averages
over a fixed set of animals. Useful applications seem to be where daily object
creation rate is quite stable by animal, or as an average over populations
of animals. Thus, usually we are assuming that even if the composition of
the population changes some over time, $b(t)$ is not meaningfully affected
by these changes. The same assumption is made about $S(t, x)$ as it might
be influenced by changes in composition of the animals present in t_0 to t_*.
We will not know about these changes from the distance sampling data.

11.10.3.1 *Steady state modelling*
The literature has considered the steady state case for which $N(t) = N$
and rates are time constant: $b(t) = b$, $h(t) = h$ (no time effects). This
model of $h(\cdot)$ is equivalently considered as an age-effects only model with
constant-age hazard, since age $a = x - t$. Now eqn (11.192) becomes

$$M(x) = \int_{-\infty}^{x} Nb \exp\{-(x - t)h\} \, dt$$

$$= \frac{Nb \exp\{-xh\}}{h} \int_{-\infty}^{x} h \exp(th) \, dt$$

$$= \frac{Nb}{h} \tag{11.193}$$

independent of x.

Alternatively, we can change variables to integrate on age:

$$M(x) = \int_{-\infty}^{x} Nb \exp\{-(x - t)h\} \, dt = \int_{0}^{\infty} Nb \exp(-ah) \, da = \frac{Nb}{h}.$$

Thus, this model is equivalently considered as one with a time-constant
hazard or as the simplest model of age-effects only on object survival.

There are units of time associated with b and h; for example, we
may define them to be per-day rates. Let S be the survival probability for one time unit. Then $S = \exp(-h)$, so that $h = -\log_e(S)$ and
$N = M(-\log_e(S))/b$. Note that in this case, object age at disappearance
('death'), A_d, has a negative exponential distribution with expected survival time units as $E(A_d) = 1/h$, so that $N = M/[bE(A_d)]$. This result does
not seem to generalize to an arbitrary hazard function. It does generalize
to any pure age-effects model for $h(\cdot)$.

11.10.3.2   *The general case*

In general we are estimating a weighted time-averaged $N(x)$ that will be an average number of animals per day, on the study region, over a not quite precise time interval. We assume that we are estimating $M(t_*)$. Hence

$$\bar{N} = \frac{M(t_*)}{\int_{-\infty}^{t_*} b(t) S(t, t_*)\, dt} = \frac{\int_{-\infty}^{t_*} N(t) b(t) S(t, t_*)\, dt}{\int_{-\infty}^{t_*} b(t) S(t, t_*)\, dt} \qquad (11.194)$$

Equation (11.194) makes clear mathematically the time-averaged nature of $\bar{N}$ that indirect distance sampling can estimate. Examination of, or trying examples with, this equation should make it clear that there is no unique time point we can associate with $\bar{N}$ (hence with $\widehat{N}$). We can only relate $\widehat{N}$ to a time interval, and by convention to some time point of our choosing, as noted above. It should also be intuitively clear from the context itself and from eqn (11.194) that for given functions $b(t)$ and $S(t, x)$, there is not a unique function $N(t)$ that produces a given $\bar{N}$. Unless the biology of the study affirms $N(t) \equiv N$ in time interval t_0 to t_*, we must interpret $\widehat{N}$ as average animal use per day on the study region in this time interval.

11.10.3.3   *Constant object production rate modelling*

It is often thought that average per animal dung or nest production is quite stable per day for a population of animals. Thus in indirect distance sampling it is common to assume $b(t) = b$. The inference equation then becomes

$$\bar{N} = \frac{M(t_*)}{b \int_{-\infty}^{t_*} S(t, t_*)\, dt}. \qquad (11.195)$$

We can change the variable of integration to be age a at fixed time t_*. Nominally, the result is

$$\int_{-\infty}^{t_*} S(t, t_*)\, dt = \int_{0}^{\infty} S(t_* - a, t_*)\, da \qquad (11.196)$$

but we need to consider implications about $h(\cdot)$. Thus, this form implies

$$S(t_* - a, t_*) = \exp\left\{ -\int_{0}^{a} h(z + t)\, dz \right\} \equiv \exp\left\{ -\int_{0}^{a} h(z - a + t_*)\, dz \right\}. \qquad (11.197)$$

Thus in general the hazard function may depend upon age as well as absolute time and in complex ways. The simple notation of $S(t_* - a, t_*)$ hides these potential complications and hence must not be taken to mean that survival is a function of age only, although that is one possibility.

The function $S(t_* - a, t_*)$ is 1 at $a = 0$; it is 0 at $a = \infty$. Also, it can be monotonic decreasing as a increases. In particular if $h(\cdot)$ depends only on age, not on either t or t_*, then

$$S(t_* - a, t_*) = \exp\left\{ -\int_0^a h(z)\,dz \right\} \tag{11.198}$$

and hence $S(t_* - a, t_*)$ is monotonic decreasing in a. Therefore, for the case where $h(\cdot)$ does not depend on 'birth' time t, just age, we can represent $S(t_* - a, t_*) \equiv 1 - F(a)$ where $F(\cdot)$ is a cdf. (This is not the only such case—just the only one we can be certain of.) Moreover, for an object still existing (i.e. 'alive') at time t_*, we may take its age as a random variable (say A_a) with cdf $P(A_a \leq a) = F(a) = 1 - S(t_* - a, t_*)$, independent of t or t_*. Age would be a random variable with this cdf if we sampled extant objects at time t_* at random. A subsample near a transect line should be such a random sample, although we will not know the object age.

In general we can have $S(t_* - a, t_*) \equiv 1 - F(a \,|\, t_*)$ for $F(\cdot \,|\, t_*)$ a cdf if $S(t_* - a, t_*)$ is monotonic decreasing in age. This can occur even if survival is not purely a function of age, for example,

$$S(t_* - a, t_*) = \exp\left[-\int_0^a h(z)\exp\{m[t(z)]\}\,dz \right], \tag{11.199}$$

where $t(z)$ is real time and $m(\cdot)$ is a function, such as of environmental covariates. It is also possible that in such a complex case $S(t_* - a, t_*)$ is not monotonic in age.

Assume the above $F(a \,|\, t_*)$ is a proper cdf. Then for the strictly positive random variable age at time t_*, A_a, it is known that

$$E(A_a \,|\, t_*) = \int_0^\infty [1 - F(a \,|\, t_*)]\,da = \int_0^\infty S(t_* - a, t_*)\,da = \int_{-\infty}^{t_*} S(t, t_*)\,dt \tag{11.200}$$

(Parzen 1960: 211). Hence, in this case of $b(t) = b$ and $S(t_* - a, t_*)$ monotonic in age, we have

$$\bar{N} = \frac{M(t_*)}{b E(A_a \,|\, t_*)}. \tag{11.201}$$

This equation does not do us much good, unless we know object ages, which we will not in the course of the usual distance sampling. We would need to identify a sample of newly created ('fresh') objects at different times before t_* and visit them at t_* to estimate the mean age of extant objects at time t_* (Laing *et al.* 2003). We do not consider here estimation of $S(t, t_*)$ (considered by Barnes and Barnes 1992; Plumptre and Harris 1995), or direct estimation of $E(A_a \,|\, t_*)$ (considered by Laing *et al.* 2003).

In the case of pure age effects,

$$S(t, t_*) = \exp\left\{ -\int_0^a h(z)\, dz \right\} \tag{11.202}$$

with $a = t_* - t$, and $E(A_d)$ does not depend on t or t_*. Moreover in this case $E(A_a \,|\, t_*) \equiv E(A_a)$ does not depend on t_* and in fact $E(A_d) = E(A_a)$. Thus if object disappearance is governed purely by age effects (hence, no time-dependent environmental covariates), a valid formula is

$$\bar{N} = \frac{M(t_*)}{b E(A_d)}. \tag{11.203}$$

This equation is not valid in all circumstances. In particular, it may fail if the expectations of A_d and A_a depend on time. Only eqn (11.194) is valid in general, and if $b(t) \equiv b$, then eqn (11.195) is valid, although eqn (11.201) might be fairly safe to use in practice.

11.10.3.4 *Retrospective vs prospective estimation of object age*
It is of interest to ask what is the relationship between $E(A_a \,|\, t_*)$ and $E(A_d \,|\, t_0)$ ($t_0 < t_*$). The basic link is that both depend on the survival function $S(t, t+u)$. However, extant mean age at a given time is retrospective and age at death is prospective from some given time (Laing *et al.* 2003). We can determine a general answer for the case of $S(t_* - a, t_*)$ monotonic in a. We start by noting that

$$E(A_d \,|\, t_0) = \int_{t_0}^{\infty} S(t_0, t)\, dt = \int_0^{\infty} S(t_0, t_0 + a)\, da = \int_0^{\infty} [1 - F(a \,|\, t_0)]\, da \tag{11.204}$$

because the probability of living longer than time t, given the object is alive at t_0, is $1 - F(t \,|\, t_0) = \Pr\{T > t \,|\, t_0\} = S(t_0, t)$.

After some algebra, we have

$$E(A_d \,|\, t_0) = \int_{t_0}^{\infty} S(t_0, t)\, dt = \int_{t_0}^{t_*} S(t_0, t)\, dt + S(t_0, t_*) E(A_d \,|\, t_*) \tag{11.205}$$

and

$$E(A_a \,|\, t_*) = \int_{-\infty}^{t_*} S(t, t_*) \, dt = S(t_0, t_*) E(A_a \,|\, t_0) + \int_{t_0}^{t_*} S(t, t_*) \, dt.$$

$$(11.206)$$

Assume that the two types of expected ages give similar values. Then by taking t_0 and t_* far enough apart that $S(t_0, t_*) \approx 0$, we get as a good approximation

$$E(A_d \,|\, t_0) \approx \int_{t_0}^{t_*} S(t_0, t) \, dt \qquad (11.207)$$

and

$$E(A_a \,|\, t_*) \approx \int_{t_0}^{t_*} S(t, t_*) \, dt. \qquad (11.208)$$

Pragmatically, then, equivalence of $E(A_d \,|\, t_0)$ and $E(A_a \,|\, t_*)$ depends on whether this type of 'forward' vs 'backward' integration of $S(t, x)$ give the same result.

Heuristically, practical equivalence of $E(A_a \,|\, t_*)$ and $E(A_d \,|\, t)$ averaged over t in t_0 to t_* seems to only require that $h(t)$ shows no time trends in t_0 to t_*. We say this based on some numerical investigation of the matter, for example, by letting $h(t) = h + m(t)$ where $m(t)$ averages 0 over t_0 to t_*. Conversely, the two expectations do differ if $m(t)$ shows a time trend, such as $m(t) = c\,t$.

11.10.4 Conclusions

If the steady state assumption holds, the problem of converting from object abundance to population abundance is greatly simplified. However, when it fails, the bias arising from a steady state model might be substantial. Seasonal changes in object creation (e.g. defecation) rates, object survival rates, and animal distribution all violate the steady state assumption (McClanahan 1986).

Most surveys to date have used prospective methods for estimating survival rates of objects. The above results show that this is satisfactory provided there is no time trend in the hazard function $h(\cdot)$. However, seasonal effects such as varying levels of rainfall (Barnes *et al.* 1997; Nchanji and Plumptre 2001; Barnes and Dunn 2002) or varying temperature can give rise to strong time trends in $h(\cdot)$ during the lifetime of objects present during the distance sampling survey. In this case, retrospective methods should be used (Laing *et al.* 2003). Another reason to favour the retrospective method is that, if object survival is related to object size (e.g. larger dung piles decaying more slowly), then the objects present at the time of the distance sampling survey are a size-biased sample of the objects deposited. The retrospective method is unaffected by this size-biased selection, but the prospective method generates bias, for example, because too few large, long-lived objects and too many small, short-lived objects are represented in the sample (Laing *et al.* 2003).

In practice, the distance sampling survey may take several weeks to complete. In this case, it may be necessary to stratify the analysis in time, to ensure that appropriate rates are applied in each time stratum.

11.11   Goodness of fit tests and q–q plots

In Buckland *et al.* (2001: 69–71), three methods of assessing the fits of models were given: the χ^2 goodness of fit test, AIC (and the second-order criterion AICc), and the likelihood ratio test. In addition, the fitted pdf of detection distances may be plotted on a histogram of observed distances. A disadvantage of both AIC and the likelihood ratio test is that they measure relative fit only; even the best model may give a poor fit, and these statistics would be unable to identify this. The χ^2 test measures absolute fit, but if the distance data are ungrouped, the choice of distance intervals to group the data into is subjective. Further, the fit of the model at zero distance from the line or point is most crucial for estimating density, and the test gives no extra weight to this region. Plots of the fitted pdf superimposed on the histogram are useful, but it is difficult to assess whether the fit is adequate, and the choice of cutpoints for the histogram bars affects the apparent fit of the model.

11.11.1   *Quantile–quantile plots*

An alternative to the plot of the fitted pdf on a histogram of the data is the quantile–quantile or q–q plot. It has the strong advantage that ungrouped data do not need to be grouped. Instead, we evaluate the fitted cdf at each observation, under our assumed model, and order the resulting values starting with the smallest. Thus if we denote the fitted cdf by $\widehat{F}(y)$, then we evaluate $\widehat{F}(y_i)$ for each recorded distance y_i, $i = 1, \ldots, n$. Denote the ith ordered value of $\widehat{F}(y_i)$ by $x_{(i)}$. (Note that, if covariates are not included in the model for the detection function, this is equivalent to ordering the detection distances to give $y_{(i)}$ say, then evaluating $x_{(i)} = \widehat{F}(y_{(i)})$; if covariates are included, it is not equivalent in general, as the ordering of observations may change.) The values $x_{(i)}$ are then plotted against the corresponding values of the empirical distribution function (edf). The edf $S(x)$ is defined as

$$S(x) = \frac{i}{n} \text{ for } x_{(i)} \leq x < x_{(i+1)} , \quad i = 0, 1, \ldots, n, \tag{11.209}$$

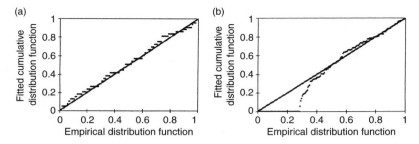

Fig. 11.9. Examples of q–q plots. In the left-hand plot, the plotted points all
fall close to the 1–1 line, indicating a good fit. In the right-hand plot, the first few
points depart appreciably from the 1–1 line, indicating an excess of observations
at zero distance.

where $x_{(i)}$ is the ith smallest value, with $x_{(0)} = -\infty$ and $x_{(n+1)} = +\infty$.
Note that this function steps up a distance $1/n$ at each observation. For the
q–q plot, it is convenient to take the edf corresponding to an observation
to be the average of $S(x)$ just before and just after the jump. That is,
$S(x_{(i)}) = (i - 0.5)/n$ for $i = 1, \ldots, n$. Also, if two observations are tied, one
is arbitrarily assigned to the first step and the other to the second; thus
if the smallest values in the sample were two zeros, one would be assigned
$i = 1$ with $S(x_{(1)}) = 0.5/n$, and the other would be assigned $i = 2$ with
$S(x_{(2)}) = 1.5/n$. This aids identification of departures of the edf from the
fitted cdf.

Distance 4.1 generates q–q plots. Figure 11.9 shows two examples. In
the plot on the left, the line through the points never goes far from the
1–1 line (the line of slope one that passes through the origin), indicating a
good fit, whereas the line through the points in the right-hand plot initially
goes along the x-axis, indicating that the data contain a number of zero
distances, which the model is unable to fit. (This is likely to be a rounding
problem, arising from poor field procedures; in this instance, we would not
want to fit this 'spike' (Buckland *et al.* 2001: 133). The q–q plot nevertheless highlights a failure of the assumption that distances are recorded
without error.)

In the left-hand plot of Fig. 11.9, the 'steps' indicate some rounding of
data, with some observations having exactly the same value on the y-axis.
Nevertheless, it is clear that the model provides a good fit. Similarly, in the
right-hand plot, the fitted cdf and the edf are clearly very different near
zero distance. Most plots are less clear cut, so we need a test of whether
the two functions differ significantly. In the following sections, we provide
such tests.

Another approach is to use the bootstrap. Suppose B bootstrap
resamples of the data are generated, for example, by resampling transect

lines or points (Buckland *et al.* 2001: 82–4). Fit the chosen model to each resample, and calculate the edf and fitted cdf values required for the q–q plot for each resample. Conceptually, all the bootstrapped q–q plots could be plotted together, to show variability. In practice, we would want to extract percentiles, so that the plot is interpretable. This can be done by taking the ordered evaluations of the fitted cdf corresponding to the original sample, and at each value, extracting the corresponding bootstrapped values of the edf. (Note that the edf is defined for the whole range of x in eqn (11.209), not just the points at which observations occur; this is required here since the bootstrap resamples will not have the same set of x values as the original sample.) The B bootstrapped values of the edf are then ordered, and percentiles extracted. For example, to plot 95% point-wise confidence limits, extract the $0.025(B+1)$st and $0.975(B+1)$st largest values. For $B = 999$, we would therefore order the 999 edf values from the resamples, and extract the 25th smallest and 25th largest values. Repeat for all values from the cdf fitted to the original sample. This provides an 'envelope' around the plotted line; if the 1–1 line (passing through the origin and through the point $(1,1)$) is entirely within this envelope, then the model provides a good fit. If it is not, there is some evidence of lack of fit; lack of fit near zero is more problematic than lack of fit near one. (We could instead condition on edf values, and extract percentiles from the bootstrapped cdf values in the same way; the two approaches lead to identical intervals, but the latter approach is problematic when there are covariates in the detection function, as we would need to evaluate the fitted cdf's after integrating over the values of the covariates.)

11.11.2   *Kolmogorov–Smirnov test*

A more formal test for comparing an edf with a cdf is the Kolmogorov–Smirnov test. If we denote the cdf by $F(x) = P(X \leq x)$, then it tests the null hypothesis $H_0 :\ F(x) = F_0(x)$ for all x against the alternative that the two functions differ for at least some value of x. In practice, we replace $F(x)$ by its estimate $\widehat{F}(x)$, and the null hypothesis is that our assumed model is the true model for our data. The test statistic is the largest absolute difference between the fitted cdf $\widehat{F}(x)$ and the edf $S(x)$, denoted D_n (Fig. 11.10). Critical values for the test are readily obtained from statistical software or tables. Distance 4.1 calculates the test statistic when data are ungrouped, together with the corresponding p-value:

$$p = 2 \sum_{i=1}^{\infty} (-1)^{i-1} \exp(-2ni^2 D_n^2) \qquad (11.210)$$

(Gibbons 1971: 81).

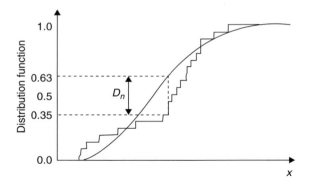

Fig. 11.10. The step function is an example of an edf, while the curve illustrates a typical fitted cdf. The Kolmogorov–Smirnov test statistic is the largest absolute difference between the edf and cdf, indicated by an arrow here, and equal to $D_n = 0.63 - 0.35 = 0.28$. The Cramér-von Mises family of tests by contrast is based on differences between the two functions over their entire range.

11.11.3   *Cramér-von Mises test*

The Cramér-von Mises test has the same null hypothesis as the Kolmogorov–Smirnov test, and is also based on differences between the edf and fitted cdf. However, the statistic is based on differences between the two functions over their entire range, rather than on just the largest difference, so that the test tends to have greater power. A convenient form for calculating the test statistic is

$$W^2 = \frac{1}{12n} + \sum_{i=1}^{n} \left[\widehat{F}(x_{(i)}) - \frac{i - 0.5}{n} \right]^2 \qquad (11.211)$$

with notation as before. Distance 4.1 calculates this test statistic and the corresponding *p*-value for ungrouped data.

11.11.4   *The Cramér-von Mises family of tests*

The Cramér-von Mises test statistic can equivalently be expressed as

$$W^2 = n \int_{-\infty}^{\infty} \left[\widehat{F}(x) - S(x) \right]^2 d\widehat{F}(x). \qquad (11.212)$$

This suggests a generalization, in which a weighting function is introduced, so that some parts of the range of x are given greater weight than others:

$$Q = n \int_{-\infty}^{\infty} \left[\widehat{F}(x) - S(x) \right]^2 \psi(x) \, d\widehat{F}(x). \qquad (11.213)$$

The usual use of the weighting function is to give greater weights to the tails of the distribution, such as the Anderson–Darling test statistic, for which $\psi(x) = 1/(F(x)[1 - F(x)])$. However, in distance sampling, it is the fit near zero distance that matters most, since estimated animal density is proportional to the estimated pdf at zero distance ($\hat{f}(0)$ or $\hat{h}(0)$). Distance 4.1 therefore includes a test and corresponding p-value from the Cramér–von Mises family with $\psi(x) = \cos(\pi x/2w)$, which puts maximum weight at $x = 0$. This test should have more power than the Cramér–von Mises test to detect departures from the fitted function near zero, and be more robust to departures at larger distances, where observations have little influence on $\hat{f}(0)$ or $\hat{h}(0)$.

11.12 Pooling robustness

Burnham *et al.* (1980) introduced the important concept of pooling robustness. They noted that detectability depends on many factors other than just distance from the line or point. Some of these factors are internal, such as age, sex, or size of the animal, or cluster size if animals occur in clusters. Others are external, such as weather, visibility, habitat, observer, or observation platform. Because data are rarely adequate to allow stratification by more than one or two factors, Burnham *et al.* (1980) note the importance of using models that yield unbiased (or nearly unbiased) estimates when distance data collected under variable conditions and on animals with varying intrinsic detectabilities are pooled. We explore here the issue of pooling robustness.

In Chapter 3, we provide methods that allow covariates to be included in the model for detectability. These methods reduce the reliance on pooling robustness, although they do not remove it altogether, as we cannot be sure that we have the right covariates, entering our model in the right form. The methods also allow the user to explore how detectability varies as covariates vary. Another advantage of modelling the covariates is when separate estimates are needed of different components of the population, such as males and females. If we just pool the data across the sexes when detectability varies by sex, pooling robustness provides an estimate of total population size with low bias (unless one sex is much more detectable than the other—see below), but the separate estimates of males and females may be appreciably biased. Addition of sex as a factor with two levels overcomes this difficulty. However, the results of this section show that, if only an overall abundance estimate is required, standard methods without covariates are satisfactory under rather mild conditions, provided heterogeneity in detectability is not too extreme.

Burnham $et\ al.$ (1980: 45) consider partitioning line transect data into r strata, chosen so that there is little heterogeneity with each stratum. The pooling robustness criterion is then defined as

$$n\hat{f}(0) = \sum_{j=1}^{r} n_j \hat{f}_j(0), \tag{11.214}$$

where n_j is the number of detections in stratum j, $\hat{f}_j(0)$ is the corresponding fitted pdf of distances of detected objects from the line evaluated at zero distance, n is the total number of detections, and $\hat{f}(0)$ is the pdf fitted to the pooled distance data, evaluated at zero distance.

In the following, we assume that animals occur singly; if they do not, then each reference to 'animal' below should be read as 'cluster'.

Using the notation of Chapter 3, y is distance from the line or point, $\underline{z}$ are the covariates associated with an animal, $g(y, \underline{z})$ is the probability of detection of an animal at distance y from the line or point and with covariate values $\underline{z}$, and $f(y, \underline{z})$ is the corresponding joint density function.

We make two key assumptions here:

1. $g(0, \underline{z}) = 1$ for all $\underline{z}$.
2. $\pi(y, \underline{z}) = \pi(y)\pi(\underline{z})$, the product of the marginals.

Random line or point placement, or more usually random placement of a systematic grid of lines or points, ensures independence of y and $\underline{z}$, so that the second assumption holds.

From eqn (3.2),

$$f(y, \underline{z}) = \frac{g(y, \underline{z})\pi(y)\pi(\underline{z})}{\iint g(y, \underline{z})\pi(y)\pi(\underline{z})\,dyd\underline{z}} \tag{11.215}$$

The marginal density of y is therefore

$$f(y) = \frac{\int g(y, \underline{z})\pi(y)\pi(\underline{z})\,d\underline{z}}{\iint g(y, \underline{z})\pi(y)\pi(\underline{z})\,dyd\underline{z}}. \tag{11.216}$$

The probability that an animal at distance y is detected, unconditional on $\underline{z}$, is

$$g(y) = \int g(y, \underline{z})\pi(\underline{z})\,d\underline{z}. \tag{11.217}$$

Hence eqn (11.216) simplifies to

$$f(y) = \frac{g(y)\,\pi(y)}{\int g(y)\,\pi(y)\,dy}. \tag{11.218}$$

For line transect sampling, given random line placement, $\pi(y) = 1/w$, so that

$$f(y) = \frac{g(y)}{\int g(y)\,dy}, \tag{11.219}$$

whereas for point transect sampling, we have $\pi(y) = 2y/w^2$ and

$$f(y) = \frac{y\,g(y)}{\int y\,g(y)\,dy}. \tag{11.220}$$

These are exactly the results that apply in the absence of heterogeneity. This is sufficient to show that the two sides of eqn (11.214) have the same large-sample expectation; the pooling robustness criterion can be assumed to hold provided we have a sufficiently flexible model for $g(y)$ to fit the observed distance data.

The implications of these results are perhaps surprising. For example, the training of shipboard observers often includes a warning that they should maintain a constant search effort. If observers search less efficiently in areas of low density through boredom, and more efficiently when they are in areas of high density, then the effective strip width is systematically greater in high-density areas and less elsewhere. This was thought to generate bias in abundance estimates, whereas the above results show that it does not—so long as they maintain $g(0) = 1$ in both high and low density areas. We would like observers to search with high efficiency throughout, as a larger number of detections gives higher precision, but abundance estimation is not compromised if efficiency is variable.

In Section 11.4.1, the hazard-rate model $g(y) = 1 - \exp[-(y/\sigma)^{-b}]$ was derived from homogeneous models for the sighting process. In Fig. 11.2, composite detection functions were shown when there is heterogeneity in detection probabilities. Although the hazard-rate detection function has a flat shoulder followed by a steep fall-off, the composite detection function has a rounded shoulder, more like that of the half-normal model. This illustrates that, to utilize the model-robustness properties of line transect estimators, it is important to use flexible models that can change shape when data arising from different detection functions are pooled.

If heterogeneity is sufficiently extreme, it becomes difficult to model adequately the detection function close to zero. For example, in some shipboard surveys, one or more observers search through $20\times$ or $25\times$ bin-oculars, while another observer 'guards the trackline' by searching with the naked eye. If $g(0)$ is appreciably less than one for the main observers, the detection function may appear to have a shoulder, but with a spike of observations at around zero distance from the line (Fig. 2.1 of Buckland *et al.* 2001: 31). Another example would be a species of songbird, for which males are almost certain to be detected out to perhaps $50\,\text{m}$ or more, whereas females beyond perhaps $5\,\text{m}$ may avoid detection. This degree of heterogeneity cannot be reliably modelled, resulting in bias.

Given assumption 1, that $g(0, \underline{z}) = 1$, then $g(0) = \int g(0, \underline{z})\, \pi(\underline{z})d\underline{z} = 1$. If $g(0, \underline{z})$ is a function of $\underline{z}$, then the model robustness criterion fails, and we must model the heterogeneity to avoid bias (Chapter 6).

Link (2003) noted the unidentifiability that can occur in estimating population size using mark-recapture, when there is heterogeneity in the capture probabilities. In the last sentence of that article, he stated that 'similar difficulties are to be anticipated in the use of distance sampling models ..., in which inference is based on a density value at zero'. The results of this section demonstrate that this is not true, provided $g(0, \underline{z}) = 1$ for all $\underline{z}$. In mark-recapture, to estimate population size, a model is needed that allows estimation of the number of animals that are never caught, and different models may give very different estimates. Mark-recapture data are discrete counts of animals with different capture histories, and from these, we must attempt to estimate the frequency of the 'never caught' capture history. By contrast, in line transect sampling, we seek to estimate $f(0)$, the pdf of detection distances at zero distance, by fitting a smooth, continuous function to $f(y)$. Under rather mild assumptions (principally that $g(0, \underline{z}) = 1$ and that $g'(0, \underline{z}) = 0$ for all $\underline{z}$), the method works well. Similar arguments apply in point transect sampling, where estimation of $h(0)$ relies on a good model for the slope of the pdf at zero distance.

References

Akaike, H. (1973) Information theory and an extension of the maximum likelihood principle, in *International Symposium on Information Theory*, 2nd edn (eds B. N. Petran and F. Csàaki), Akadèemiai Kiadi, Budapest, Hungary, pp. 267–81.

Akaike, H. (1985) Prediction and entropy, in *A Celebration of Statistics* (eds A. C. Atkinson and S. E. Fienberg), Springer-Verlag, Berlin, pp. 1–24.

Aldrin, M., Holden, M., and Schweder, T. (2003) Comment on Cowling's 'spatial methods for line transect surveys'. *Biometrics*, **59**, 186–8.

Alho, J. M. (1990) Logistic regression in capture–recapture models. *Biometrics*, **46**, 623–35.

Alpizar-Jara, R. (1997) *Assessing Assumption Violation in Line Transect Sampling*, Ph. D. Thesis, North Carolina State University.

Alpizar-Jara, R. and Pollock, K. H. (1996) A combination line transect and capture recapture sampling model for multiple observers in aerial surveys. *Environmental and Ecological Statistics*, **3**, 311–27.

Alpizar-Jara, R. and Pollock, K. H. (1999) Combining line transect and capture-recapture for mark-resighting studies, in *Marine Mammal Survey and Assessment Methods* (eds G. W. Garner, S. C. Amstrup, J. L. Laake, B. F. J. Manly, L. L. McDonald, and D. G. Robertson), Balkema, Rotterdam, pp. 99–114.

Anderson, D. R. (2001) The need to get the basics right in wildlife field studies. *Wildlife Society Bulletin*, **29**, 1294–7.

Anderson, D. R., Burnham, K. P., White, G. C., and Otis, D. L. (1983) Density estimation of small-mammal populations using a trapping web and distance sampling methods. *Ecology*, **64**, 674–80.

Anderson, D. R., Burnham, K. P., Lab, B. C., Thomas, L., Corn, P. S., Medica, P. A., and Marlow, R. W. (2001) Field trials of line transect methods applied to estimation of desert tortoise abundance. *Journal of Wildlife Management*, **65**, 583–97.

Anderson, D. R. and Pospahala, R. S. (1970) Correction of bias in belt transects of immotile objects. *Journal of Wildlife Management*, **34**, 141–6.

Anderson, M. J. and Thompson, A. A. (in press) Multivariate control charts for ecological and environmental monitoring. *Ecological Applications*.

Atkinson, A. J., Yang, B. S., Fisher, R. N., Ervin, E., Case, T. J., Scott, N., and Shaffer, H. B. (2003) *MCB Camp Pendleton Arroyo Toad monitoring protocol*, Western Ecological Research Center, U.S. Geological Survey.

Augustin, N. H., Mugglestone, M. A., and Buckland, S. T. (1998) The role of simulation in modeling spatially-correlated data. *Environmetrics*, **9**, 175–96.

Baranov, T. I. (1918) On the question of the biological basis of fisheries. *Reports of the Division of Fish Management and Scientific Study of the Fishing Industry*, **1**, 81–128.

Barlow, J. (1999) Trackline detection probability for long-diving whales, in *Marine Mammal Survey and Assessment Methods* (eds G. W. Garner, S. C. Amstrup, J. L. Laake, B. F. J. Manly, L. L. McDonald, and D. G. Robertson), Balkema, Rotterdam, pp. 209–21.

Barlow, J., Oliver, C. W., Jackson, T. D., and Taylor, B. L. (1988) Harbor porpoise, *Phocoena phocoena*, abundance estimation for California, Oregon, and Washington: aerial surveys. *Fishery Bulletin*, **86**, 433–44.

Barndorff-Nielsen, O. E. (1986) Inference on the full or partial parameters based on the standardized log likelihood ratio. *Biometrika*, **73**, 307–22.

Barnes, R. F. W. (2001) How reliable are dung counts for estimating elephant numbers? *African Journal of Ecology*, **39**, 1–9.

Barnes, R. F. W., Asamoah-Boateng, B., Naada Majam, J., and Agyei-Ohemeng, J. (1997) Rainfall and the population dynamics of elephant dung-piles in the forests of southern Ghana. *African Journal of Ecology*, **35**, 39–52.

Barnes, R. F. W. and Barnes, K. L. (1992) Estimating decay rates of elephant dung-piles in forest. *African Journal of Ecology*, **30**, 316–21.

Barnes, R. F. W. and Dunn, A. (2002) Estimating forest elephant density in Sapo National Park (Liberia) with a rainfall model. *African Journal of Ecology*, **40**, 159–63.

Barry, S. C. and Welsh, A. H. (2001) Distance sampling methodology. *Journal of the Royal Statistical Society Series B*, **63**, 31–53.

Beavers, S. C. and Ramsey, F. L. (1998) Detectability analysis in transect surveys. *Journal of Wildlife Management*, **62**, 948–57.

Besbeas, P., Freeman, S. N., and Morgan, B. J. T. (in press) The potential of integrated population modelling. *Australian and New Zealand Journal of Statistics*.

Besbeas, P., Freeman, S. N., Morgan, B. J. T., and Catchpole, E. A. (2002) Integrating mark-recapture-recovery and census data to estimate animal abundance and demographic parameters. *Biometrics*, **58**, 540–7.

Besbeas, P., Lebreton, J.-D., and Morgan, B. J. T. (2003) The efficient integration of abundance and demographic data. *Applied Statistics*, **52**, 95–102.

Borchers, D. L. (1996) *Line Transect Estimation with Uncertain Detection on the Trackline*, Ph. D. thesis, University of Cape Town.

Borchers, D. L. (1999) Composite mark-recapture line transect surveys, in *Marine Mammal Survey and Assessment Methods* (eds G. W. Garner, S. C. Amstrup, J. L. Laake, B. F. J. Manly, L. L. McDonald, and D. G. Robertson), Balkema, Rotterdam, pp. 115–26.

Borchers, D. L., Buckland, S. T., Goedhart, P. W., Clarke, E. D., and Hedley, S. L. (1998a) Horvitz–Thompson estimators for double-platform line transect surveys. *Biometrics*, **54**, 1221–37.

Borchers, D. L., Buckland, S. T., and Zucchini, W. (2002) *Estimating Animal Abundance: Closed Populations*, Springer-Verlag, London.

Borchers, D. L. and Burt, M. L. (unpublished) Generalized regression methods for estimating school size from line transect data. Paper SC/54/IA23, presented to the Scientific Committee of the International Whaling Commission, 2002.

Borchers, D. L., Laake, J. L., Southwell, C., and Paxton, C. G. M. (in preparation *a*) Accommodating unmodelled heterogeneity in double-observer distance sampling surveys.

Borchers, D. L., Pike, D., Gunnlaugsson, Φ., and Vikingsson, G. A. (in preparation *b*) Maximum likelihood methods for distance sampling with measurement errors.

Borchers, D. L., Zucchini, W., and Fewster, R. M. (1998b) Mark-recapture models for line transect surveys. *Biometrics*, **54**, 1207–20.

Bowman, A. W. and Azzalini, A. (1997) *Applied Smoothing Techniques for Data Analysis: The Kernel Approach with S-Plus Illustrations*, Oxford University Press, Oxford.

Box, G., Jenkins, G. M., and Reinsel, G. (1994) *Time Series Analysis: Forecasting and Control*, 3rd edn, Prentice Hall, New York.

Bravington, M. V., Wood, S. N., and Hedley, S. L. (in preparation) Spatial modelling of line transect data with clustering.

Breiwick, J. M., Rugh, D. J., Withrow, D. E., Dahlheim, M. E., and Buckland, S. T. (unpublished) Preliminary population estimate of gray whales during the 1987/88 southward migration. Paper SC/40/PS12, presented to the Scientific Committee of the International Whaling Commission, 1988.

Brown, B. M. and Cowling, A. (1998) Clustering and abundance estimation for Neyman–Scott models and line transect surveys. *Biometrika*, **85**, 427–38.

Buckland, S. T. (1984) Monte Carlo confidence intervals. *Biometrics*, **40**, 811–17.

Buckland, S. T. (1985) Perpendicular distance models for line transect sampling. *Biometrics*, **41**, 177–95.

Buckland, S. T. (1987a) On the variable circular plot method of estimating animal density. *Biometrics*, **43**, 363–84.

Buckland, S. T. (1987b) An assessment of the performance of line transect models for fitting IWC/IDCR cruise data, 1978/79 to 1984/85. *Report of the International Whaling Commission*, **37**, 277–9.

Buckland, S. T. (1992a) Fitting density functions using polynomials. *Applied Statistics*, **41**, 63–76.

Buckland, S. T. (1992b) Maximum likelihood fitting of Hermite and simple polynomial densities. *Applied Statistics*, **41**, 241–66.

Buckland, S. T. (1992c) Effects of heterogeneity on estimation of probability of detection on the trackline. *Report of the International Whaling Commission*, **42**, 569–73.

Buckland, S. T., Anderson, D. R., Burnham, K. P., and Laake, J. L. (1993a) *Distance Sampling: Estimating Abundance of Biological Populations*, Chapman and Hall, London.

Buckland, S. T., Anderson, D. R., Burnham, K. P., Laake, J. L., Borchers, D. L., and Thomas, L. (2001) *Introduction to Distance Sampling*, Oxford University Press, Oxford.

Buckland, S. T. and Breiwick, J. M. (2002) Estimated trends in abundance of eastern Pacific gray whales from shore counts, 1967/68 to 1995/96. *Journal of Cetacean Research and Management*, **4**, 41–8.

Buckland, S. T., Breiwick, J. M., Cattanach, K. L., and Laake, J. L. (1993b) Estimated population size of the California gray whale. *Marine Mammal Science*, **9**, 235–49.

Buckland, S. T., Burnham, K. P., and Augustin, N. H. (1997) Model selection: an integral part of inference. *Biometrics*, **53**, 603–18.

Buckland, S. T., Cattanach, K. L., and Anganuzzi, A. A. (1992) Estimating trends in abundance of dolphins associated with tuna in the eastern tropical Pacific Ocean, using sightings data collected on commercial tuna vessels. *Fishery Bulletin*, **90**, 1–12.

Buckland, S. T., Goudie, I. B. J., and Borchers, D. L. (2000) Wildlife population assessment: past developments and future directions. *Biometrics*, **56**, 1–12.

Buckland, S. T., Newman, K. B., Thomas, L., and Koesters, N. B. (2004) State-space models for the dynamics of wild animal populations. *Ecological Modelling*, **171**, 157–75.

Buckland, S. T., Thomas, L., Marques, F. F. C., Strindberg, S., Hedley, S. L., Pollard, J. H., Borchers, D. L., and Burt, M. L. (2002) Distance sampling: recent advances and future directions, in *Quantitative Methods for Current Environmental Issues* (eds C. W. Anderson, V. Barnett, P. Chatwin, and A. El-Shaarawi), Springer-Verlag, London, pp. 79–97.

Buckland, S. T. and Turnock, B. J. (1992) A robust line transect method. *Biometrics*, **48**, 901–9.

Burnham, K. P. and Anderson, D. R. (1976) Mathematical models for nonparametric inferences from line transect data. *Biometrics*, **32**, 325–36.

Burnham, K. P. and Anderson, D. R. (2002) *Model Selection and Inference: a Practical Information-Theoretic Approach*, 2nd edn, Springer, New York.

Burnham, K. P., Anderson, D. R., and Laake, J. L. (1980) Estimation of density from line transect sampling of biological populations. *Wildlife Monographs*, **72**, 1–202.

Burnham, K. P. and White, G. C. (2002) Evaluation of some random effects methodology applicable to bird ringing data. *Journal of Applied Statistics*, **29**, 245–64.

Butcher, G. S., Peterjohn, B. G., and Ralph, C. J. (1993) Overview of national bird population monitoring programs and databases, in *Status and Management of Neotropical Migratory Birds* (eds D. M. Finch and P. W. Stangel), General Technical Report RM-229, U.S.D.A. Forest Service Rocky Mountain Forest and Range Experiment Station, Fort Collins, pp. 192–203.

Butterworth, D. S. (1982) A possible basis for choosing a functional form for the distribution of sightings with right angle distance: some preliminary ideas. *Report of the International Whaling Commission*, **32**, 555–8.

Butterworth, D. S., Best, P. B., and Basson, M. (1982) Results of analysis of minke whale assessment cruise, 1980/1. *Report of the International Whaling Commission*, **32**, 819–34.

Butterworth, D. S. and Borchers, D. L. (1988) Estimates of $g(0)$ for minke schools from the results of the independent observer experiment on the 1985/86 and 1986/87 IWC/IDCR Antarctic assessment cruises. *Report of the International Whaling Commission*, **38**, 301–13.

Cañadas, A., Desportes, G., and Borchers, D. L. (in preparation) Estimation of $g(0)$ and abundance of common dolphins (*Delphinus delphis*) from the NASS-95 Faroese survey.

Carretta, J. V., Forney, K. A., and Laake, J. L. (1998) Abundance of California coastal bottlenose dolphins estimated from tandem aerial surveys. *Marine Mammal Science*, **14**, 655–75.

Cassey, P. and McArdle, B. H. (1999) An assessment of distance sampling techniques for estimating animal abundance. *Environmetrics*, **10**, 261–78.

Caswell, H. (2001) *Matrix Population Models*, Sinauer Associates, Sunderland, MA.

Chen, S. X. (1996) Studying school size effects in line transect sampling using the kernel method. *Biometrics*, **52**, 1283–94.

Chen, S. X. (1998) Measurement errors in line transect surveys. *Biometrics*, **54**, 899–908.

Chen, S. X. (1999) Estimation in independent observer line transect surveys for clustered populations. *Biometrics*, **55**, 754–9.

Chen, S. X. (2000) Animal abundance estimation in independent observer line transect surveys. *Environmental and Ecological Statistics*, **7**, 285–99.

Chen, S. X. and Cowling, A. (2001) Measurement errors in line transect surveys where detectability varies with distance and size. *Biometrics*, **57**, 732–42.

Chen, S. X. and Lloyd, C. J. (2000) A nonparametric approach to the analysis of two-stage mark-recapture experiments. *Biometrika*, **87**, 633–49.

Cochran, W. G. (1977) *Sampling Techniques*, 3rd edn, Wiley, New York.

Cook, J. R. and Stefanski, L. A. (1994) Simulation-extrapolation estimation in parametric measurement error models. *Journal of the American Statistical Association*, **89**, 1314–28.

Cook, R. D. and Jacobsen, J. O. (1979) A design for estimating visibility bias in aerial surveys. *Biometrics*, **35**, 735–42.

Cooke, J. G. (1985) Estimation of abundance from surveys. Unpublished manuscript.

Cooke, J. G. (1997) An implementation of a surfacing-based approach to abundance estimation of minke whales from shipborne surveys. *Report of the International Whaling Commission*, **47**, 513–28.

Cooke, J. G. (unpublished) A modification of the radial distance method for dual-platform line transect analysis, to improve robustness. Paper SC/53/IA31 presented to the Scientific Committee of the International Whaling Commission, 2001.

Cooke, J. G. and Leaper, R. (unpublished) A general modelling framework for the estimation of whale abundance from line transect surveys. Paper SC/50/RMP21 presented to the Scientific Committee of the International Whaling Commission, 1998.

Cowling, A. (1998) Spatial methods for line transect surveys. *Biometrics*, **54**, 828–39.

Cox, D. R. (1962) *Renewal Theory*, Methuen, London.

Cox, D. R. and Snell, E. J. (1989) *Analysis of Binary Data*, 2nd edn, Chapman and Hall, London.

Crain, B. R., Burnham, K. P., Anderson, D. R., and Laake, J. L. (1979) Nonparametric estimation of population density for line transect sampling using Fourier series. *Biometrics*, **21**, 731–48.

Cressie, N. A. C. (1991) *Statistics for Spatial Data*, Wiley, New York.

Dagum, C. and Dagum, E. B. (1988) Trend, in *Encyclopedia of Statistical Sciences Volume 9* (eds S. Kotz and N. L. Johnson), Wiley, New York, pp. 321–4.

Davison, A. C. and Hinkley, D. V. (1997) *Bootstrap Methods and their Application*, Cambridge University Press, Cambridge.

Dennis, B. and Taper, M. L. (1994) Density-dependence in time-series observations of natural populations: estimation and testing. *Ecological Monographs*, **64**, 205–24.

Diggle, P. J. (1983) *Statistical Analysis of Spatial Point Patterns*, Academic Press, London.

Diggle, P. J., Tawn, J. A., and Moyeed, R. A. (1998) Model-based geostatistics. *Applied Statistics*, **47**, 299–350.

Drummer, T. D. and McDonald, L. L. (1987) Size bias in line transect sampling. *Biometrics*, **43**, 13–21.

Eberhardt, L. L. (1968) A preliminary appraisal of line transect. *Journal of Wildlife Management*, **32**, 82–8.

Eberhardt, L. L. (1978) Transect methods for population studies. *Journal of Wildlife Management*, **42**, 1–31.

Efford, M. G. (in press) Density estimation in live-trapping studies. *Oikos*.

Efford, M. G., Dawson, D. K., and Robbins, C. S. (in press) DENSITY: software for analysing capture-recapture data from passive detector arrays. *Animal Biodiversity and Conservation*.

Efford, M. G., Warburton, B., Coleman, M. C., and Barker, R. J. (in preparation) A field test of two methods for density estimation.

Efron, B. and Tibshirani, R. J. (1993) *An Introduction to the Bootstrap*, Chapman and Hall, London.

Eggert, L. S., Eggert, J. A., and Woodruff, D. S. (2003) Censusing elusive animals: the forest elephants of Kakum National Park, Ghana. *Molecular Ecology*, **12**, 1389–402.

Evans-Mack, D., Raphael, M. G., and Laake, J. L. (2002) Probability of detecting marbled murrelets at sea: effects of single versus paired observers. *Journal of Wildlife Management*, **66**, 865–73.

Fancy, S. G. (1997) A new approach for analyzing bird densities from variable circular-plot counts. *Pacific Science*, **51**, 107–14.

Farnsworth, G. L., Pollock, K. H., Nichols, J. D., Simons, T. R., Hines, J. E., and Sauer, J. R. (2002) A removal model for estimating detection probabilities from point-count surveys. *The Auk*, **119**, 414–25.

Fernández, C. and Niemi, A. (in preparation) Bayesian spatial point process modelling of line transect data.

Fewster, R. M., Buckland, S. T., Siriwardena, G. M., Baillie, S. R., and Wilson, J. D. (2000) Analysis of population trends for farmland birds using generalized additive models. *Ecology*, **81**, 1970–84.

Franklin, A. B., Anderson, D. R., and Burnham, K. P. (2002) Estimation of long-term trends and variation in avian survival probabilities using random effects models. *Journal of Applied Statistics*, **29**, 267–87.

Freeman, S. N., Noble, D. G., Newson, S. E., and Baillie, S. R. (2003) Modelling bird population changes using data from the Common Birds Census and the Breeding Bird Survey. *British Trust for Ornithology Research Report*, **303**, 1–44.

Freilich, J., Burnham, K. P., Collins, C. M., and Garry, C. A. (2000) Factors affecting population assessments of desert tortoises. *Conservation Biology*, **14**, 1479–89.

Gates, C. E., Marshall, W. H., and Olson, D. P. (1968) Line transect method of estimating grouse population densities. *Biometrics*, **24**, 135–45.

Geissler, P. H. and Sauer, J. R. (1990) Topics in route regression analysis, in *Survey Designs and Statistical Methods for the Estimation of Avian Population Trends* (eds J. R. Sauer and S. Droege), U.S. Fish and Wildlife Service Biological Report 90(1), pp. 54–7.

Gerrodette, T. (1987) A power analysis for detecting trends. *Ecology*, **68**, 1364–72.

Gerrodette, T. (1991) Models for power of detecting trends: a reply to Link and Hatfield. *Ecology*, **72**, 1889–92.

Gibbons, J. D. (1971) *Nonparametric Statistical Inference*, McGraw-Hill, New York.

Goldsmith, B. (ed.) (1991) *Monitoring for Conservation and Ecology*, Chapman and Hall, London.

Gotway, C. A. and Stroup, W. W. (1997) A generalized linear model approach to spatial data analysis and prediction. *Journal of Agricultural, Biological, and Environmental Statistics*, **2**, 157–78.

Graham, A. and Bell, R. (1989) Investigating observer bias in aerial survey by simultaneous double-counts. *Journal of Wildlife Management*, **53**, 1009–16.

Hagen, G. and Schweder, T. (1995) Point clustering of minke whales in the northeastern Atlantic, in *Whales, Seals, Fish and Man* (eds A. S. Blix, L. Walløe, and Ø. Ulltang), Elsevier Science B. V., Amsterdam, pp. 27–33.

Hain, J. H. W., Ellis, S. L., Kenney, R. D., and Slay, C. K. (1999) Sightability of right whales in coastal waters of the southeastern United States with implications for the aerial monitoring program, in *Marine Mammal Survey and Assessment Methods* (eds G. W. Garner, S. C. Amstrup, J. L. Laake, B. F. J. Manly, L. L. McDonald, and D. G. Robertson), Balkema, Rotterdam, pp. 191–207.

Halley, J. M. (1996) Ecology, evolution and 1/f noise. *Trends in Ecology and Evolution*, **11**, 33–7.

Hammond, P. S. (1990) Capturing whales on film: estimating cetacean population parameters from individual recognition data. *Mammal Review*, **20**, 17–22.

Hansen, M. M. and Hurwitz, W. N. (1943) On the theory of sampling from finite populations. *Annals of Mathematical Statistics*, **14**, 333–62.

Harvey, A. C. (1989) *Forecasting, Structural Time Series Models and the Kalman Filter*, Cambridge University Press, Cambridge.

Hastie, T. J. and Tibshirani, R. J. (1990) *Generalized Additive Models*, Chapman and Hall, London.

Hayes, R. J. and Buckland, S. T. (1983) Radial-distance models for the line-transect method. *Biometrics*, **39**, 29–42.

Hayne, D. W. (1949) An examination of the strip census method for estimating animal populations. *Journal of Wildlife Management*, **13**, 145–57.

Hedley, S. L. (2000) *Modelling Heterogeneity in Cetacean Surveys*, Ph. D. thesis, University of St Andrews.

Hedley, S. L. and Buckland, S. T. (in press) Spatial models for line transect sampling. *Journal of Agricultural, Biological, and Environmental Statistics*.

Hedley, S. L., Buckland, S. T., and Borchers, D. L. (1999) Spatial modelling from line transect data. *Journal of Cetacean Research and Management*, **1**, 255–64.

Hiby, A. R. (1985) An approach to estimating population densities of great whales from sighting surveys. *IMA Journal of Mathematics Applied in Medicine and Biology*, **2**, 201–20.

Hiby, A. R. and Hammond, P. S. (1989) Survey techniques for estimating abundance of cetaceans, in *The Comprehensive Assessment of Whale Stocks: The Early Years* (ed. G. P. Donovan), International Whaling Commission, Cambridge, pp. 47–80.

Hiby, L. (1999) The objective identification of duplicate sightings in aerial survey for porpoise, in *Marine Mammal Survey and Assessment Methods* (eds G. W. Garner, S. C. Amstrup, J. L. Laake, B. F. J. Manly, L. L. McDonald, and D. G. Robertson), Balkema, Rotterdam, pp. 179–89.

Hiby, L. and Lovell, P. (1998) Using aircraft in tandem formation to estimate abundance of harbour porpoise. *Biometrics*, **54**, 1280–9.

Hiby, L., Ward, A., and Lovell, P. (1989) Analysis of the North Atlantic sightings survey 1987: aerial survey results. *Report of the International Whaling Commission*, **39**, 447–55.

Hjort, J. G. and Ottestad, P. (1933) The optimum catch. *Hvalrådets Skrifter, Oslo*, **7**, 92–127.

Hobbs, R. C., Rugh, D. J., Waite, J. M., Breiwick, J. M., and DeMaster, D. P. (unpublished) Preliminary estimate of the abundance of gray whales in the 1995/96 southbound migration. Paper SC/48/AS9, presented to the Scientific Committee of the International Whaling Commission, 1996.

Hobson, K. A., Rempel, R. S., Greenwood, H., Turnbull, B., and Van Wilgenburg, S. L. (2002) Acoustic surveys of birds using electronic recordings: new potential from an omnidirectional microphone system. *Wildlife Bulletin*, **30**, 709–20.

Högmander, H. (1991) A random field approach to transect counts of wildlife populations. *Biometrical Journal*, **33**, 1013–23.

Högmander, H. (1995) *Methods of Spatial Statistics in Monitoring Wildlife Populations*, Ph. D. thesis, University of Jyväskylä.

Holgate, P. (1964) The efficiency of nearest neighbour estimators. *Biometrics*, **20**, 647–9.

Horvitz, D. G. and Thompson, D. J. (1952) A generalization of sampling without replacement from a finite universe. *Journal of the American Statistical Association*, **47**, 663–85.

Huggins, R. M. (1989) On the statistical analysis of capture experiments. *Biometrika*, **76**, 133–40.

Huggins, R. M. (1991) Some practical aspects of a conditional likelihood approach to capture experiments. *Biometrics*, **47**, 725–32.

Innes, S., Heide-Jørgensen, M. P., Laake, J. L., Laidre, K. L., Cleator, H. J., Richard, P., and Stewart, R. E. A. (2002) Surveys of belugas and narwhals in the Canadian High Arctic in 1996. NAMMCO Scientific Publications, **4**, 169–90.

James, F. C., McCulloch, C. E., and Wiedenfeld, D. A. (1996) New approaches to the analysis of population trends in land birds. *Ecology*, **77**, 13–27.

Jamieson, L. E. and Brooks, S. P. (2003) State space models for density dependence in population ecology, in *Bayesian Statistics VII* (eds J. M. Bernardo, M. J. Bayarri, J. O. Berger, A. P. Dawid, D. Heckerman, A. F. M. Smith, and M. West), Oxford University Press, Oxford, pp. 565–75.

Jessen, R. J. (1942) Statistical investigation of a farm survey for obtaining farm facts. *Iowa Agricultural Station Research Bulletin*, **304**, 54–9.

Johnson, D. S. and Hoeting, J. A. (2003) Autoregressive models for capture-recapture data: a Bayesian approach. *Biometrics*, **59**, 340–9.

Kelker, G. H. (1940) Estimating deer populations by a differential hunting loss in the sexes. *Proceedings of Utah Academy of Science, Arts and Letters*, **17**, 6–69.

Klavitter, J. L. (2000) *Survey Methodology, Abundance and Demography of the Endangered Hawaiian hawk: Is delisting warranted?*, M. S. thesis, University of Washington.

Klavitter, J. L., Marzluff, J. M., and Vekasy, M. S. (2003) Abundance and demography of the Hawaiian hawk: is delisting warranted? *Journal of Wildlife Management*, **67**, 165–76.

Koopman, B. O. (1980) *Search and Screening: General Principles with Historical Applications*, Pergamon Press, New York.

Krzysik, J. A., Woodman, A. P., and Hagan, M. (1995) A field evaluation of four methods for estimating desert tortoise densities. *Proceedings of the Desert Tortoise Council*, **1995**, 92.

Laake, J. L. (1978) *Line Transect Estimators Robust to Animal Movement*, M. S. thesis, Utah State University.

Laake, J. L. (1999) Distance sampling with independent observers: reducing bias from heterogeneity by weakening the conditional independence assumption, in *Marine Mammal Survey and Assessment Methods* (eds G. W. Garner, S. C. Amstrup, J. L. Laake, B. F. J. Manly, L. L. McDonald, and D. G. Robertson), Balkema, Rotterdam, pp. 137–48.

Laake, J. L., Calambokidis, J. C., Osmek, S. D., and Rugh, D. J. (1997) Probability of detecting harbor porpoise from aerial surveys: estimating $g(0)$. *Journal of Wildlife Management*, **61**, 63–75.

Laake, J. L., Rugh, D. J., Lerczak, J. A., and Buckland, S. T. (unpublished) Preliminary estimates of population size of gray whales from the 1992/93 and 1993/94 shore-based surveys. Paper SC/46/AS7, presented to the Scientific Committee of the International Whaling Commission, 1994.

Laing, S. E., Buckland, S. T., Burn, R. W., Lambie, D., and Amphlett, A. (2003) Dung and nest surveys: estimating decay rate. *Journal of Applied Ecology*, **40**, 1102–11.

Lawton, J. H. (1988) More time means more variation. *Nature*, **334**, 563.

Lin, X. H. and Zhang, D. W. (1999) Inference in generalized additive mixed models by using smoothing splines. *Journal of the Royal Statistical Society Series B*, **61**, 381–400.

Lincoln, F. C. (1930) Calculating waterfowl abundance on the basis of banding returns. *United States Department of Agriculture Circular*, **118**, 1–4.

Link, W. A. (2003) Nonidentifiability of population size from capture–recapture data with heterogeneous detection probabilities. *Biometrics*, **59**, 1123–30.

Link, W. A. and Hatfield, J. S. (1990) Power calculations and model selection for trend analysis: a comment. *Ecology*, **71**, 1217–20.

Link, W. A. and Nichols, J. D. (1994) On the importance of sampling variance to investigations of temporal variation in animal population size. *Oikos*, **69**, 539–44.

Link, W. A. and Sauer, J. R. (1997a) Estimation of population trajectories from count data. *Biometrics*, **53**, 488–97.

Link, W. A. and Sauer, J. R. (1997b) New approaches to the analysis of population trends in land birds: comment. *Ecology*, **78**, 2632–4.

Loader, C. (1999) *Local Regression and Likelihood*, Springer, New York.

Lukacs, P. M. (2001) *Estimating Density of Animal Populations using Trapping Webs: Evaluation of Web Design and Data Analysis*, M. S. thesis, Colorado State University.

Lukacs, P. M. (2002) WebSim: simulation software to assist in trapping web design. *Wildlife Society Bulletin*, **30**, 1259–61.

Lukacs, P. M., Anderson, D. R., and Burnham, K. P. (in preparation *a*) Evaluation of trapping web designs.

Lukacs, P. M., Burnham, K. P., and Anderson, D. R. (in preparation *b*) The use of recaptures to estimate density using the trapping web.

Mack, Y. P. and Quang, P. X. (1998) Kernel methods in line and point transect sampling. *Biometrics*, **54**, 606–19.

Mackenzie, M. L., Donovan, C. R., and McArdle, B. H. (in preparation) Regression spline mixed models: a forestry example.

Manly, B. F. J., McDonald, L. L., and Garner, G. W. (1996) Maximum likelihood estimation for the double-count method with independent observers. *Journal of Agricultural, Biological, and Environmental Statistics*, **1**, 170–89.

Marques, F. F. C. (2001) *Estimating Wildlife Distribution and Abundance from Line Transect Surveys conducted from Platforms of Opportunity*, Ph. D. thesis, University of St Andrews.

Marques, F. F. C. and Buckland, S. T. (2003) Incorporating covariates into standard line transect analyses. *Biometrics*, **59**, 924–35.

Marques, F. F. C., Buckland, S. T., Goffin, D., Dixon, C. E., Borchers, D. L., Mayle, B. A., and Peace, A. J. (2001) Estimating deer abundance from line transect surveys of dung: sika deer in southern Scotland. *Journal of Applied Ecology*, **38**, 349–63.

Marques, T. A. (in press) Predicting and correcting bias caused by measurement error in line transect sampling using multiplicative error models. *Biometrics*.

Marsh, H. and Sinclair, D. F. (1989) Correcting for visibility bias in strip transect aerial surveys of aquatic fauna. *Journal of Wildlife Management*, **53**, 1017–24.

McClanahan, T. R. (1986) Quick population survey method using faecal droppings and a steady state assumption. *African Journal of Ecology*, **24**, 37–9.

McCullagh, P. and Nelder, J. A. (1989) *Generalized Linear Models*, 2nd edn, Chapman and Hall, London.

McLaren, I. A. (1961) Methods of determining the numbers and availability of ring seals in the eastern Canadian Arctic. *Arctic*, **14**, 162–75.

McLuckie, A. M., Harstad, D. L., Marr, J. W., and Fridell, R. A. (2002) Regional desert tortoise monitoring in the upper Virgin River recovery unit, Washington County, Utah. *Chelonian Conservation and Biology*, **4**, 380–6.

Melville, G. J. and Welsh, A. H. (2001) Line transect sampling in small regions. *Biometrics*, **57**, 1130–7.

Meyer, R. and Millar, R. B. (1999) Bayesian stock assessment using a state-space implementation of the delay difference model. *Canadian Journal of Fisheries and Aquatic Sciences*, **56**, 37–52.

Millar, R. B. and Meyer, R. (2000) Non-linear state space modelling of fisheries biomass dynamics by using Metropolis-Hastings within-Gibbs sampling. *Applied Statistics*, **49**, 327–42.

Mills, L. S., Citta, J. J., Lair, K. P., Schwarz, M. K., and Tallmon, D. A. (2000) Estimating animal abundance using noninvasive DNA sampling: Promise and pitfalls. *Ecological Applications*, **10**, 283–94.

Montgomery, D. C. (1996) *Introduction to Statistical Quality Control*, 3rd edn, Wiley, New York.

Moran, P. A. P. (1951) A mathematical theory of animal trapping. *Biometrika*, **38**, 307–11.

Morgan, B. J. T. and Freeman, S. N. (1989) A model with first-year variation for ring-recovery data. *Biometrics*, **45**, 1087–101.

NAMMCO (1997) *Annual Report 1996: Report of the Scientific Committee*, North Atlantic Marine Mammal Commission, Tromsø.

Nchanji, A. C. and Plumptre, A. (2001) Seasonality in elephant dung decay and implications for censusing and population monitoring in south-western Cameroon. *African Journal of Ecology*, **39**, 24–32.

Newman, K. B. (2000) Hierarchic modeling of salmon harvest and migration. *Journal of Agricultural, Biological, and Environmental Statistics*, **5**, 430–55.

Newman, K. B., Fernández, C., Buckland, S. T., and Thomas, L. (in preparation) Inference for state-space models for wild animal populations.

Newman, K. B. and Lindley, S. T. (in preparation) Modelling the population dynamics of Sacramento winter run chinook salmon.

Nichols, J. D., Hines, J. E., Sauer, J. R., Fallon, F. W., Fallon, J. E., and Heglund, P. J. (2000) A double-observer approach for estimating detection probability and abundance from point counts. *The Auk*, **117**, 393–408.

Norvell, R. E., Howe, F. P., and Parrish, J. R. (2003) A seven-year comparison of relative-abundance and distance-sampling methods. *The Auk*, **120**, 1013–28.

Okamura, H., Kitakado, T., Hiramatsu, K., and Mori, M. (2003) Abundance estimation of diving animals by the double-platform line transect method. *Biometrics*, **59**, 512–20.

Otis, D. L., Burnham, K. P., White, G. C., and Anderson, D. R. (1978) Statistical inference from capture data on closed animal populations. *Wildlife Monographs*, **62**, 1–135.

Otto, M. C. and Pollock, K. H. (1990) Size bias in line transect sampling: a field test. *Biometrics*, **46**, 239–45.

Palka, D. L. (1993) *Estimating Density of Animals when Assumptions of Line Transect Survey are Violated*, Ph. D. thesis, University of California San Diego.

Palka, D. (1995) Abundance estimate of the Gulf of Maine harbor porpoise, in *Biology of the Phocoenids* (eds A. Bjørge and G. P. Donovan), International Whaling Commission, Cambridge, pp. 27–50.

Palka, D. and Pollard, J. (1999) Adaptive line transect survey for harbor porpoises, in *Marine Mammal Survey and Assessment Methods* (eds G. W. Garner, S. C. Amstrup, J. L. Laake, B. F. J. Manly, L. L. McDonald, and D. G. Robertson), Balkema, Rotterdam, pp. 3–11.

Pannekoek, J. and van Strien, A. (1996) *TRIM (TRends and Indices for Monitoring data)*, research paper 9634, Statistics Netherlands, Voorburg, The Netherlands.

Parmenter, R. R., Yates, T. L., Anderson, D. R., Burnham, K. P., Dunnum, J. L., Franklin, A. B., Friggens, M. T., Lubow, B. C., Miller, M., Olson, G. S., Parmenter, C. A., Pollard, J., Rexstad, E., Shenk, T. M., Stanley, T. R., and White, G. C. (2003) Small-mammal density estimation: a field comparison of grid-based versus web-based density estimators. *Ecological Monographs*, **73**, 1–26.

Parzen, E. (1960) *Modern Probability Theory and its Applications*, Wiley, New York.

Patil, S. A., Burnham, K. P., and Kovner, J. L. (1979) Nonparametric estimation of plant density by the distance method. *Biometrics*, **35**, 597–604.

Perryman, W. L., Donahue, M. A., Laake, J. L., and Martin, T. E. (1999) Diel variation in migration rates of eastern Pacific gray whales measured with thermal imaging sensors. *Marine Mammal Science*, **15**, 426–45.

Peterman, R. M. (1990) Statistical power analysis can improve fisheries research and management. *Canadian Journal of Fisheries and Aquatic Sciences*, **47**, 2–15.

Petersen, C. G. J. (1896) The yearly immigration of young plaice into the Limfjord from the German Sea. *Report of the Danish Biological Station*, **6**, 1–48.

Pettersson, M. (1998) Monitoring a freshwater fish population: statistical surveillance of biodiversity. *Environmetrics*, **9**, 139–50.

Pledger, S. (2000) Unified maximum likelihood estimates for closed capture-recapture models using mixtures. *Biometrics*, **56**, 434–42.

Plumptre, A. J. and Harris, S. (1995) Estimating the biomass of large mammalian herbivores in a tropical montane forest—a method of fecal counting that avoids assuming a steady-state system. *Journal of Applied Ecology*, **32**, 111–20.

Pollard, J. H. (2002) *Adaptive Distance Sampling*, Ph. D. thesis, University of St Andrews.

Pollard, J. H. and Buckland, S. T. (1997) A strategy for adaptive sampling in shipboard line transect surveys. *Report of the International Whaling Commission*, **47**, 921–31.

Pollard, J. H., Palka, D., and Buckland, S. T. (2002) Adaptive line transect sampling. *Biometrics*, **58**, 862–70.

Pollock, K. H. and Kendall, W. L. (1987) Visibility bias in aerial surveys: a review of estimation procedures. *Journal of Wildlife Management*, **51**, 502–10.

Pollock, K. H., Nichols, J. D., Simons, T. R., Farnsworth, G. L., Bailey, L. L., and Sauer, J. R. (2002) Large scale wildlife monitoring studies: statistical methods for design and analysis. *Environmetrics*, **13**, 105–19.

Quang, P. X. (1991) A nonparametric approach to size-biased line transect sampling. *Biometrics*, **47**, 269–79.

Quang, P. X. and Becker, E. F. (1997) Combining line transect and double count sampling techniques for aerial surveys. *Journal of Agricultural, Biological, and Environmental Statistics*, **2**, 230–42.

Ramsey, F. L., Scott, J. M., and Clark, R. T. (1979) Statistical problems arising from surveys of rare and endangered forest birds. *Proceedings of the 42nd Session of the International Statistical Institute*, 471–83.

Ramsey, F. L., Wildman, V., and Engbring, J. (1987) Covariate adjustments to effective area in variable-area wildlife surveys. *Biometrics*, **43**, 1–11.

R Development Core Team (2003) *R: a Language and Environment for Statistical Computing*, R Foundation for Statistical Computing, Vienna. www.R-project.org.

Reilly, S. B., Rice, D. W., and Wolman, A. A. (1980) Preliminary population estimate for the California gray whale based upon Monterey shore censuses, 1967/68 to 1978/79. *Report of the International Whaling Commission*, **30**, 359–68.

Reilly, S. B., Rice, D. W., and Wolman, A. A. (1983) Population assessment of the gray whale, *Eschrichtius robustus*, from California shore censuses, 1967–80. *Fishery Bulletin*, **81**, 267–81.

Rexstad, E. A. and Debevec, E. M. (in preparation) Metric of environmental change using measures of process variation.

Reynolds, R. T., Scott, J. M., and Nussbaum, R. A. (1980) A variable circular-plot method for estimating bird numbers. *Condor*, **82**, 309–13.

Rice, J. A. and Wu, C. O. (2001) Nonparametric mixed effects models for unequally sampled noisy curves. *Biometrics*, **57**, 253–9.

Robinette, W. L., Loveless, C. M., and Jones, D. A. (1974) Field tests of strip census methods. *Journal of Wildlife Management*, **38**, 81–96.

Rorabaugh, J. C., Palermo, C. L., and Dunn, S. C. (1987) Distribution and relative abundance of the flat-tailed horned lizard (*Phrynosoma mcallii*) in Arizona. *The Southwestern Naturalist*, **32**, 103–9.

Rosenstock, S. S., Anderson, D. R., Giesen, K. M., Leukering, T., and Carter, M. F. (2002) Landbird counting techniques: current practices and an alternative. *The Auk*, **119**, 46–53.

Rugh, D. J., Breiwick, J. M., Dahlheim, M. E., and Boucher, G. C. (1993) A comparison of independent, concurrent sighting records from a shore-based count of gray whales. *Wildlife Society Bulletin*, **21**, 427–37.

Samuel, M. D., Garton, E. O., Schlegel, M. W., and Carson, R. G. (1987) Visibility bias during aerial surveys of elk in north central Idaho. *Journal of Wildlife Management*, **51**, 622–30.

Schipper, M., den Hartog, J., and Meelis, E. (1997) Sequential analysis of environmental monitoring data: optimal SPRTs. *Environmetrics*, **8**, 29–41.

Schweder, T. (1974) *Transformations of Point Processes: Applications to Animal Sighting and Catch Problems, with Special Emphasis on Whales*, Ph. D. thesis, University of California Berkeley.

Schweder, T. (1977) Point process models for line transect experiments, in *Recent Developments in Statistics* (eds J. R. Barba, F. Brodeau, G. Romier, and B. Van Cutsem), North-Holland Publishing Company, New York, USA, pp. 221–42.

Schweder, T. (1990) Independent observer experiments to estimate the detection function in line transect surveys of whales. *Report of the International Whaling Commission*, **40**, 349–55.

Schweder, T. (1996) A note on a buoy-sighting experiment in the North Sea in 1990. *Report of the International Whaling Commission*, **46**, 383–5.

Schweder, T. (1997) Measurement error models for the Norwegian minke whale survey in 1995. *Report of the International Whaling Commission*, **47**, 485–8.

Schweder, T. (1999) Line transecting with difficulties; lessons from surveying minke whales, in *Marine Mammal Survey and Assessment Methods* (eds G. W. Garner, S. C. Amstrup, J. L. Laake, B. F. J. Manly, L. L. McDonald, and D. G. Robertson), Balkema, Rotterdam, pp. 149–66.

Schweder, T., Hagen, G., Helgeland, J., and Koppervik, I. (1996) Abundance estimation of northeastern Atlantic minke whales. *Report of the International Whaling Commission*, **46**, 391–408.

Schweder, T. and Høst, G. (1992) Integrating experimental data and survey data to estimate $g(0)$: a first approach. *Report of the International Whaling Commission*, **42**, 575–82.

Schweder, T., Høst, G., and Øien, N. (1991) A note on the bias in capture-recapture type estimates of $g(0)$ due to the fact that whales are diving. *Report of the International Whaling Commission*, **41**, 397–9.

Schweder, T., Skaug, H. J., Dimakos, X. K., Langaas, M., and Øien, N. (1997) Abundance of northeastern Atlantic minke whales, estimates for 1989 and 1995. *Report of the International Whaling Commission*, **47**, 453–83.

Schweder, T., Skaug, H. J., Langaas, M., and Dimakos, X. K. (1999) Simulated likelihood methods for complex double-platform line transect surveys. *Biometrics*, **55**, 678–87.

Scott, J. M. and Ramsey, F. L. (1981) Length of count period as a possible source of bias in estimating bird densities, in *Estimating Numbers of Terrestrial Birds. Studies in Avian Biology No. 6* (eds C. J. Ralph and J. M. Scott), Cooper Ornithological Society, pp. 409–13.

Seber, G. A. F. (1979) Transects of random length, in *Sampling Biological Populations* (eds R. M. Cormack, G. P. Patil, and D. S. Robson), International Co-operative Publishing House, Fairland, MD, USA, pp. 183–92.

Seber, G. A. F. (1982) *The Estimation of Animal Abundance and Related Parameters*, Macmillan, New York.

Sen, A. R., Tourigny, J., and Smith, G. E. J. (1974) On the line transect sampling method. *Biometrics*, **30**, 329–41.

Severini, T. A. (2000) *Likelihood Methods in Statistics*, Oxford University Press, Oxford.

Shenk, T. M., White, G. C., and Burnham, K. P. (1998) Sampling-variance effects on detecting density dependence from temporal trends in natural populations. *Ecological Monographs*, **68**, 445–63.

Simonoff, J. S. (1996) *Smoothing Methods in Statistics*, Springer, New York.

Skaug, H. J. and Schweder, T. (1999) Hazard models for line transect surveys with independent observers. *Biometrics*, **55**, 29–36.

Smith, T. D., Allen, J., Clapham, P. J., Hammond, P. S., Katona, S., Larsen, F., Lein, J., Mattila, D., Palsbøl, P. J., Sigurjónsson, J., Stevick, P. T., and Øien, N. (1999) An ocean-basin-wide mark-recapture study of the north Atlantic Humpback whale (*Megaptera novaeangliae*). *Marine Mammal Science*, **15**, 1–32.

Southwell, C., de la Mare, B., Underwood, M., Quartararo, F., and Cope, K. (2002) An automated system to log and process distance sight-resight aerial survey data. *Wildlife Society Bulletin*, **30**, 394–404.

Southwell C., de la Mare, B., Borchers, D. L., Paxton, C. G. M., and Burt, M. L. (in preparation). The effect of unmodelled heterogeneity in double-platform aerial surveys of pack-ice seals.

Steel, R. G. D. and Torrie, J. H. (1980) *Principles and Procedures of Statistics: A Biometrical Approach*, 2nd edn, McGraw-Hill Inc., Singapore.

Steidl, R. J., Hayes, J. P., and Schauber, E. (1997) Statistical power in wildlife research. *Journal of Wildlife Management*, **61**, 270–9.

Steidl, R. J. and Thomas, L. (2001) Power analysis and experimental design, in *Design and Analysis of Ecological Experiments*, 2nd edn (eds S. M. Scheiner and J. Gurevitch), Oxford University Press, New York, pp. 14–36.

Steinhorst, R. K. and Samuel, M. D. (1989) Sightability adjustment methods for aerial surveys of wildlife populations. *Biometrics*, **45**, 415–25.

Stoyan, D. (1982) A remark on the line transect method. *Biometrical Journal*, **24**, 191–5.

Strindberg, S. (2001) *Optimized Automated Survey Design in Wildlife Population Assessment*, Ph. D. thesis, University of St Andrews.

Strindberg, S. and Buckland, S. T. (in press) Zigzag survey designs in line transect sampling. *Journal of Agricultural, Biological, and Environmental Statistics*.

Swartz, S. L., Jones, M. L., Goodyear, J., Withrow, D. E., and Miller, R. V. (1987) Radio-telemetric studies of gray whale migration along the California coast: a preliminary comparison of day and night migration rates. *Report of the International Whaling Commission*, **37**, 295–9.

Taylor, B. L. and Gerrodette, T. (1993) The uses of statistical power in conservation biology: the Vaquita and Northern Spotted Owl. *Conservation Biology*, **7**, 134–46.

ter Braak, C. J. F., van Strien, A. J., Meijer, R., and Verstrael, T. J. (1994) Analysis of monitoring data with many missing values: which method? in *Bird Numbers 1992. Distribution, Monitoring and Ecological Aspects. Proceedings of the 12th International Conference of the International Bird Census Committee and European Ornithological Atlas Committee* (eds W. Hagemeijer and T. Verstrael), SOVON, Beek-Ubbergen, The Netherlands, pp. 663–73.

Thomas, L. (1996) Monitoring long-term population change: Why are there so many analysis methods? *Ecology*, **77**, 49–58.

Thomas, L. (1997) Retrospective power analysis. *Conservation Biology*, **11**, 276–80.

Thomas, L., Buckland, S. T., Newman, K. B., and Harwood, J. (in press) A unified framework for modelling wildlife population dynamics. *Australian and New Zealand Journal of Statistics*.

Thomas, L. and Juanes, F. (1996) The importance of statistical power analysis: an example from animal behaviour. *Animal Behaviour*, **52**, 856–9.

Thomas, L., Laake, J. L., Derry, J. F., Buckland, S. T., Borchers, D. L., Anderson, D. R., Burnham, K. P., Strindberg, S., Hedley, S. L., Marques, F. F. C., Pollard, J. H., and Fewster, R. M. (1998) *Distance 3.5*, Research Unit for Wildlife Population Assessment, University of St Andrews.

Thomas, L., Laake, J. L., Strindberg, S., Marques, F. F. C., Buckland, S. T., Borchers, D. L., Anderson, D. R., Burnham, K. P., Hedley, S. L., Pollard, J. H., and Bishop, J. R. B. (2003) *Distance 4.1. Release 1*, Research Unit for Wildlife Population Assessment, University of St Andrews. www.ruwpa.st-and.ac.uk/distance/

Thomas, L. and Martin, K. (1996) The importance of analysis method for breeding bird survey population trend estimates. *Conservation Biology*, **10**, 479–90.

Thompson, S. K. (1990) Adaptive cluster sampling. *Journal of the American Statistical Association*, **85**, 1050–9.

Thompson, S. K. (2002) *Sampling*, 2nd edn, Wiley, New York.

Thompson, S. K. and Seber, G. A. F. (1994) Detectability in conventional and adaptive sampling. *Biometrics*, **50**, 712–24.

Thompson, S. K. and Seber, G. A. F. (1996) *Adaptive Sampling*, Wiley, New York.

Trenkel, V. M., Elston, D. A., and Buckland, S. T. (2000) Fitting population dynamics models to count and cull data using sequential importance sampling. *Journal of the American Statistical Association*, **95**, 363–74.

Turner, F. B. and Medica, P. A. (1982) The distribution and abundance of flat-tailed horned lizard (*Phrynosoma mcallii*). *Copeia*, **4**, 815–23.

Underwood, F. M. (2004) *Design-based Adaptive Monitoring Strategies for Wildlife Population Assessment*, Ph. D. thesis, University of St Andrews.

Urquhart, N. S. and Kincaid, T. M. (1999) Designs for detecting trend from repeated surveys of ecological resources. *Journal of Agricultural, Biological, and Environmental Statistics*, **4**, 404–14.

Urquhart, N. S., Paulsen, S. G., and Larsen, D. P. (1998) Monitoring for policy-relevant regional trends over time. *Ecological Applications*, **8**, 246–57.

US Fish and Wildlife Service (2003) *Waterfowl Population Status, 2003*, US Department of the Interior, Washington D.C.

van der Meer, J. (1997) Sampling design of monitoring programmes for marine benthos: a comparison between the use of fixed versus randomly selected stations. *Journal of Sea Research*, **37**, 167–79.

Walton, I. (1653) *The Compleat Angler*, reproduced by Everyman's Library, London, 1953.

Warner, R. M. (1998) *Spectral Analysis of Time-series Data*, Guilford Press, New York.

White, G. C. (2000) Population viability analysis: data requirements and essential analyses, in *Research Techniques in Animal Ecology: Controversies and Consequences* (eds L. Boitani and T. K. Fuller), Columbia University Press, New York, pp. 288–331.

Williams, B. K., Nichols, J. D., and Conroy, M. J. (2002) *Analysis and Management of Animal Populations: Modeling, Estimation, and Decision Making*, Academic Press, San Diego.

Wood, S. N. (2000) Modelling and smoothing parameter estimation with multiple quadratic penalties. *Journal of the Royal Statistical Society Series B*, **62**, 413–28.

Wood, S. N. (2003) Thin plate regression splines. *Journal of the Royal Statistical Society Series B*, **65**, 95–114.

INDEX